An Introduction to Electrical Science

An Introduction to Electrical Science

Adrian Waygood

Routledge
Taylor & Francis Group

NEW YORK AND LONDON

First edition published 2013
by Routledge
2 Park Square, Milton Park, Abingdon, Oxon OX14 4RN

Simultaneously published in the USA and Canada
by Routledge
711 Third Avenue, New York, NY 10017

Routledge is an imprint of the Taylor & Francis Group, an informa business

British Library Cataloguing in Publication Data
A catalogue record for this book is available from the British Library

Library of Congress Cataloging in Publication Data
Waygood, Adrian.
   An introduction to electrical science / Adrian Waygood.
      pages cm
   1. Electric currents, Alternating.   2. Electrons.   3. Electrical engineering.   I. Title.
   TK1141.W39 2013
   621.3—dc23   2012033453

ISBN: 978-0-415-81002-9 (pbk)
ISBN: 978-0-203-44109-1 (ebk)

Typeset in Times
by RefineCatch Limited, Bungay, Suffolk

Dedication

To Anita, for having put up with a complete lack of conversation as I sat in front of my iMac for hours on end, and to my friends Greg Collins, George Brain and Anne Wafaa who, with others, provided encouragement long before I embarked on this particular project.

Contents

Foreword

I feel honoured to have been asked to provide a short forward to Adrian Waygood's book on electrical basics. I know him well and worked closely with him in the Electrical Programs at the Northern Alberta Institute of Technology – NAIT. Adrian was highly respected by his colleagues, an effective instructor, and very popular with his students. We all felt a sense of loss when, in December 1990, he chose to move on in order to expand his experience.

For as long as I've known him, Adrian has reinforced his lectures with his own superb printed materials. I, and other instructors, have used many of his handouts for years. In fact, I used several just this year to assist me teaching basic electricity to employees of a local start-up, new-technology, lighting company.

I am thrilled to see the chapters of this new and exciting textbook exhibit the same engaging style. The conversational nature of the presentation engages learners and encourages them to read on. It appears he has never forgotten what it can be like for a learner to travel down this unfamiliar path. He does everything possible to remove the usual frustrations and make learning rewarding and enjoyable.

Topics are presented and developed in a logical and easy-to-follow sequence. The author strives to ensure no 'holes' are left in the concepts, and no assumptions made about prior knowledge. One of the most frustrating statements in printed learning materials is 'It can be shown that. . .' which, in my opinion, leaves the reader with a feeling of having missed something. You won't find that in this book.

Outstanding strengths of the book include:

- Ahead of each chapter, learning outcomes are presented clearly and thoroughly. The learner is made aware of exactly what he or she will be able to do upon completion of the chapter.
- At the beginning of each chapter an overview paints a broad picture of what is about to be covered enabling the learner to see the material in context.
- The conventions employed on the book are identified at appropriate points and well explained.
- Technical terms are well defined and explained. Definitions are highlighted.
- Step-by-step explanations are accompanied with numerous diagrams to assist the learner in understanding and organizing information.
- Worked examples are plentiful, timely, and clear.
- Equations are developed from first principles allowing the reader to understand them rather than simply accept and memorize.
- Periodic summaries are provided in text boxes, highlighting important concepts as they are developed.

I am pleased to see Adrian's work pulled together to create this book. This is a thorough and comprehensive, yet easy to understand introduction to electricity and circuit fundamentals. It is an excellent resource for apprentices and tradesmen as well as aspiring technicians and technologists.

Greg Collins
Chair (Retired)
Electrical Engineering Technology
Northern Alberta Institute of Technology – NAIT
Edmonton, Alberta, Canada

Introduction

The original intention of this book was to provide an introduction to electrical science for apprentices and tradespersons in the electrotechnology industry, and for others who require an understanding of electricity for their work. However, it is to be hoped that others, such as hobbyists, will also find the book interesting and useful.

While there are a great many books which support the regulations and practical aspects of training in the electrotechnology industry, by necessity these books are unable to cover electrical science in any depth. There is a tendency, therefore, for many otherwise excellent tradespersons to have a weak understanding of the science behind their chosen trade.

Furthermore, even a superficial examination of the questions and answers on websites, such as *WikiAnswers,* reveal a great many misconceptions on the subject of electricity, such as 'Ohm's Law applies to all circuits' (it doesn't!), 'the end of a compass needle that points North is a south pole' (it isn't!), 'there are three categories of electric circuit' (there are four!), and so on!

So these are the two areas that this book tries to address: an improvement in the knowledge and understanding of electrical science, and an attempt to eliminate many of the misconceptions that people have about this subject.

I have tried my best to make each topic interesting by providing, where appropriate, the historical context behind the various 'rules' and 'laws' discussed, as I believe that there is far more to electrical science than rote-learned rules and equations. By making it interesting, I hope to pass on to the reader my own enthusiasm for the subject, in the same way as I tried to do for my former students.

Wherever possible, I have used a step-by-step approach to some of the more complicated topics. For example, in chapters on alternating current theory, I have used several illustrations to show how each phasor diagram is constructed, rather than just providing a completed phasor diagram. I have also tried to emphasise the need to *develop*, rather than memorise, equations as the key to understanding topics such as alternating current.

Adrian Waygood
August 2012

Online resources
There is also a companion website for this book featuring multiple choice questions, further written questions and an extra chapter on electrical measuring instruments. The website can be accessed via the following link: www.routledge.com/cw/waygood

Chapter 1

SI system of measurements

On completion of this chapter, you should be able to

1   explain the term 'SI base units'.
2   list the seven SI base units.
3   explain the term 'SI derived units'.
4   recognise SI prefixes.
5   apply correct SI symbols.
6   use correct SI prefixes.
7   apply SI conventions when writing SI units.

Introduction

In 1948, the **General Conference of Weights and Measures (CGPM)** charged an international committee, the CIPM*, to *'study the establishment of a complete set of rules for (metric) units of measurement'*.

> ***CIPM (Comité international des poid et mesures)** is an International Committee for Weights and Measurements comprising eighteen individuals, each from a different member state, whose principal task is to promote worldwide uniformity in units of measurement. The Committee achieves this either by direct action, or by submitting proposals to the General Conference on Weights and Measures.

The outcome of this study was a rational system of metric units termed 'SI'.

The abbreviation **SI** stands for *Système Internationale d'Unités* and this system of measurements has been adopted internationally by the scientific and engineering communities, as well as by businesses for the purpose of international trade. Whereas *most* countries now use SI exclusively, *some* countries – most notably the United States and, to a lesser extent, the United

Kingdom – still make wide use of non-metric units, especially for day-to-day use.

Earlier versions of the metric system include the **'cgsA'** ('centimetre, gram, second, ampere') and the **'mksA'** ('metre, kilogram, second, ampere') systems. SI is largely based on the 'mksA' system.

SI comprises *two* classes of units:

- base units
- derived units.

Base units

There are *seven* **base units** from which *all* other SI units are derived. These are shown in Table 1.1.

Table 1.1

Quantity	SI unit	SI symbol
length	metre	m
mass*	kilogram	kg
time	second	s
electric current	ampere	A
temperature	kelvin	K
luminous intensity	candela	cd
amount of substance	mole	mol

* The kilogram is a little confusing because it is the only base unit with a prefix (kilo). It has been suggested that the name 'kilogram' should be replaced as the unit for mass, but this is likely to cause more confusion than necessary.

An Introduction to Electrical Science, Waygood, ISBN 9780415810029, 2013. © Taylor & Francis

Derived units

Those SI units which are *not* base units are called **derived units**.

Derived units are formed by combining base units – for example, the '**volt**' is defined as *'the potential difference between two points such that the energy used in conveying a charge of one coulomb from one point to the other is one joule'*.

So the **volt** is defined in terms of the **coulomb** and the **joule**. The **coulomb** (see page 5), in turn, is defined in terms of the **ampere** and the **second** (both base units). The **joule** is defined in terms of the **newton** and the **metre** (a base unit). Finally, the **newton** is defined in terms of the **kilogram**, the **metre** and the **second** (all base units).

So, by 'deconstructing' the **volt**, we find that it is ultimately derived from a combination of each of the base units underlined, in Figure 1.1 – i.e. the **ampere**, the **second**, the **kilogram** and the **metre**.

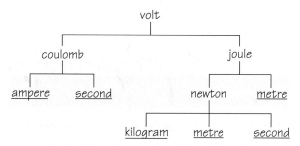

Figure 1.1

Most derived units have been given **special names** in honour of famous physicists whose research has contributed to our knowledge of the quantity concerned – for example, as we have learnt, the derived unit for potential difference is the '**volt**', which is simply a special name given to a '**joule per coulomb**', and is named after the Italian nobleman and professor of physics, Count Alessandro Volta (1745–1827).

Table 1.2 lists **SI derived units** with special names that you will meet in this text.

Non-SI metric units

Not all *metric* units are SI units, although many may be '*used alongside*' SI units. These include the commonly used units shown in Table 1.3.

Table 1.2

Quantity	Symbol	SI unit	SI symbol
capacitance	C	farad	F
capacitive reactance	X_C	ohm	Ω
conductance	G	siemens	S
electric charge	Q	coulomb	C
force	F	newton	N
frequency	F	hertz	Hz
impedance	Z	ohm	Ω
inductance	L	henry	H
inductive reactance	X_L	ohm	Ω
potential difference	E, U, V	volt	V
power	P	watt	W
pressure	P	pascal	Pa
resistance	R	ohm	Ω
magnetic flux	Φ	weber*	Wb
magnetic flux density	B	tesla	T
magnetomotive force	H	ampere**	A
mutual inductance	M	henry***	H
self inductance	L	henry	H
work, energy	W	joule	J

Notes:
*The weber is pronounced 'vay-ber'.
**Often spoken as 'ampere turn'.
***The plural of 'henry' is 'henrys', not 'henries'.

Table 1.3

Quantity	Unit	Symbol
energy	watt hour	W·h
mass	tonne	t*
volume	litre	L or ℓ **
rotation	revolutions per second	r/s
temperature	degree Celsius***	°C
time	minute (60 s); hour; day; year	min, h, d, a

Notes:
*The unit of mass, the *tonne*, is pronounced, or spoken, as 'metric ton'.
**Since a lower-case 'ell' (l) can be confused with the number 1, we shall use a capital 'ell' (L), in common with North American SI practice.
***The division intervals are identical for the both the *Celsius* and *kelvin* scales. However, 0°C corresponds to 273.15 K, and 100°C corresponds to 373.15 K.

Multiples and sub-multiples

Frequently, we have to deal with very large, or very small, quantities. For example, the resistance of insulation is measured in millions of ohms, while the resistance of a conductor is measured in thousandths of an ohm.

To avoid having to express very large or very small values in this way, we use, instead, **multiples** and **sub-multiples**. These are indicated by assigning a *prefix* to the SI unit. The more common are listed in Table 1.4.

Table 1.4

Multiplication factor	Power of ten	Prefix	Symbol
1 000 000 000 000 ×	10^{12}	tera	T
1 000 000 000 ×	10^{9}	giga	G
1 000 000 ×	10^{6}	mega	M
1 000 ×	10^{3}	kilo	k
0.001 ×	10^{-3}	milli	m

Multiplication factor	Power of ten	Prefix	Symbol
0.000 001 ×	10^{-6}	micro	μ
0.000 000 000 001 ×	10^{-12}	pico	p

Examples
• 10 000 000 watts can be written as $10×10^6$ W, or as 10 MW
• 33 000 volts can be written as $33×10^3$ V, or as 33 kV
• 0.025 amperes can be written as $25×10^{-3}$ A, or as 25 mA
Note: the correct spelling for one-millionth of an ohm, is 'microhm', *not* 'microohm' or 'micro-ohm'.

Note that SI prefixes employ the '**Engineering System**' – i.e. powers of ten increase or decrease by a factor of *three*. Accordingly, units such as 'centimetre', should *not* be used when working in SI.

We *cannot* insert multiples or sub-multiples into equations. For example, we must *always* convert microwatts, milliwatts, kilowatts, megawatts, etc., into **watts** whenever we insert that quantity into an equation.[12]

To do this:

Table 1.5

to convert...	into...	multiply by...
picowatts	watts	$× 10^{-12}$
microwatts	watts	$× 10^{-6}$
milliwatts	watts	$× 10^{-3}$
kilowatts	watts	$× 10^{3}$
megawatts	watts	$× 10^{6}$
gigawatts	watts	$× 10^{9}$
terawatts	watts	$× 10^{12}$

Although, in the above example, we have used watts, this applies of course to *any* SI unit.

SI conventions

SI specifies *how* its units of measurement should be written. These rules, or **conventions**, apply to the units themselves, to their symbols and to their associated numerals.

You should be aware of the following conventions.

Rules for writing SI units

1 SI units and their symbols are never italicised:
  e.g. ampere, not *ampere*
    mV, not *mV*

2 When written in full, units are *never* capitalised.
  e.g. watt, not Watt
    ampere, not Ampere

3 SI symbols are written in lower-case, *unless they are named after someone*, in which case they are capitalised.
  e.g. symbol for metre: m
    symbol for ampere: A (after André-Marie Ampère)

4 SI symbols are symbols, not abbreviations, so are *not* punctuated with full stops (periods).
  e.g. 230 V, not 230 V.
    13 A, not 13 A.

5 There is no plural form of an SI symbol:
  e.g. 500 kg, not 5000 kgs
    40 W, not 40 Ws

6 Numerals are always followed by the *symbol* for a unit:
  e.g. 400 V, not 400 volts
    10 kW, not 10 kilowatts

7 Written numbers are always followed by a written unit:
  e.g. Twelve volts, not twelve V

8 A *space* is always placed between a number and the unit symbol:
  e.g. 5000 W, not 5000W
    275 kV, not 275kV

9 A hyphen *may* be used (optionally) between a number and the unit symbol, when the combination is used as an adjective:
  e.g. 'A 66-kV power line.' or 'A 66 kV power line'
    'A 13-A socket.' or 'A 13 A socket'

10 Compound derived unit symbols are separated by a point placed above the line:
  e.g. SI unit for apparent power: V·A (volt ampere)
    SI unit for resistivity: Ω·m (ohm metre)

11 No space is placed between an SI unit or symbol and its multiplier:
  e.g. kilowatt, not kilo-watt or kilo watt
    kW, not k-W or k W
  Special case for ohms:
    microhm, not microohm or micro-ohm
    kilohm, not kiloohm or kilo-ohm

12 *Spaces*, not commas, are used as thousand separators with large numbers:
  e.g. 11 000 V, not 11,000 V
    15 000.000 075, not 15,000.000075

The space is *optional* for four digits: 1500 mW *or* 1 500 mW

13 Square and cubic measurements are written as exponents:
  e.g. m^2 (square metres) not sq m.
  e.g. m^3 (cubic metres) not cu m.

Proposed new definitions for SI base units

A subcommittee of the **International Committee for Weights and Measures (CIPM)** has proposed revised definitions of the SI base units, for consideration at the 25th **General Conference of Weights and Measures (CGPM)** scheduled to be held in 2014.

The proposal is that three of the base units, the metre, second and candela, will essentially remain unchanged, except for the phrasing of their definitions, whereas the remaining four will each undergo fundamental changes, as a result of allocating exact values to five fundamental constants:

- speed of light – based on the metre per second
- Planck's constant – based on the joule second
- elementary charge – based on the coulomb
- Boltzmann constant – based on the joule per kelvin
- Avogadro constant – based on the reciprocal of the mole

Of particular interest to those of us who work in the electricity industry, is a proposed major redefinition of the **ampere**. The proposed new definition removes its current dependency on the kilogram and the metre and bases it, instead, on fixing the numerical value of the elementary charge (the amount of charge on a single electron) to a figure yet to be agreed.

If this change goes ahead, then it seems that the ampere may be defined in terms of charge (but *not* the coulomb) and time.

Definitions of electrical SI units

ampere (symbol: **A**)

The **ampere** is defined as '*the constant current that, if maintained in two straight parallel conductors of infinite length and negligible cross-sectional area and placed one metre apart in a vacuum, would produce* *between them a force equal to 2 × 10^{-7} newtons per unit length'.*

coulomb (symbol: **C**)

The **coulomb** is defined as *'the charge transported through any cross-section of a conductor in one second by a constant current of one ampere'.*

volt (symbol: **V**)

The **volt** is defined as *'the potential difference between two points such that the energy used in conveying a charge of one coulomb from one point to the other is one joule'.*

joule (symbol: **J**)

The **joule** is defined as *'the work done when the point of application of a force of one newton is displaced one metre in the direction of that force'.*

ohm (symbol: **Ω**)

The **ohm** is defined as *'the electrical resistance between two points of a conductor, such that when a constant potential difference of one volt is applied between those points, a current of one ampere results'.*

newton (symbol: **N**)

The **newton** is defined as *'the force which, when applied to a mass of one kilogram, will give it an acceleration of one metre per second per second'.*

watt (symbol: **W**)

The **watt** is defined as *'the power resulting when one joule of energy is dissipated in one second'.*

farad (symbol: **F**)

The **farad** is defined as *'the capacitance of a capacitor, between the plates of which there appears a difference in potential of one volt, when it is charged to 1 coulomb.*

weber* (symbol: **Wb**)

The **weber** is defined as *'the magnetic flux that, linking a circuit of one turn, produces a potential difference of one volt when it is reduced to zero at a uniform rate in one second'.*
(*pronounced 'vay-ber')

tesla (symbol: **T**)

The **tesla** is defined as *'one weber of magnetic flux per square metre of circuit area'.*

henry (symbol: **H**)

The **henry** is defined as *'the self- or mutual-inductance of a closed loop if a current of one ampere gives rise to a magnetic flux of one weber'.*

Misconceptions

The Metric System and the SI system are the same thing

While the SI system is certainly 'metric', it is incorrect to think that the metric system is 'SI'! The metric system existed long before the introduction of SI. Different versions of the metric system include the **cgsA** system, whose base units included the centimetre, gram, second and ampere. It also included the **mksA** system, which shared the same base units as SI.

Celsius and litre are SI units

Celsius and litre are 'metric' units and in common use, but they are *not* SI units. They are classified as units that 'may be used alongside' the SI system. The SI unit for temperature is the kelvin and the SI unit for volume is the cubic metre.

The centimetre is an SI unit

For engineering, the SI system prefers those prefixes based on powers of ten raised to multiples of 3. These include: micro (10^{-6}), milli (10^{-3}), kilo (10^3) and mega (10^6). In SI, other prefixes are considered to be 'non-preferred'.

The metre, kilogram, second and ampere are 'fundamental' SI units

SI doesn't use the expression 'fundamental units' to describe these units. The correct expression is 'base units'.

Finally...

Now that you have completed this chapter, are you able to achieve the objectives or learning outcomes listed at the beginning of this chapter?

Ask yourself, 'Can I …'

1 explain the term 'SI base units'.
2 list the seven SI base units.
3 explain the term 'SI derived units'.
4 recognise SI prefixes.
5 apply correct SI symbols.
6 use correct SI prefixes.
7 apply SI conventions when writing SI units.

Chapter 2

The 'electron theory' of electricity

On completion of this chapter, you should be able to

1 explain the difference between
   a elements
   b compounds.
2 describe the structure of Bohr's model of an atom.
3 list the electric charges associated with
   a protons
   b electrons.
4 state the laws of attraction and repulsion between electric charges.
5 explain the term 'free electron'.
6 explain the significance of an atom's valence shell.
7 explain what is meant by 'ionisation', and 'positive' and 'negative ions'.
8 explain the difference between conductors and insulators.
9 describe an electric current in metals, in terms of the flow of free electrons.
10 describe how current flow in non-metallic conductors differs from that in metallic conductors.
11 specify the direction of
   a electron flow current
   b conventional current.
12 list examples of and describe applications for
   a practical conductors
   b practical insulators.
13 explain what is meant by the term 'displacement current'.

Introduction

Mankind has known about the existence of **electricity** for thousands of years but, for most of that time, electricity was regarded as either something to be feared (e.g. lightning storms), or simply as a curiosity – with its various phenomena regarded as little more use than to provide entertainment for the curious.

We should understand that, these days, the term 'electricity' is used to describe a branch of science in the same way as, for example, we use the word 'chemistry', and it is wrong to use it as though it were a quantity, such as charge or current. It is quite meaningless, these days, to ask, for example, *'How much "electricity" does a residence consume?'*. Having said that, until well into the 1960s, it was common for textbooks to describe a *'quantity of electricity'* to mean what we now call a *'quantity of electric charge'*.

By the nineteenth century, despite not really understanding the true nature of 'electricity', scientists such as **Michael Faraday** in England, **Joseph Henry** in the United States, **Georg Ohm** in Germany, and many others, were establishing 'rules' regarding its *behaviour*, based on the results of their practical experiments.

However, it wasn't until the gradual accumulation of knowledge on the **structure of the atom**, from the

An Introduction to Electrical Science, Waygood, ISBN 9780415810029, 2013. © Taylor & Francis

late nineteenth century onwards, that the secrets of electricity finally began to be revealed.

The problem with **atoms**, of course, is that we can't 'see' them! And without being able to see them, it's difficult to visualise how they behave. For something to be visible, it must be capable of reflecting light, and atoms are very much smaller than the wavelength of light and, so, are quite incapable of reflection!

So if we can't *see* atoms, how, then, can we possibly know what they 'look' like and understand how they behave?

The answer, of course, is that we have no idea of what they look like and we still don't fully understand how they behave!

In the much-acclaimed BBC television documentary series *'The Ascent of Man'*, the presenter, physicist Dr **Jacob Bronowski**, summed up this situation by saying,

When it comes to atoms, language is not for describing facts, but for creating images. What lies below the visible world is always imaginary; there is no other way to talk about the invisible . . .

He then continued,

When we step through the gateway of the atom, we are in a world which our senses cannot experience. Things are put together in a way we cannot know; we (can) only try to picture it by ***analogy***.

What Dr Bronowski was saying is that it is *impossible* to describe the structure and behaviour of an atom *as it actually is*, because it is simply *beyond our understanding;* all we can do is try to create an *anology* (a 'likeness'), by comparing it with something that we *can* describe and understand. We call this analogy a '**model**'.

> A '**model**' provides us with a way of imagining what an atom *might* look like, based upon its apparent behaviour and, from that behaviour, predicting how it would behave under different circumstances. A model doesn't necessarily represent the way an atom *really* is, because it is invisible and its actual behaviour is probably beyond our understanding anyway.

One of the earliest models of the atom resembled a mass, within which its various sub-atomic particles were randomly dispersed – rather like the ingredients of a Christmas pudding. But, over time, this particular model gradually changed as scientists from all over the world sought to refine that model in light of the results of their continuing experiments.

For example, in 1911, the New Zealander **Sir Ernest Rutherford** suggested that an atom consisted mainly of 'empty space', with a heavy 'nucleus', around which electrons moved in different orbits – similar to the way in which the planets circle the Sun.

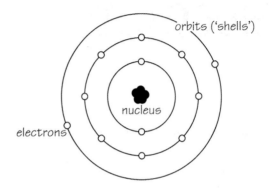

Figure 2.1

However, this 'planetary' model, didn't quite account for the way in which the electrons appeared to behave. The planets in our solar system are continuously losing the energy they need to maintain themselves in their elliptical orbits.

This causes those orbits to get a little smaller over time – eventually, trillions of years into the future (hopefully!), they will spiral into the Sun and become destroyed! But this *didn't* seem to be the case with electrons; their orbits weren't getting any smaller, which suggested that electrons simply couldn't be 'gradually' losing energy!

Scientists were intrigued about what prevented an electron from gradually losing its energy and eventually spiralling into its nucleus. The eventual answer to this riddle resulted in a completely new way at looking at physics: something called 'Quantum Mechanics'.

Amongst other things, Quantum Mechanics explains that an electron's energy can only change in *discrete* (*distinct*) *amounts*. These discrete 'packets' of energy, are called '**quanta**'.

The radius of an electron's orbit (called a 'shell') depends on the energy level of that electron; the higher its energy level, the greater the radius of its orbit. If its level of energy cannot change gradually, then neither

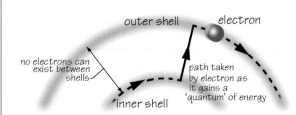

Figure 2.2

can the radius of its orbit. So, if the energy levels of electrons can only exist in discrete amounts ('quanta'), then their orbits can only exist at discrete distances from an atom's nucleus. So Quantum Mechanics explains that electrons cannot exist *between* these orbits any more than cars can park between the floors of a multi-storey car park!

To move from an inner orbit or shell, to an outer one, an electron must gain a quantum (discrete amount) of energy, as illustrated in Figure 2.2. Similarly, to move from an outer shell to an inner shell, an electron *must* lose that quantum of energy. So electrons *cannot* 'gradually' gain or lose energy in order to change their orbits.

In 1913 the Danish physicist, **Neils Bohr** (who was to go on to play a major part in developing the atomic bombs which brought about the surrender of Japan in 1945) combined Rutherford's model of the atom with Plank's theory of 'quanta', and published his scientific paper *On the Constitution of Atoms and Molecules*, in which he proposed what is considered to be the basis of the model of the atom that we will now examine.

Bohr's model of the atom

Bohr's model of the atom consists of a **nucleus**, surrounded by up to seven orbits called **electron shells** (labelled *k, l, m, n, o, p* and *q*). The electron capacity of each shell being determined by the expression, $2n^2$ – where *n* represents the sequence of the shell, counting outwards from the nucleus. So, the innermost ('*k*') shell has a capacity of 2, the next shell has a capacity of 8, the next a capacity of 18, and so on.

An atom's electron shells begin with a '*k*' because the originator of this system of labelling, Charles Barkla, believed that there were undiscovered

shells closer to the nucleus, and wanted to be able to maintain the sequence once they had been discovered. It turned out that he was wrong!

Bohr's model of the atom, of course, had its limitations, which prevented it from explaining later discovered behaviours, and other physicists have continued to develop that model further. In 1932, for example, James Chadwick proposed that the atom's nucleus was *not* a single heavy mass but, instead, made up of *two* types of 'elementary particle': positively charged **protons**, and **neutrons** which carry no charge at all. And it was the number of *protons* within the nucleus that determined all of nature's elements: hydrogen with a single proton, helium with two, and so on, up to uranium with 92 protons. Since those days, a number of man-made elements have also been created, extending the number of different types of atom to well over 100.

But there remained many questions that still needed answering. And, today, physicists working for organisations such as **CERN** (the **European Organisation for Nuclear Research**) are continuing to unravel the true nature of the atom, and have either discovered, or believe in the existence of, new sub-atomic particles that contribute to explaining its mysteries; these include many with strange names such as '*up quarks*', '*down quarks*', '*neutrinos*', '*pions*', '*muons*', '*kaons*', etc. It's hard to keep up!

In July 2012, CERN scientists claimed what is considered to be the biggest-ever breakthrough in research into the atom: apparent confirmation of the existence of what had been dubbed the 'God particle' (more accurately, the 'Higgs boson'). This sub-atomic particle, the existence of which had been long predicted by a Scottish physicist, Peter Higgs, is believed to be responsible for giving matter its mass – it ultimately being what holds the universe together!

Today's scientists no longer think of the atom in terms of Bohr's model – the present model is still evolving, but bears very little resemblance to his model of electrons travelling around a nucleus!

However *we* are not physicists, and Bohr's model of the atom is perfectly adequate to help us understand the nature and behaviour of electricity. So, for the remainder of this chapter, we will concentrate on learning a little more about Bohr's model, and how some atoms provide the charge carriers necessary for the phenomenon we call an **electric current**.

Electrons and electricity

Anything that has mass and occupies volume, we call 'matter'. And matter is made up of tiny particles, called **atoms**.

Matter that consists *entirely of identical atoms* is termed an **element**, and there are 92 naturally occurring elements, together with others that are man-made. Examples of elements include hydrogen, helium, oxygen, carbon, copper, uranium, etc.

Matter that consists of *a combination of different atoms* is termed a **compound**, and these combinations of atoms we call **molecules**.

Water, for example, is a compound made up of molecules, each of which comprises two atoms of hydrogen (symbol: H), together with one atom of oxygen (symbol: O) – hence its chemical symbol: 'H_2O'.

The structure of a molecule determines whether an element or a compound is a solid, a liquid or a gas. With *solids*, molecules form rigid, crystal-like, structures. With *liquids*, the molecules are not bound together as rigidly, resulting in the fluid nature of a liquid. And with *gases*, the molecules drift apart from each other, and disperse to fill their container. Some elements and compounds can exist as solids, liquids *and* gases (e.g. ice/water/steam), according to their temperature.

The atom

As we have learnt, Bohr's model of the atom resembles a miniature solar system, with a **nucleus** (corresponding to our sun), surrounded by tiny particles called **electrons** (corresponding to the planets) travelling in different orbits called 'shells'.

Unlike the solar system, however, which is governed by *gravitational* forces, the behaviour of an atom is governed by *electric* forces: that is, by the attraction and repulsion between **electric charges**.

These electric forces are *enormously* greater than the gravitational forces which exist within an atom. For example, in a hydrogen atom the *electric* force between a proton and electron is 2.3×10^{39} times as great as the corresponding *gravitational* force! Unfortunately, we cannot even begin to imagine such huge figures!

Electric charges have been arbitrarily assigned as being either **positive** or **negative**, and behave according to a universal law which states that

- *like* charges *repel* each other, while
- *unlike* charges *attract* each other.

Protons are *positively charged*, whereas electrons are *negatively charged*. Accordingly, electrons are held in their orbits by their attraction towards the protons within the nucleus, which is then exactly balanced by a centrifugal reaction which acts in the opposite direction.

Protons are fixed within their nucleus, and all elements are defined by their number of protons.

The simplest element, hydrogen, has just one proton in its nucleus; oxygen has eight; copper has twenty-nine; etc., as listed in the *Periodic Table of Elements*. The naturally occurring element with the greatest number of protons is *uranium* which has 92 protons in its nucleus.

So **electrons** are *negatively charged* particles which whizz around the nucleus within several fixed three-dimensional orbits, called '**shells**'. As already explained, the energy level of each electron determines which shell it occupies. Electrons with the lowest energy level occupy the shell nearest to the nucleus, while electrons with the greater energy level occupy the shell further away from the nucleus.

Ernest Rutherford was absolutely right when he believed that most of the volume occupied by an atom was simply 'empty space'. To put this into perspective, the diameter of Bohr's atom is *at least* 100 000 times the diameter of its nucleus! So the atom in Figure 2.1 is *nothing* like its true scale; in fact, if we were to represent the nucleus by printing it as a 1-mm dot in the middle of this page, then the page itself would really need to be 100-m wide to represent the diameter of the atom!

To emphasise the amount of 'empty space' there is within an atom, if it was possible to remove all the empty space in the atoms that comprise, say, the Empire State Building, in New York, then its volume would probably be reduced to something the size of an orange pip – yet its mass would remain unchanged, at thousands of tonnes!

The amount of negative charge on *one* electron is identical to the amount of positive charge on *one* proton. Under normal circumstances, atoms are *neutral* and, so, for every *positively* charged proton, there must be a corresponding *negatively* charged electron (in other words, overall, *the two charges must cancel, or neutralise, each other*).

In Bohr's model of the atom, the most complex atoms have as many as *seven* shells. The outermost shell in any atom is called its **valence shell**, and it is *this shell that determines an atom's electrical (and chemical) properties*. So, for our purposes, we can now ignore the complex structure of the inner shells.

Regardless of its electron *capacity*, the valence shell can actually only *support* up to eight electrons. So, for example, if the valence shell of a particular atom has a 'capacity' of, say, 18, once it is *actually* occupied by eight electrons, no further electrons can be added. Instead, a *new* valence shell is formed and, once this new valence shell has acquired eight electrons, the previous valence shell will then be able to continue to build up its capacity to 18.

So the valence shell has a maximum capacity of just *eight* electrons. If the valence shell has less than four electrons, then that shell is considered to be *unstable* – by which we mean its electrons are loosely secured within that shell and can easily break away from the atom to become what are then termed 'free electrons'.

In Table 2.1, we list a number of **metallic elements**. For each element, if you examine the column for the outermost shell, you will find that it contains *less than four electrons*. Each of these metallic elements, therefore, can release electrons from their valence shells which then drift haphazardly from atom to atom within the element.

We call these haphazardly drifting electrons '**free electrons**'. Elements with large numbers of free electrons are mainly metallic elements, like those listed in Table 2.1, and are called **conductors**. As we shall learn, conductors provide the free electrons necessary to support electric current.

It would be a mistake to assume that, from Table 2.1, because aluminium has three valence electrons, then it must be a better conductor than, say, copper with just one valence electron. What matters is the *overall number of free electrons*, and this depends on the density of free electrons within the conductor. And copper has a greater density of free electrons than a corresponding volume of aluminium, making copper the better conductor. In fact, the best conductor is silver, followed very closely by copper.

If the valence shell contains *more than four electrons,* then that shell is said to be *stable* and its electrons are held tightly within that shell. The resulting scarcity (in relative terms!) of free electrons causes this type of element to behave as an **insulator**.

Although many *elements* behave as insulators, most *practical* insulators are actually manufactured from *compounds* such as plastics, rubber, glass, ceramics, etc.

Ionisation and ions

When an atom's valence shell temporarily loses or gains an electron, it acquires an electric charge due to the imbalance between its number of electrons and the number of protons contained within its nucleus. We call a charged atom an **ion**, and the process of losing or gaining an electron is called **ionisation**.

If an otherwise neutral atom *loses* an electron (so there are now more protons than electrons), then

Table 2.1

Element	Symbol	Atomic number	Actual occupancies of shells (capacities shown in brackets)						
			k (2)	l (8)	m (18)	n (32)	o (50)	p (72)	q (98)
aluminium	Al	13	2	8	3				
copper	Cu	29	2	8	18	1			
silver	Ag	47	2	8	18	18	1		
mercury	Hg	80	2	8	18	32	18	2	

The 'electron theory' of electricity   **11**

it acquires an overall positive charge and, so, is called a **positive ion** and tends to attract a nearby free electrons – thus becoming neutral again. So within *conductors,* both positive ions and free electrons have a very short lifespan! Immediately an electron breaks away from an atom to become a free electron, that atom becomes a positive ion and attracts a nearby free electron, thus becoming neutral once more.

If an otherwise neutral atom *gains* an electron (so there are more electrons than protons), then it acquires an overall negative charge, and is called a **negative ion** and tends to repel nearby free electrons.

Electric current

Practical conductors, then, are mainly metallic elements with an abundance of free electrons which move extremely rapidly from one atom to another in *haphazard* and *random* directions. You can, if you like, imagine the free electrons forming a sort of negatively charged, rapidly vibrating 'cloud' that fills in the voids between fixed atoms.

In Figure 2.3, we see a length of metal conductor within which free electrons (represented by arrowed dots) move chaotically from atom to atom in random directions.

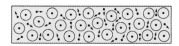
chaotic electron movement

Figure 2.3

Suppose we now apply an **external positive charge** to one end of this length of conductor, and an **external negative charge** at the other end (don't worry, at this stage, *where* these external charges might come from or *how* they are produced; we'll deal with that later).

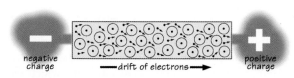
negative charge   —drift of electrons→   positive charge

Figure 2.4

Free electrons, being negatively charged, will be *repelled* by the external negative charge, and *attracted* towards the positive external charge. So, while the free electrons *still* continue to move frantically and haphazardly from atom to atom, *there will now be a general* **tendency** *for the free electrons to drift from*

the negative end of the conductor towards the positive end. This **drift** of free electrons through a metallic conductor is termed an **electric current**.

> In a metallic conductor, an electric current is a drift of free electrons from its negative end, towards its positive end.

Figure 2.4 is actually rather misleading because it suggests that the electrons all move in a relatively orderly 'flow' along the entire length of the conductor when under the influence of the external charges. This *isn't* really the case! What *actually* happens is that the electrons continue with their frantic and chaotic movement, but there is a *very gradual tendency* for them to *drift* towards the positive end of the conductor.

This is illustrated in Figure 2.5. The solid line represents the typical chaotic movement of an individual electron when there are *no* external charges applied to the conductor. The broken line, on the other hand, shows the same electron when external charges *are* applied. As you can see, it *still* follows a frantic and chaotic path, but *its finish point will be a little further towards the positive end of the conductor.* The new finish point is a mere fraction of the diameter of a single atom.

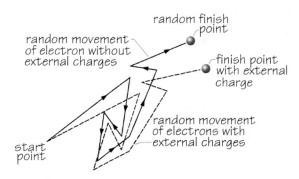
random finish point
random movement of electron without external charges
finish point with external charge
start point
random movement of electrons with external charges

Figure 2.5

So, in *both* cases then, *electrons continue to move chaotically and at great speed* **within** *the conductor.* But, under the influence of external charges, there is a very *gradual drift along* the conductor.

So, while the *effect* of this drift of free electrons, due to their mutual repulsion (like charges repel), is practically *instantaneous* along the entire length of the conductor, the progress of *individual* free electrons along the conductor is *very, v-e-r-y,* slow. So slow, in fact, that *an individual free electron is unlikely to complete its journey through the length of*

a small movement here...   ...is felt immediately...   ...here.

Figure 2.6

a torch's lamp filament during the lifetime of that torch's battery!

The easiest way of visualising this, is to compare this behaviour with that of a row of coupled railway goods wagons, as shown in Figure 2.6: a very small force applied to the wagon at one end of the row will cause a near-instantaneous movement of the wagon at the far end of the row, even though the individual wagons will have moved only a very short distance.

Why conductors are neutral

It's very important to understand that, despite having enormous quantities of free electrons, *a conductor normally remains electrically* **neutral**. This is because, for every single electron (whether 'free' or fixed in an orbit) within the conductor, there is *always* a corresponding proton within the fixed atoms.

So what happens when a free electron exits the conductor at the end connected to an external positive charge? Well, in order to maintain the conductor's neutrality, another free electron is instantaneously drawn into the conductor from the end connected to the external negative charge.

Non-metallic conductors

So far, we have only discussed conduction within *solid metallic conductors*.

But not all conductors are metallic and solid!

For example, some *liquids* are excellent conductors (conducting liquids are called 'electrolytes'). Unlike solids, the atoms in electrolytes are loosely bound together, and are themselves free to move around. So, current flow through electrolytes is *not* due to a drift of electrons, but due to a drift of *ions* (charged atoms) which takes place when the otherwise neutral atoms release free electrons into the external (metallic) circuit.

> In liquid conductors (electrolytes), electric current is a drift of ions (charged atoms).

In electrolytes, a current is not only a drift of ions but, often, a drift of both positive *and* negative ions which move *in opposite directions to each other at the same time!*

In most cases, of course, we are mainly concerned with current flow through *metal* conductors, so we will not spend any further time studying current flow in electrolytes in this particular chapter. Instead, we'll wait until we examine *cells and batteries*, in a later chapter.

However, *it is important to understand that current flow is* **not** *confined just to a flow of free electrons in metals, but can also be a flow of ions in liquids.*

As electrons and ions are generically termed '**charge carriers**', it would be more accurate, therefore, to describe an electric current as *a drift of* **charge carriers.** This definition of current will then apply to solid, liquid or gaseous conductors.

> *An electric current is defined as 'a drift of charge carriers'.*

Strictly speaking, this drift of charges is termed a '**conduction current**', because there is *another* type of current, termed a '**displacement current**', which we shall examine next.

Displacement current in insulators or dielectrics

As we have learnt, there are relatively few free electrons available as charge carriers in insulators so, for most practical purposes, within these materials the drift of free electrons is insignificantly small. We call this tiny drift of electrons a '**leakage current**'.

However, whenever we apply external charges across a sample of insulator, something very interesting happens to the fixed atoms. The atoms themselves cannot move but, under the influence of external charges, *the shape of the orbits of their electrons can!* So, as illustrated in Figure 2.7, the orbits of the electrons within a sample of insulation become distorted, or 'stretched', towards the positive external charge. We say that the atoms have become 'polarised'. The greater the difference between the external charges, the greater the amount of distortion.

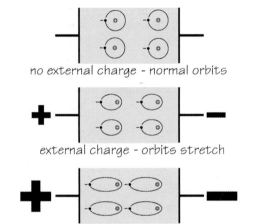

no external charge - normal orbits

external charge - orbits stretch

greater external charge - orbits stretch further

Figure 2.7

This 'stretching' of the electron orbits represents a *momentary* current, which we call a 'displacement current'. So, a displacement current *only* takes place *during* any change to the magnitude of the external charges. If an alternating voltage were applied across an insulating material, then it would be accompanied by a continuous displacement current that would vary in both magnitude and direction, just like that voltage.

A '**displacement current**', then, is associated with insulators, and is a 'momentary' current due to the distortion of the fixed atoms' electron orbits under the influence of external electric charges, whereas a '**conduction current**' is a current due to the movement of free charges (electrons, in the case of metal conductors) through a conductor, or as a tiny 'leakage current' through an insulator.

As we shall learn in a later chapter on *capacitors and capacitance*, displacement currents are very important when dealing with the insulators or dielectrics used,

for example, in the manufacture of circuit devices called capacitors.

'Electron flow' versus 'conventional current' flow

By 'direction of current' in metallic conductors, we mean the direction in which current *passes through the load*. In other words, it is its direction through the external circuit, *never* within the source of potential difference.

During the eighteenth century, the great American scientist and statesman **Benjamin Franklin** (1706–1790), along with others, believed that an electric current was some sort of mysterious 'fluid' that flowed inside a conductor from a high-pressure area to low-pressure area. He naturally labelled high pressure as being 'positive' pressure, and low pressure as being 'negative' pressure and so he believed that an electric current flowed from *'positive to negative'* – i.e. in a direction *opposite* to that of the drift of free electrons!

Franklin's mistaken theory on current direction was, unfortunately, reinforced during the following century, as result of experiments in electrolysis conducted by the English scientist **Michael Faraday** (1791–1867). Electrolysis is a method of depositing ('plating') metal on an electrode immersed in an electrolyte. Faraday noticed during his experiments that metal was removed from the positive electrode and deposited on the negative electrode, from which he, too, concluded that current moved from positive to negative, although he rejected Franklin's idea that it was a 'fluid', in favour of it being a 'field'.

Over the following years, various rules (e.g. to determine the direction of magnetic fields) were devised, based on the mistaken belief that current in metallic conductors flowed from positive to negative. So, as strange as it might seem and despite today's knowledge about current in metal conductors being a flow of free electrons, Franklin's current direction *is still widely used as a convention in a great many textbooks*, and is known as *'Franklinian'* or, more commonly, **conventional flow**.

Electron flow:
A drift of free electrons, from *negative to positive*.
Conventional flow:
'Current' direction from *positive to negative*.

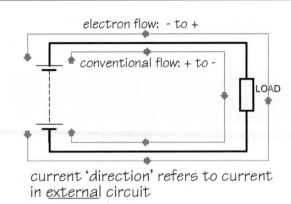

current 'direction' refers to current in <u>external</u> circuit

Figure 2.8

As you will learn later in this book, knowing which direction that a current is flowing is *essential* to understanding the laws of electromagnetism and electromagnetic induction.

So, for consistency with most other textbooks, we have reluctantly adopted **conventional flow**, and *this is the direction that will be used generally throughout this book.*

Having said this, however, there will be occasions when using conventional flow will complicate our understanding of a topic so, on those occasions, we will revert to using electron flow and make it clear that we are doing so.

Practical conductors and insulators

There is no such thing as a 'perfect' conductor or a 'perfect' insulator ('dielectric'). In fact, if we were to make a list of conductors and insulators, we could arrange the list in such a way that the best conductors (the 'worst insulators') appear towards the *top* of the list and the best insulators (the 'worst conductors') appear towards the *bottom* of that list.

Even the best insulators contain impurities, and these impurities contribute a relatively small number of free electrons, which enable tiny currents to flow. Furthermore, thermal activity causes free electrons to be generated within most insulators.

So, in practice, there is simply *no such thing as a 'perfect' insulator*, and the worst enemy of any insulation is high temperature. At high temperatures, even the best insulator will break down and conduct.

In practical terms, the very *best* metallic conductors is **silver** which, because of its cost, is only used for special applications such as relay contacts and printed circuit boards. The most *commonly used* conductors are **copper** and **aluminium**; copper comes a close second to silver and is a better conductor than aluminium but far more expensive – which is why (together with its lower mass) aluminium is preferred for high-voltage transmission and distribution line conductors.

Most practical insulators are compounds rather than elements and include toughened **glass** and **ceramics** (used to insulate overhead power lines), certain **gases** (used to insulate high-voltage busbars), **mineral oils** (used to insulate high-voltage transformers and circuit breakers), or **oil-impregnated paper** and **plastics** (used to insulate electric cables).

Probably the most common insulator is dry air. Thousands of kilometres of overhead lines use bare conductors insulated simply by the surrounding air.

It's important to emphasise that an insulator doesn't 'oppose' an electric current. Rather, it simply *doesn't have sufficient charge carriers available to support an electric current.*

So, what do we mean by 'insufficient' charge carriers? There is a misconception by many students that insulators have 'very few' free electrons. In fact, *it's all relative* – for example, polystyrene is a very good insulator indeed, yet it contains around 60×10^6 electrons per cubic millimetre! Now, that might sound like an absolutely enormous figure but, compared to copper, which contains around 85×10^{18} electrons per cubic millimetre, that figure is actually very small indeed! To put it another way, copper has around 1.42×10^{12} times as many free electrons, per cubic millimetre, compared to polystyrene!

Summary

In this chapter, we have examined Bohr's model of the atom and the part electrons play as charge carriers in metal conductors. In fact, we have gone into a little more detail than is really necessary to understand electric current. So, in this section, we are going to summarise the essential requirement for understanding this topic.

- **Atoms** consist of a nucleus, containing positively charged **protons**, and **neutrons** that carry no charge; the nucleus is surrounded by negatively charged **electrons**, which travel around the nucleus in orbits called 'shells'.

- The amount of positive charge on each proton is identical to the amount of negative charge on each electron. As atoms are normally neutral, it follows that **the number of electrons normally equal the number of protons**.

- If a normally neutral atom temporarily loses an electron, it acquires a positive charge and is called a **positive ion**. If an atom temporarily gains an electron, then it acquires a negative charge and is called a **negative ion**. The process of losing or gaining electrons is called **ionisation**.

- **Like charges repel** and **unlike charges attract**. So, electrons are held within their shells by their attraction to the positively charged protons within the nucleus.

- An **element** is made up of identical atoms determined by the number of protons that their nucleii contain. The simplest element is hydrogen, whose atoms contain just one proton. The most complex 'natural' element is uranium, whose atoms contain 92 protons.

- A **compound** is made up of two or more different elements, whose atoms combine to form molecules. An example of a compound is water, each molecule of which contains two atoms of hydrogen and one of oxygen (H_2O).

- Electron **shells** exist at fixed distances from the nucleus. An electron's energy level determines which shell it will occupy. The further a shell is from the nucleus, the greater the energy level of those electrons that occupy that shell.

- As the energy level of each electron exists in discrete amounts, called 'quanta', electrons cannot exist between shells.

- Shells can only contain specific numbers of electrons – the innermost shell can only contain two electrons, the next eight, etc. (determined by the equation $2n^2$ – where n represents the number of the shell, working from the inner to the outer).

- The outermost shell, called the **valence shell**, determines the electrical characteristic of the atom and *cannot contain more than eight electrons* (*regardless* of its $2n^2$ capacity).

- If the valence shell has less than four electrons, then those electrons are weakly held by their atom and can leave their shell to become **free electrons**.

- Materials with a large number of free electrons are called **conductors**.

- If the valence shell has more than four electrons, then those electrons are strongly held by their atom and *cannot* leave their shell to become **free electrons**.

- Materials with relatively few free electrons (compared to conductors) cannot support conduction and are called **insulators**.

- If an external negative charge is applied to one end of a metallic conductor, and an external positive charge to the other end, the free electrons within the conductor will be repelled by the external negative charge and attracted by the external positive charge and **drift** towards the positive end of the conductor.

- The velocity of electron drift within an insulator is *v-e-r-y* slow!

- This drift of free electrons in a metallic conductor is termed an **electric current** or, more accurately, a 'conduction current'.

- In conducting liquids (electrolytes), an electric current is usually due to a drift of charged atoms, or **ions**.

- In general, then, an electric current is best defined as 'a drift of electric charges'.

- 'Displacement currents', as opposed to 'conduction currents', are associated with insulators or dielectrics. These are momentary currents resulting from the temporary distortion of the electron orbits, under the influence of changes to external charges.

- Before the discovery of atoms, an electric current in metal conductors was thought to be a fluid that 'flowed' from a high-pressure (positive) area to a low-pressure (negative) area. This direction is still used in a great many textbooks and is called '**conventional flow**' to distinguish it from 'electron flow'.

- In common with those textbooks, this book will assume conventional flow, *except where indicated*.

Misconceptions

Atoms look like miniature solar systems
No one has ever seen an atom and no one ever will. Atoms are so complex that they cannot be described in terms that laymen can understand. Scientists are constantly learning new things about the way atoms behave and are discovering more and more new particles within the atom.

Our concept of an atom resembling a tiny solar system is nothing more than a 'model' – in other words, we are trying to describe something we *can't* fully understand in terms of something we *can* understand.

Electrons are tiny particles that orbit the atom's nucleus

Again, this is only a model to help us visualise what 'might' be going on inside the atom. In reality, this is unlikely to be the case. Electrons behave both as charged particles *and* as waves. Sometimes, scientists find it convenient to think of them as charged particles; at other times, they find it convenient to think of them as waves. In reality, they could be neither but something else completely!

An electric current is always a flow of free electrons

This is only true in the case of metallic conductors, such as copper and aluminium. This is not necessarily the case in semiconductors, liquids and gases! A far better definition of current is that it is a 'flow of charges'.

Current flows at the speed of light

While the *effect* of current within a conductor may be detected more or less instantaneously, individual electrons drift along *very* slowly. Research suggests that an individual electron will not travel the length of a flashlight's filament within the lifetime of that flashlight's battery!

Conductors have lots of free electrons, therefore they must be negatively charged

Although conductors do have large numbers of free electrons, for every free electron, there is a corresponding proton within the atoms or positive ions. So conductors don't have an overall charge; they are neutral.

Insulators 'block' current flow

Insulators don't 'block' current flow; they simply don't have sufficient charge carriers to *support* current flow.

Insulators contain few free electrons

Insulators actually contain billions of free electrons per cubic millimetre but, compared to conductors, this figure is relatively small and certainly insufficient to support current flow.

'Conventional flow' is a flow of positive charges in the opposite direction to electrons.

No. 'Conventional flow' isn't a flow of anything. It's simply a 'direction', mistakenly chosen, for current, from positive to negative.

Finally . . .

Now that you have completed this chapter, are you able to achieve the objectives or learning outcomes listed at the beginning of this chapter?

Ask yourself, 'Can I . . .'

1 explain the difference between
   a elements
   b compounds.
2 describe the structure of Bohr's model of an atom.
3 list the electric charges associated with
   a protons
   b electrons.
4 state the laws of attraction and repulsion between electric charges.
5 explain the term 'free electron'.
6 explain the significance of an atom's valence shell.
7 explain what is meant by 'ionisation', and 'positive' and 'negative ions'.
8 explain the difference between conductors and insulators.
9 describe an electric current in metals, in terms of the flow of free electrons.
10 describe how current flow in non-metallic conductors differs from that in metallic conductors.
11 specify the direction of
    a electron flow current
    b conventional current.
12 list examples of and describe applications for
    a practical conductors
    b practical insulators.
13 explain what is meant by the term 'displacement current'.

Chapter 3

Electric current

On completion of this chapter, you should be able to

1 list the three effects of an electric current.
2 specify the SI unit of measurement of electric current.
3 specify which of the three effects is used to define the SI unit of electric current.
4 state the relationship between electric current and electric charge.
5 state the SI unit of measurement of electric charge.
6 solve simple problems on the relationship between electric current and electric charge.

Measuring electric current

Earlier, we learnt that the general definition for **electric current** is *'a drift of electric charges'*.

In the case of metal conductors, it's a drift of free electrons. In other materials, such a conducting liquids ('electrolytes'), it's a drift of ions (charged atoms): it can even be a drift of positive ions in one direction, and negative ions in the opposite direction *at the same time!*

There is nothing obvious to indicate the presence of an electric current in a metal conductor or in any other material; after all, we most certainly won't be able to *see* any charges! Even if it were possible to see these charges, then all we would see would be their chaotic movement; we certainly wouldn't be able to perceive any drift in a particular direction because, as we have learnt, this drift is far too slow!

So the presence of a current can *only* be detected by observing one or more of the *three* **effects** produced by that current. These are the current's

* **heating** effect
* **chemical** effect
* **magnetic** effect.

Heating effect – an electric current drifting through a conductor causes that conductor's temperature to rise. We make practical use of this effect with incandescent lamps, electric heaters, etc. It is also responsible for the operation of fuses, which melt in response to rises in temperature due to excessively high currents. This effect is also responsible for wasteful energy *losses*, due to heat transfer away from a conductor into its surroundings.

Chemical effect – an electric current can be responsible for chemical reactions. This can be useful, and we make use of this effect in electrolysis (electroplating) – but it may also be harmful, as it is also responsible for some types of corrosion.

Magnetic effect – an electric current produces a magnetic field, which surrounds the conductor through which that current is drifting. We make use of the forces resulting from the interactions between magnetic fields to drive motors, operate relays, etc.

The unit of current: the ampere

The SI unit of measurement of electric current is the **ampere** (symbol: **A**), which, as we have learnt, is one of the seven 'base units' of the SI system.

The ampere could be defined in terms of *any* of the three effects described above. And, indeed, since 1947, the definition of the ampere is based on the **force** resulting from the *magnetic effect* of an electric current, as follows:

> The **ampere** (symbol: **A**) is defined as *'that constant current which, when maintained in two straight parallel conductors of infinite length and of negligible circular cross-sectional area, and placed one metre apart in a vacuum, would produce between them a force equal to 2×10^{-7} newton per metre of length'*.

An Introduction to Electrical Science, Waygood, ISBN 9780415810029, 2013. © Taylor & Francis

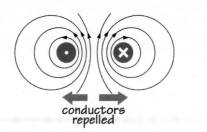
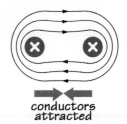

conductors
repelled

conductors
attracted

Figure 3.1

While it's not necessary for you to memorise this definition, you should certainly understand how important it is, for it provides the link between the invisible world of electricity and the highly visible world of mechanics!

The 'force' referred to in this definition is due to the attraction or repulsion (depending in the relative current directions, as seen in Figure 3.1) between two parallel conductors, due to the interaction of the magnetic fields set up around the currents drifting through those conductors.

In Figure 3.1, the dot indicates conventional current drifting *towards* you and the cross represents conventional current drifting *away* from you, and the lines represent the shape of the resulting magnetic fields.

The reason *why* magnetic fields should produce such forces will be discussed in detail in the later chapter on *electromagnetism*. For now, however, you need to accept that a force is indeed set up by the interaction between magnetic fields, and it is the resulting strength of this force that allows us to define the ampere.

> As pointed out in the chapter on the *SI system of measurements*, the ampere is one of the SI base units whose definition may be changed in the future, under proposals by the CIPM.

Electric charge

Until now, we have described charges as being either 'positive' or 'negative', without making any attempt to assign any values to them. So how *do* we measure electric charge?

Well, the smallest quantity of charge must be the amount of negative charge possessed by an *individual* electron which, of course, is *exactly* the same as the amount of positive charge possessed by an *individual* proton.

As these amounts of charge are so incredibly tiny, we need a much larger, and more practical, unit of measurement for electric charge, rather than measuring it in terms of individual electrons – in much the same way in which we measure sugar by the kilogram, *not* by counting its individual granules. We also need to be able to define this new unit in terms of units we have already defined (which is why the SI system is termed a 'rationalised' system of measurement!).

As current is a drift of electric charge, we can define charge in terms of *the amount of charge transported, per second, by a current of one ampere*, and the name we give to this amount of charge is the **coulomb** (symbol: **C**), named in honour of the French academic, Charles de Coulomb (1736–1806).

> The **coulomb** (symbol: **C**), then, is defined as *'the quantity of electric charge transported, per second, by a current of one ampere'*.

We mentioned earlier that the amount of charge on an individual electron is incredibly small; well, to put this in perspective, a coulomb equates to the amount of charge possessed by **6.24×10^{18} electrons** – that's 624 followed by *sixteen* zeros! So, when a current of one ampere is drifting through a conductor, that is the number of electrons which are being transported past a given point in that conductor every second!

If there are 6.24×10^{18} electrons per coulomb, then an individual electron must possess a charge of 160×10^{-21} C.

It's *not* necessary to memorise the definition of a coulomb but, simply, to appreciate that there *is* a relationship between electric charge and electric current. This relationship is written as follows:

$$Q = I\,t$$

where:

Q = quantity of charge, in coulombs (symbol: C)

I = electric current, in amperes (symbol: A)

t = time, in seconds (symbol: s)

If we rearrange this equation, making current the subject, then . . .

$$I = \frac{Q}{t}$$

. . . we see that electric current 'corresponds' to '*the rate of drift of electric charge*'. Unfortunately, some textbooks take this statement one step further, and incorrectly 'define' the ampere as a 'coulomb per second'.

While it is certainly true that an ampere 'corresponds' to an 'ampere per second', this is *not* the SI definition of the ampere, as you *cannot* define a base unit (the ampere) in terms of a derived unit (the coulomb)!

It must be clearly understood that the **ampere** *cannot* be defined as a 'coulomb per second', because you *cannot* define an SI base unit (the ampere) in terms of a derived unit (the coulomb).

Unfortunately, this is precisely what some learned institutions have done in some of their learning materials! It is also very common for North American textbooks to 'define' the ampere as a 'coulomb per second'!

Worked example 1   What quantity of current is drifting if a charge of 25 mC drifts past a point in a circuit every 5 min?

Solution   **Important!** Don't forget, we must first convert millicoulombs (mC) to coulombs (C), and minutes (min) to seconds (s):

$$I = \frac{Q}{t}$$
$$= \frac{(25 \times 10^{-3})}{(5 \times 60)}$$
$$= 83.33 \times 10^{-6}\, \text{A (Answer)}$$

Worked example 2   How many electrons will be transported past a given point in a circuit, when a current of 3 A drifts for 10 s?

Solution
$$Q = It$$
$$= 3 \times 10$$
$$= 30\, \text{C}$$

Since 1 C = 6.24×10^{18} electrons, then:

$$Q = 30 \times (6.24 \times 10^{18})$$
$$= 187.20 \times 10^{18}\, \text{electrons (Answer)}$$

Summary

- The unit of electric current is the **ampere** (symbol: **A**).
- The ampere is defined in terms of the *force* between two, straight, parallel, current-carrying conductors due to the magnetic fields that surround those currents.
- The amount of negative charge transported, per second, by a current of one ampere is one **coulomb** (symbol: **C**).
- The **coulomb** is equivalent to the amount of charge on 6.24×10^{18} electrons.

Misconceptions

The ampere is defined in terms of the rate of drift of electric charge

No! The ampere is defined in terms of the magnetic effect of an electric current – the force between two, parallel, current-carrying conductors, due to their magnetic fields. While the ampere certainly *corresponds* to the rate of drift of electric charge, expressed in coulombs per second, this is *not* how it is defined.

The ampere describes the speed of an electric current

The 'speed' of an electric current has nothing to do with its unit of measurement. Electric charge drifts v-e-r-y slowly, regardless of the amount of current.

Finally . . .

Now that you have completed this chapter, are you able to achieve the objectives or learning outcomes listed at the beginning of this chapter?

Ask yourself, 'Can I . . .'

1   list the three effects of an electric current.
2   specify the SI unit of measurement of electric current.
3   specify which of the three effects is used to define the SI unit of electric current.
4   state the relationship between electric current and electric charge.
5   state the SI unit of measurement of electric charge.
6   solve simple problems on the relationship between electric current and electric charge.

Online resources

The companion website to this book contains further resources relating to this chapter. The website can be accessed via the following link:

www.routledge.com/cw/waygood

Chapter 4

Potential and potential difference

On completion of this chapter, you should be able to

1   explain the need for external negative and positive potentials to cause current flow in a conductor.
2   describe simple electric field patterns.
3   explain the terms, potential and potential difference in terms of charge movement within electric fields.
4   explain what is meant by 'charge separation'.
5   state the SI unit of measurement of potential difference.
6   briefly explain the differences between each of the following terms:
    a   potential difference
    b   voltage
    c   electromotive force
    d   potential.

Introduction

In an earlier chapter, we learnt that if we applied an external *negative* charge to one end of a *metal* conductor, and an external *positive* charge to the other end of that conductor, an electric current – *a drift of free electrons* – will take place along that conductor.

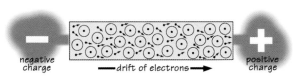

negative charge   ——drift of electrons——▶   positive charge

Figure 4.1

We describe the external negative charge as having a **negative 'potential'**, and the external positive charge as having a **positive 'potential'**. The difference between these potentials, therefore, is called a **'potential difference'**, and it is this which provides the *'driving force'* for current flow.

So, for current to flow between two points,

1   we need a **conducting path** between those two points.
2   there must be a **potential difference** between the two points.

But what exactly do we mean by **'potential'** and **'potential difference'**, and where do they come from? In order to understand these terms, we need a basic understanding of **electric fields**.

Electric fields

The area surrounding an electric charge, in which the effects of that charge may be observed, is termed an **'electric field'**.

An electric field is graphically represented using lines of force, called **'electric flux'**. It must be clearly understood that these electric flux lines are *imaginary* – in other words, the lines themselves don't actually exist, but are used simply to provide us with a 'model' (a visual representation) of an electric field in exactly the same way as *magnetic* lines of force are used to represent a magnetic field (as we shall learn later).

An Introduction to Electrical Science, Waygood, ISBN 9780415810029, 2013. © Taylor & Francis

These flux lines emanate perpendicularly from an electric charge and spread out in all directions towards infinity, as illustrated in Figure 4.2, with individual flux lines repelling adjacent flux lines. Although they are represented in just two dimensions, it must be understood that they actually extend in all *three* dimensions.

It is conventional to allocate a *direction* to electric flux lines using arrow heads and, by common agreement, this direction is determined by the direction in which an isolated, mobile, **positive** charge would move if placed within the field. As a mobile positive charge would be repelled by another positive charge (like charges repel) and attracted towards a negative charge (unlike charges attract), that direction is *along the lines of electric flux, towards a negative charge* – as illustrated by the arrow heads, for a single negative charge, in Figure 4.2.

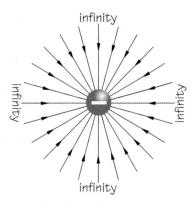

Figure 4.2

If point charges of *opposite* polarity are located near to each other, then these electric flux lines link the two

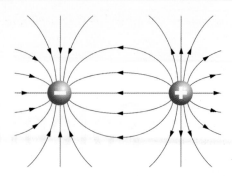

Figure 4.3

charges as well as extending to infinity, as shown in Figure 4.3.

Imagine, now, moving an electron from infinity, along one of these lines of magnetic flux, *towards a fixed negative charge*, as illustrated in Figure 4.4. As 'like poles repel', **work** must be done to overcome the force of repulsion due to the fixed negative charge, in order to move that electron the distance from infinity to point **B**. The work done in moving the electron results in an increase in that electron's **potential energy** or, simply, its '**potential**'.

The absolute potential at *any* point along a line of electric flux is defined in terms of the work done in moving a negative charge (in Figure 4.4, a single electron but, in practice, a negative charge equal to one coulomb) *from infinity to that particular point (point B)*.

Unfortunately, this is *not* a very practical definition for potential – after all, 'infinity' is hardly 'accessible'!

So, instead, we choose an *accessible*, but arbitrary, point of reference – such as point **A** in Figure 4.4 – and then find the work done in transporting the charge *from* that point *to* point **B**. In other words,

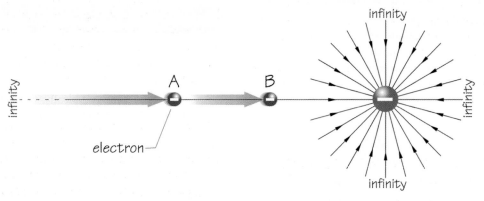

Figure 4.4

we determine the potential at point **B** *with respect to point A*.

Or, to put it another way, it allows us to determine the **potential difference** between points **A** and **B**.

> The **potential difference** between two points in an electric field is defined in terms of the *work done* in transporting *electric charge* between those two points.

The term '**voltage**' is synonymous with 'potential difference' (but *not* potential) – i.e. they both mean *exactly* the same thing! The symbol for potential difference or voltage is E, U or V – depending on context (more on this later).

We shall be examining the differences between potential and potential difference in more detail later in this chapter.

The volt

We are now in a position to define the SI unit of both potential and potential difference, which is the **volt**, named in honour of the Italian physicist, Count Alessandro Volta (1745–1827).

> The **volt** (symbol: **V**) is defined as *'the potential difference between two points such that the energy used in conveying a charge of one coulomb from one point to the other is one joule'*.

Again, you do *not* have to memorise this definition, however we *will* need to refer back to it when we discuss energy, work and power, in a later chapter.

In practice, potential differences can vary enormously. For example, a simple AA disposable battery (or, more accurately, 'cell') will provide a potential difference of just 1.5 V, whereas electricity transmission voltages (in the UK) can be as high as 400 kV.

The preceding definition may be expressed in the form of an equation:

$$E = \frac{W}{Q}$$

where:

E = potential-difference, in volts (symbol: V)

W = work, in joules (symbol: J)

Q = electric charge, in coulombs (symbol: C)

Worked example 1   The work done by a generator in separating a charge of 20 C is 50 kJ. What is the resulting potential difference across its terminals?

Solution   **Important!** Don't forget, we must first convert the kilojoules to joules.

$$E = \frac{W}{Q}$$
$$= \frac{(50 \times 10^3)}{20}$$
$$= 2.5 \times 10^3$$
$$= 2500 \text{ V (Answer)}$$

Creating a potential difference through charge separation

So we now know what we mean by a potential difference and how it is measured. But *how* do we obtain a potential difference in practice?

All materials, including conductors, are usually electrically **neutral** because, under normal circumstances, their atoms contain equal numbers of protons and electrons whose equal, but opposite, charges act to neutralise each other.

In order to acquire a charge, an object must either gain or lose electrons – thereby acquiring an excess or a deficiency of negative charge. For example, if an object has more electrons than protons, it is *negatively charged;* if it has more protons than electrons, then it is *positively charged.*

However, it's *not* necessary for two objects to be *literally* negatively and positively charged for a potential difference to exist between them. For example, if two objects are both negatively charged, but one is *less* negatively charged than the other, then a potential difference will appear between them also.

For example, if object **A** is *less* negative than object **B**, then we can say that object **A** is *'positive with respect to object B'*. Or, if you prefer, *'object B is 'negative with respect to object A'*.

In practice, *this is by far the most common situation we encounter in any circuit,* and is practically *always* the case for electrodes in cells and batteries: with the battery's so-called 'positive' electrode actually being negatively charged, but *less negatively charged* than (or 'positive with respect to') the 'negative' electrode.

The process by which this can be made to happen is called **charge separation**.

There are a great many ways in which charge separation can be achieved, but let's look at just *one* method by considering a simple battery or, more accurately, an electrochemical **cell** (a 'battery' is a number of 'cells' connected together).

A detailed description on how a simple electrochemical cell separates charges in order to create a potential difference is covered in a later chapter on *cells and batteries* but, for now, a simplified description will suffice.

There are many different types of cell but, in its simplest form, a cell consists simply of two dissimilar conductors (e.g. zinc and copper), called 'plates' or 'electrodes', immersed in a conducting liquid (e.g. dilute sulfuric acid) called an '**electrolyte**'.

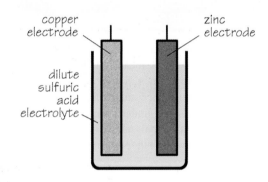

Figure 4.5

When the zinc electrode is inserted into the electrolyte, it reacts chemically with the electrolyte, and starts to dissolve. As zinc dissolves, positively charged zinc ions are released into the electrolyte, leaving electrons behind to accumulate on the zinc electrode – which, therefore, acquires a negative charge.

This action continues until the zinc electrode acquires sufficient negative charge to prevent any further positive ions from escaping.

A similar chemical reaction occurs at the copper electrode, with positive copper ions dissolving into the electrolyte, leaving electrons behind to accumulate on the copper electrode. The reaction of copper, however, is far less vigorous than it is for zinc, and the amount of negative charge acquired by the copper electrode is significantly *less* than the amount of negative charge acquired by the zinc electrode. We say that the copper electrode, therefore, is *'positive with respect to the zinc*

electrode', and is named the 'positive electrode' (or 'positive plate') while the zinc is named the 'negative electrode' (or 'negative plate').

The difference between these two amounts of negative charge results in a potential difference of about 1.1 V appearing between the two electrodes.

This is a gross over-simplification of the chemical process that is *actually* taking place within the cell, but is more than adequate, at this stage, to explain how a chemical cell *separates charges* and *provides a potential difference*. As already mentioned, we will have a more in-depth examination of cells and batteries in a later chapter.

The *open-circuit* potential difference created by the charge separation process is called the **electromotive force (e.m.f.**, symbol: E) of the source. However, when a load current flows, the open-circuit potential difference will reduce somewhat. We will discuss e.m.f. in greater detail in a later chapter.

There are a great many *other* methods of separating charges, including the use of **light** (photovoltaic cells), **pressure** (piezoelectricity), **heat** (thermocouples) and, of course, **magnetism** (generators). The most important of these various methods is magnetism, and we will learn how a generator uses magnetic fields to separate charges later in the book.

How charges gain and lose potential

Now, let's put together what we have learnt about charge separation and electric fields, to find out about the **gain and loss of potential** that takes place in a circuit.

This time, we'll use **gravity** as an analogy to describe what is going on.

Whenever an object is lifted vertically, *against the force of gravity*, then that object is said to *acquire* potential energy. When that object is allowed to *fall under the influence of gravity*, it *gives up* that potential energy in the form of kinetic energy.

Now, let's turn our simple battery on its side, just so we can continue with the 'gravity' analogy.

Figure 4.6 represents a chemical cell turned sideways, with its electric field shown between its two electrodes. To the left, we see a negative charge moving vertically upwards, through the electrolyte*, along one of the flux lines, and *against* the force of repulsion offered by the negative electrode. The work done in moving this negative charge increases the potential of that charge, in much the same way as raising a mass *against* the force of gravity acquires potential energy. The rise in the potential of this negative charge is

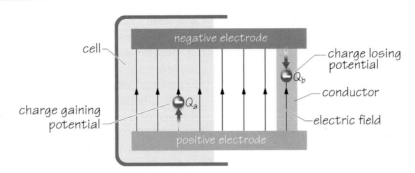

Figure 4.6

acquired at the expense of the energy available through the cell's chemical reaction.

At the same time, another negative charge is being repelled by the negative electrode and is moving through an external conductor that links the two electrodes (for the sake of clarity, no load is shown) – to the right of Figure 4.6. As this negative charge moves 'down' through the conductor, along the flux line towards the positive electrode, the potential that this negative charge acquired moving through the electrolyte, is given up (usually by doing work on the external conductor), in much the same way as a mass gives up its potential energy as it falls under the influence of gravity.

Again, this is a simplification* of what is *actually* taking place, chemically, within the cell itself (as we have learnt, it's *ions* that move through the electrolyte, not electrons), but the principle applies: the potential *acquired* by *any* charge (electrons, ions, whatever) as it moves between electrodes within the cell is then *lost* as it moves through the external conductor.

> *As we have learnt, in the case of a cell, electrons don't actually move through the electrolyte but, rather, ions do. But to reduce the complexity of this topic, we have assumed that an electron moving between the positive and negative electrodes is *equivalent* to the actual movement of ions within the electrolyte.

Voltage drop

Yet another term that we must thoroughly understand is '**voltage drop**'. This is, perhaps, best explained using yet another analogy. Think about a simple central-heating **system** that uses hot water flowing through, say, three radiator panels.

In order for the water to flow we need a **difference in pressure** across the entire heating system, which is provided by a pump.

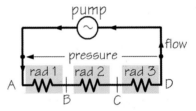

Figure 4.7

The pump produces a difference in pressure across *all three radiators* and causes the hot water to flow through the entire *system*. But at the same time, for water to flow through each *individual* radiator, there must be *a difference in pressure between its inlet and its outlet* – in other words, as well as the difference in pressure across *all three* radiators (i.e. A–D), there must also be individual pressure differences across radiator 1 (i.e. A–B), *and* across radiator 2 (i.e. B–C), *and* across radiator 3 (i.e. C–D). And, obviously, *the sum of these individual pressure differences must equal the pressure difference across the system:*

i.e. **pressure**$_{(A-D)}$ = **pressure**$_{(A-B)}$ + **pressure**$_{(B-C)}$ + **pressure**$_{(C-D)}$

Now, let's compare this heating system with an electric circuit. Instead of a pump, we have a cell or battery and, instead of radiators, we have three lamps.

The cell produces a potential difference across *all three lamps* which causes current to flow through the circuit. However, for current to flow through each

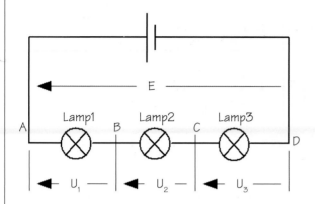

Figure 4.8

individual lamp, there must also be *a difference in potential across each lamp* too – in other words, as well as the potential difference across *all* three lamps (*E*), there must also be individual potential differences across lamp 1 (i.e. U_1), *and* across lamp 2 (i.e. U_2) and across lamp 3 (i.e. U_3). And, obviously, *the sum of these individual potential differences must equal the electromotive force across the complete circuit:*

$$\text{i.e. } E = U_1 + U_2 + U_3$$

These individual potential differences that appear across individual circuit components are known as **'voltage drops'**. We will learn more about this, in much greater detail, in a later chapter on *electric circuits*.

Potential v potential difference (voltage)

Earlier in this chapter, we learnt that the absolute **potential** at any point within an electric field is the energy required to move a charge of one coulomb from *infinity* to that point. But we also learned that this is an impractical means of defining the potential at any particular point. In practice, the potential at any point can be measured from *any* convenient reference point we care to choose. By general consent, for most practical circuits, the 'zero-reference' point for measuring potential is usually **earth**.

This doesn't mean that the earth is *literally* at 'zero potential', it simply means measurements of potential are made *with respect to* (or compared to the potential of) earth.

It's *very* important that you understand the difference between 'potential' and 'potential difference' – as you

will learn later, it's particularly important for anyone dealing with earthed systems, where it is quite common to measure *potentials* with respect to earth. It's also important for vehicle electricians, who regularly have to measure potentials with respect not to 'true earth', but to 'chassis earth' (where the negative terminal of the battery is connected to the common metal parts or 'chassis' of a vehicle).

To help us reinforce our understanding of the difference between potential and potential difference, we will use yet another analogy which, this time, compares 'potential' and 'potential difference' with the terms 'height' and 'difference in height', where

- 'potential' is equivalent to '**height**'
- 'potential difference' is equivalent to '**difference in height**'.

Whenever we talk about the 'height' of an object, we have to measure it from some agreed **datum point**. For example, we normally measure the height of a mountain, using *sea level* as its datum point. So, the heights of two points, *A* and *B*, on a mountain are normally measured from (or 'with respect to') sea level.

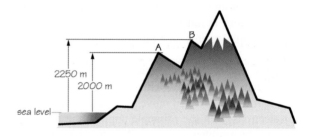

Figure 4.9

Suppose the *height* of point *A* is 2000 m from sea level, and the *height* of point *B* is 2250 m from sea level. We can then say that the *difference in height* between points *A* and *B* is 250 m (i.e. 2250 –2000).

If we had chosen a completely different datum point (say, for example, we chose point *A*) then the *heights* of points *A* and *B* would be completely different, with the height of point *A* now being zero metres, and the height of point *B* being *plus* 250 m. On the other hand, if we had chosen point *B* as the datum point, then the height of point *B* would be zero metres, and the height of point *A* would be *minus* 250 m. The *difference in heights* between points *A* and *B* would *always* remain the same. In other words, 'height' is *relative* (i.e. it depends from where it is measured), whereas 'difference in height' is *absolute*.

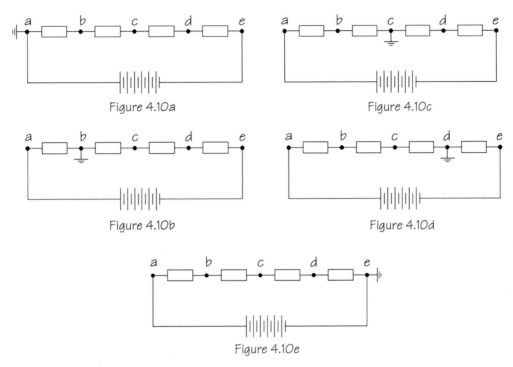

Figure 4.10a

Figure 4.10c

Figure 4.10b

Figure 4.10d

Figure 4.10e

Figure 4.10

Let's compare this use of the terms 'height' and 'difference in height' with 'potential' and 'potential difference'.

Like 'height', 'potential' exists at a *particular point* in a circuit and is measured from some agreed datum point. The datum point could be *anywhere* (another point in the same circuit, for example), but the most commonly used datum point in electrical-distribution systems is the **earth**.

The general mass of the earth is considered to be a conductor and, by *convention*, is allocated a potential of zero volts. Again, it's important to understand that the *actual* potential of earth isn't necessarily zero; but this doesn't matter, it's simply zero by convention! For example, in an electrical installation, the neutral conductor is connected to earth and, so, behaves as the point of reference for measurements of potential made at any point along the line conductor.

'Potential difference' means the difference between the potentials at two separate points in a circuit. So, whereas 'potential' exists at a *single* point and its value depends upon where the reference is located, 'potential difference' is measured between *two* points in a circuit, and *is independent of the datum point*. Furthermore, we can describe potential as having 'positive' or

'negative' polarity relative to the chosen datum point. We *cannot* allocate charge polarity to a potential *difference*.

So, 'potential' is *relative*, whereas 'potential difference' is *absolute*.

Let's look at the examples in Figure 4.10. Each represents a circuit supplied by a 200-V supply. Let's assume that each resistor has the same value, in which case (as we shall learn later) the voltage drop across each resistor will be identical: 50 V.

Note how, in this example, the potentials at point *a*, *b*, *c*, *d* and *e*, depend upon where the earth reference is, whereas the potential difference across each resistor remains 50 V regardless of the position of the earth connection.

Figure 4.10a:

- the potential at point *a* with respect to earth is **zero** (because its *directly connected* to earth)
- the potential at point *b* with respect to earth is **– 50 V**
- the potential at point *c* with respect to earth is **– 100 V**
- the potential at point *d* with respect to earth is **– 150 V**

- the potential at point *e* with respect to earth is – 200 V

(note that each potential is 'negative', because each point is located between the earthed point and the negative terminal of the battery).

Figure 4.10b:

- the potential at point *a* with respect to earth is + 50
- the potential at point *b* with respect to earth is zero
- the potential at point *c* with respect to earth is – 50 V
- the potential at point *d* with respect to earth is – 100 V
- the potential at point *e* with respect to earth is – 150 V

Figure 4.10c:

- the potential at point *a* with respect to earth is +100
- the potential at point *b* with respect to earth is + 50 V
- the potential at point *c* with respect to earth is zero
- the potential at point *d* with respect to earth is – 50 V
- the potential at point *e* with respect to earth is – 100 V

Figure 4.10d:

- the potential at point *a* with respect to earth is + 150
- the potential at point *b* with respect to earth is +100 V
- the potential at point *c* with respect to earth is + 50 V
- the potential at point *d* with respect to earth is zero

- the potential at point *e* with respect to earth is – 50 V

Figure 4.10e:

- the potential at point *a* with respect to earth is + 200
- the potential at point *b* with respect to earth is +150 V
- the potential at point *c* with respect to earth is + 100 V
- the potential at point *d* with respect to earth is + 50 V
- the potential at point *e* with respect to earth is zero

You will note that the polarity of the various potentials in the above examples change, *according to where the earth is connected in the circuit*. So it's important to understand that these 'polarities' are not *absolute*, but are *relative to the other points within the circuit*. For example, a point that is labelled '+' isn't necessarily positive in the sense that there are less electrons than protons at that particular point; it simply means that it is 'less negative' than another point.

So, while the potential at any point varies, depending on the position of its point of reference (in the above examples, the earth), the potential difference across each resistor remains at 50 V. You can go ahead and confirm this for yourself if you wish, by simply subtracting two adjacent potentials in any of the above examples – the answer will *always* be 50 V.

Summary

'Voltage' is simply another word for '**potential difference**'; i.e. they are synonyms. '**Voltage**' is *not* another term for '**potential**'.

Table 4.1

potential difference	The difference in potentials between any two points in a circuit. Symbol: E or U – depending on context.
voltage	A synonym for 'potential difference'. Symbol: E or U – depending on context.
electromotive force	The potential difference produced, internally, by a battery, generator, etc., and which appears across its terminals *when it is not supplying a load*. Symbol: E.
voltage drop	The potential difference across an individual circuit component, such as a resistor, responsible for current flow through that component. Symbol: U_1, U_2, etc.
potential	Potential exists at a single point in a circuit, and is measured relative to another randomly chosen point (in practice, often earth). Potential is either negative or positive with respect to the point of reference. Symbol: U.

- It is *correct* to say, 'The voltage *across* a resistor is so-many volts'.
- It is *incorrect* to say, 'The voltage *at a point* is so-many volts'; instead, we should say, 'The *potential* at a point....'.
- It is *correct* to say, 'The voltage *between* line conductor and earth is 230 V'.
- It is *incorrect* to say, 'The *voltage* of the line conductor with respect to earth is 230 V'; instead, we should say, 'The *potential* of the line conductor with respect to earth is 230 V'.

Traditionally, in the English-speaking world, the symbol for **potential difference**, or **voltage**, has always been '*V*', and the symbol for **electromotive force** has been '*E*' – these are what you will see in most electrical science textbooks.

However, the ***IET Wiring Regulations*** have adopted the European standard (formerly a German standard) symbol, '*U*', in place of '*V*', and that is what we have done throughout this book.

Finally ...

Now that you have completed this chapter, are you able to achieve the objectives or learning outcomes listed at the beginning of this chapter?

Ask yourself, 'Can I ...'

1 explain the need for external negative and positive potentials to cause current flow in a conductor.
2 describe simple electric field patterns.
3 explain the terms, potential and potential difference in terms of charge movement within electric fields.
4 explain what is meant by 'charge separation'.
5 state the SI unit of measurement of potential difference.
6 briefly explain the differences between each of the following terms:
   a potential difference
   b voltage
   c electromotive force
   d potential.

Online resources

The companion website to this book contains further resources relating to this chapter. The website can be accessed via the following link:
www.routledge.com/cw/waygood

Chapter 5

Resistance

On completion of this chapter, you should be able to

1 explain the term 'resistance'.
2 state the SI unit of measurement of resistance.
3 list the factors that affect the resistance of a material.
4 state the effect upon resistance of varying the
   a length of a material.
   b cross-sectional area of a material.
5 define the term 'resistivity'.
6 state the SI unit of measurement of resistivity.
7 solve simple problems, based on the equation:
   $R = \rho \dfrac{l}{A}$.
8 explain how the above equation relates to conductor insulation.
9 explain, in general terms, the effect of an increase in temperature upon the resistance of:
   a pure metal conductors.
   b insulators.
10 explain how resistance values are indicated on schematic symbols.

Resistance

Resistance is the natural opposition offered by *any* material to the drift of an electric current through that material. In this chapter, we will be examining the resistance of conductors and insulators. The resistance of semiconductors and fluids (liquids and gases) behaves differently, and is beyond the scope of this book.

The use of the term 'resistance', in the sense of meaning the *'opposition to the drift of current'*, is credited to the German schoolmaster, Georg Simon Ohm (1789–1854). However, the *concept* of electrical resistance predates Ohm, and was the subject of experiments by the eccentric English physicist, Henry Cavendish (1731–1810), after whom the famous 'Cavendish Laboratory' at Cambridge University is named. Long before the days of electrical measuring instruments, Cavendish studied the effects of the 'opposition to current' by different conductors by subjecting himself to a series of electric shocks – the more intense the shock, the lower the material's opposition to current! Judging from his meticulous and extensive research notes, Cavendish must have subjected himself to *thousands* of such shocks! And, rather surprisingly, his results apparently compare remarkably well with what we know today about electrical resistance.

Resistance (symbol: ***R***) is, to some extent, dependent upon the quantity of free electrons available as charge carriers within a given volume of conductor or insulator, and the opposition to the drift of those free electrons due to the fixed atomic structure and forces within those materials.

For example, conductors have very large numbers of free electrons available as charge carriers and, therefore, have low values of resistance. On the other hand, insulators have relatively few free electrons in comparison with conductors, and, therefore, have very high values of resistance.

But resistance is also the result of collisions between free electrons drifting through the conductor under the influence of the external electric field, and the stationary atoms. Such collisions represent a considerable reduction in the velocity of these electrons, with the resulting loss of their kinetic energy contributing to the rise of the conductor's temperature. So it can be said that *the consequence of resistance is heat.*

Resistance, therefore, can be considered to be a *useful* property as it is responsible for the operation of incandescent lamps, heaters, etc. On the other hand, resistance is also responsible for temperature increases in conductors which results in heat transfer *away* from

An Introduction to Electrical Science, Waygood, ISBN 9780415810029, 2013. © Taylor & Francis

those conductors into their surroundings – we call these energy *losses*, which are undesirable.

We can modify the natural resistance of any circuit by adding **resistors**. These are circuit components, which are manufactured to have specific amounts of resistance, and are used to change the resistance of a circuit for various reasons, such as to modify or to limit the current flowing through the circuit.

Conductance

You may come across the term **conductance** (symbol: *G*). Conductance is the *reciprocal* of resistance, that is:

$$G = \frac{1}{R}$$

Until the adoption of SI, the unit of measurement of conductance was the **mho** – that's 'ohm', spelt backwards (proving that some scientists do, indeed, have a sense of humour)! The SI unit of measurement for conductance, however, is the **siemens** (symbol: **S**).

Conductance is a particularly useful concept to use when we study a.c. theory in a later chapter.

The unit of resistance: the ohm

The SI unit of measurement of resistance is the **ohm** (symbol: Ω), named in honour of Georg Simon Ohm.

The **ohm** is defined as *'the electrical resistance between two points along a conductor such that, when a constant potential difference of one volt is applied between those points, a current of one ampere results'*.

In other words, an **ohm** is equivalent to a *volt per ampere*, and you should understand the significance of this very important definition, as it will become important when we study *Ohm's Law* in a later chapter.

The **resistance** of any material depends upon the following factors:

*   its **length** (symbol: *l*)
*   its **cross-sectional area** (symbol: *A*)
*   its **resistivity** (symbol: ρ, pronounced '*rho*').

Length

The resistance of a material is *directly proportional to its length*. In other words, *doubling* the length of a conductor will *double* its resistance, while *halving* its length will *halve* its resistance, etc.

Area

The resistance of a material is also *inversely proportional to its cross-sectional area*. In other words, doubling its cross-sectional area will *halve* its resistance, while halving its cross-sectional area will *double* its resistance, etc.

Important! The area of a circle is proportional to the *square of its diameter*. So doubling the diameter of a circular-section wire, will actually *quadruple* its cross-sectional area and, therefore, *reduce its resistance to a quarter*. For the same reason, halving the diameter of a wire will *quarter* its cross-sectional area and *quadruple its resistance*.

Resistivity

Knowing how the length and cross-sectional area affects resistance, we can now express these relationships, mathematically, as follows:

$$R \propto \frac{l}{A}$$

As you know, we can change a proportion sign ($\propto$) to an equals sign (=) by inserting a *constant of proportionality:*

$$R = \text{constant} \times \frac{l}{A}$$

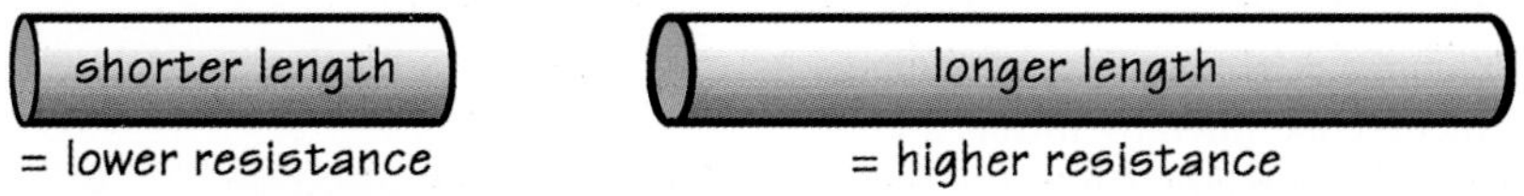

shorter length
= lower resistance

longer length
= higher resistance

Figure 5.1

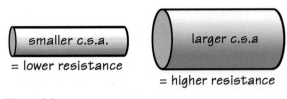

Figure 5.2

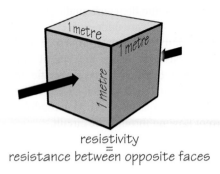

resistivity
=
resistance between opposite faces

Figure 5.3

This constant is called the **resistivity** (symbol: ρ, pronounced 'rho') of the material. So, the final equation for resistance becomes:

$$R = \rho \frac{l}{A}$$

where:

R = resistance, in ohms (symbol: Ω)

ρ = resistivity, in ohm metres (symbol: $\Omega \cdot m$)

l = length, in metres (symbol: m)

A = cross-sectional area, in square metres (symbol: m^2)

So what is the real significance of **resistivity**?

Well, the problem with *resistance* is that it depends not only on the material from which a conductor is made, but also on the physical dimensions of that conductor. As we have learnt, if we were to increase the length of a conductor, then the measured resistance would also increase; if we were to increase the cross-sectional area of the conductor, then the measured resistance would decrease.

Resistivity allows us to compare different materials' abilities to pass electrical current that is independent from these geometrical factors.

Resistivity is a *physical property* of a material, which varies from material to material, and is affected by **temperature**. *Accordingly, values of resistivity are always quoted at a particular temperature – usually, 293 K (20°C).*

> **Resistivity** is defined as *'the resistance of a unit length of a substance with a uniform cross-section'.*

In SI, the above definition corresponds to *'the resistance between the opposite faces of a one-metre cube of material'.* It's important to understand that a 'metre cube', as shown in Figure 5.3, is *not* the same thing as a 'cubic metre'. A 'metre cube' literally means a cube having sides each measuring a metre, whereas a 'cubic metre' has the same volume, but can be of *any* shape.

Resistivity, then, is a fundamental property of a material that describes how easily that material can allow the passage of an electrical current. High values of resistivity imply that the material is very resistant to the drift of current; low values of resistivity imply that the material allows current to pass very easily.

In fact, materials are classified as being a 'conductor', an 'insulator' or a 'semiconductor' (semiconductors are materials used in the manufacture of electronic components, such as diodes and transistors) *according to its range of resistivities.*

As you can see from Table 5.1, conductors represent just a *tiny* range of resistivity values, compared to those of semiconductors and insulators.

Table 5.2 lists the resistivity values (at 20°C) of some common conductors.

Table 5.1

Material	Resistivity / $\Omega \cdot m$
conductors	10^{-8}–10^{-6}
semiconductors	10^{-6}–10^{7}
insulators	10^{7}–10^{23}

Table 5.2

Material	Resistivity / $\Omega \cdot m$
silver	16.4×10^{-9}
copper	17.5×10^{-9}
aluminium	28.5×10^{-9}
tungsten	5.6×10^{-8}
carbon	3.5×10^{-5}

Worked example 1 Using Table 5.2, calculate the resistance of 1000 m of copper wire of cross-sectional area 1.5 mm².

Solution **Important!** Don't forget to convert square millimetres into square metres.

$$R = \rho \frac{l}{A}$$

$$= (17.5 \times 10^{-9-3}) \times \frac{1000}{(1.5 \times 10^{-6})}$$

$$= \frac{17.5 \times 10^{-3} \times 1000}{1.5}$$

$$= 11.7 \ \Omega \ \text{(Answer)}$$

Worked example 2 Using Table 5.2, calculate the resistance of 1000 m of aluminium wire of cross-sectional area 1.5 mm².

Solution **Important!** Don't forget to convert square millimetres into square metres.

$$R = \rho \frac{l}{A}$$

$$= (28.5 \times 10^{-9}) \times \frac{1000}{(1.5 \times 10^{-6})}$$

$$= \frac{28.5 \times 10^{-3} \times 1000}{1.5}$$

$$= 19.0 \ \Omega \ \text{(Answer)}$$

Worked example 3 *Manganin* is a metal alloy comprising copper, manganese and nickel. Its resistance remains approximately constant over a wide range of temperatures, and it is used in the manufacture of wire-wound resistors. What length of manganin wire, of diameter 1 mm, would be required to make a resistor with a resistance of 0.7 Ω? Its resistivity is 0.4 μΩ·m.

Solution **Important!** Don't forget to convert microhm metres to ohm metres, and the diameter in millimetres into an area in square metres.

$$R = \rho \frac{l}{A}$$

$$l = \frac{RA}{\rho}$$

$$= \frac{R\pi r^2}{\rho}$$

$$= \frac{0.7\pi (0.5 \times 10^{-3})^2}{0.4 \times 10^{-6}}$$

$$= \frac{0.7\pi \times 0.25 \times 10^{-6}}{0.4 \times 10^{-6}}$$

$$= 1.37 \ \text{m (Answer)}$$

Worked example 4 A circular section copper cable has a resistance of 0.5 Ω. What will be the resistance of a copper cable of the same length but of twice its diameter?

Solution We know that the resistance of a conductor is *inversely proportional* to its cross-sectional area. So, what happens to the cross-sectional area of a circular-section cable if its *diameter* is doubled?

Well, doubling its diameter will *quadruple* its cross-sectional area. So, in this example, *the cable's resistance will be reduced by a factor of 4.*

$$\text{resistance} = \frac{\text{original resistance}}{4}$$

$$= \frac{0.5}{4}$$

$$= 0.125 \ \Omega \ \text{(Answer)}$$

The resistivity of insulators is *not* necessarily constant for a particular temperature, as it is with conductors. Instead, it generally varies considerably according (amongst other things) to the insulator's purity, its surface condition and the duration of the application of a potential difference. Furthermore the resistivity of most insulators *decreases* with an *increase* in temperature.

Table 5.3 lists typical resistivities of common insulators.

Table 5.3

Material	Resistivity/Ω·m
Air	13×10^{15} to 33×10^{15}
Glass	100×10^{9} to 1×10^{15}
Porcelain	1×10^{12}
Mica	10×10^{12}
Polystyrene	100×10^{12}
Teflon	100×10^{21} to 10×10^{24}

The enormous differences between the resistivities of insulators and conductors result in *massive* differences in their values of resistance. Comparing the resistance values of conductors and insulators will reveal some truly astonishing figures – as the following worked example will illustrate.

Worked example 5

a Determine the resistance of a 25-mm length of mica, of cross-sectional area 2.5 mm², if its resistivity is 10×10^{12}.

b What length of copper conductor, having the same cross-sectional area, would have the same resistance as this sample of mica (the resistivity of copper is 17.5×10^{-7})?

Solution   For **mica**:

$$R_{mica} = \rho \frac{l}{A} = (10 \times 10^{12}) \times \frac{25 \times 10^{-3}}{2.5 \times 10^{-6}}$$

$$= 100 \times 10^{15} \ \Omega \ \text{(Answer a.)}$$

For **copper**:

$$R_{copper} = \rho \frac{l}{A}$$

$$l = \frac{R_{copper} A}{\rho} = \frac{(100 \times 10^{15}) \times (2.5 \times 10^{-6})}{17.5 \times 10^{-9}}$$

$$= 14.29 \times 10^{18} \ \text{m (Answer b.)}$$

So, from the above example, sample of mica, just 25 mm long, has the same resistance as a copper conductor, having the same cross-sectional area, but measuring a staggering **14 290 000 000 000 000 kilometres** long!

To put this in some sort of perspective, the average distance between the Earth and the Sun is just 150 000 000 km!

The above worked example dramatically illustrates the scale of difference between the resistance of conductors and the resistance of insulators.

Before we leave the subject of resistivity, we need to know that the reciprocal of resistivity is called '**conductivity**', which is measured in siemens per metre (S/m).

Resistors

The **circuit symbol** for both *resistance* (i.e. the *quantity*) and *resistors* (i.e. the *circuit component*) is the same. The European standard symbol for both **resistance** (i.e. the quantity) and for a **fixed-value resistor** (i.e. the electrical component) is show to the left, and the alternative ('non-preferred') symbol (which is the standard symbol in the United States) is shown to the right of Figure 5.4.

Figure 5.4

On schematic diagrams, the value of resistance, in **ohms**, is shown adjacent to its symbol, often in the form illustrated in Figure 5.5.

15R

Figure 5.5

In this particular example, the letter 'R' acts as a *multipler* (in this case, ×1) while its *placement*, relative to the number, represents the position of a decimal point. Similarly, the letter 'K' represents the multiplier ×1000, and the letter 'M' represents the multiplier ×1 000 000. This system for indicating resistance values is explained in Table 5.4.

Table 5.4

15R	=	15 Ω
1R5	=	1.5 Ω
R15	=	0.15 Ω
15K	=	15 kΩ
1K5	=	1.5 kΩ
K15	=	0.15 kΩ
15M	=	15 MΩ
1M5	=	1.5 MΩ
M15	=	0.15 MΩ

Fixed-value resistors are manufactured in a wide range of shapes and sizes. Inexpensive resistors are manufactured from short rods of carbon mixed with an insulating material; more expensive and accurate resistors are made from ceramic rods coated with a thin film of carbon ('cracked carbon' types) or metal ('metal film' types); resistors designed to carry larger currents are manufactured from lengths of resistance wire, wound around ceramic tubes. The physical size of a resistor determines its ability to dissipate heat and, therefore, its power rating.

A **variable resistor** has *three* terminals, enabling it to be connected in such a way as to control current or to control voltage. When connected to control *current*, it is being used as a **rheostat**; when used to control *voltage*, it is being used as a **potentiometer**. The terms 'rheostat' and 'potentiometer', then, refer to their *applications* and *not* to the variable resistors themselves (despite the common use of the word 'pot' to describe a variable resistor!).

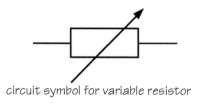

circuit symbol for variable resistor

Figure 5.6

General effect of temperature upon resistance

Resistance will vary with **temperature** because, as we have learned, a material's *resistivity* is affected by temperature. It is quite possible to calculate the resistance of a material at *any* temperature but, in this chapter, it

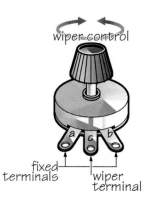

Figure 5.7

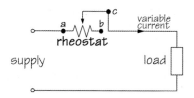

Figure 5.8

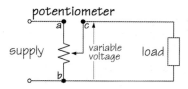

Figure 5.9

is only necessary to learn some *general* rules for pure-metal conductors, insulators, carbon, and certain alloys.

The 'hot' (operating) resistance of an incandescent lamp's tungsten filament is typically around 10–18 times greater than that of its 'cold' resistance. So this type of lamp takes a significantly greater current at the instant it is switched on, compared to its operating current – this is *not* a problem, however, as it only takes less than 0.1 s for the current to fall to, and remain at, its operating value.

Because the resistance of pure metal conductors increases with temperature, special alloys (such as *manganin* and *constantan*) whose resistance remains

Table 5.5

Pure metal conductors	an increase in temperature of a pure metal conductor will cause its resistance to *increase*.
Insulators and carbon	an increase in temperature of an insulator, and of carbon (a conductor) will cause their resistance to *decrease*.
Alloys	certain alloys, such as *constantan* (copper with 10–55% nickel) are manufactured so that an increase in temperature has very little effect upon their resistance.

approximately constant over a wide range of temperatures have been developed for applications where a constant resistance is important – e.g. precision wire-wound resistors.

Temperature increase, therefore, has an adverse effect upon insulators. As temperature increases, an insulator's resistance will decrease and, eventually, may lead to a catastrophic breakdown of its insulating properties. *Excessive temperature is the major cause of insulation failure*. It's essential, therefore, that the ventilation features of electrical devices should never be obstructed.

'A.C. resistance' due to the 'skin effect'

As we have learned, the resistance of a metal conductor is inversely proportional to the cross-sectional area of that conductor – i.e. the *lower* the cross-sectional area, the *greater* the resulting resistance.

When a direct current drifts through a conductor, free electrons are distributed uniformly across the cross-section of the conductor. However, *this isn't the case with alternating current*.

For reasons that are beyond the scope of this particular chapter, whenever an *alternating* current drifts, the free electrons tend to travel closer to the surface of the conductor – we call this the '**skin effect**'.

The 'skin effect' becomes increasingly pronounced at higher frequencies – in other words, at higher frequencies, the charge carriers tend to drift even

closer to the surface of the conductor, reducing the effective cross-sectional area even more. This is illustrated in Figure 5.10, where the grey area represents the distribution of charge carriers across a section of conductor.

The effect of the 'skin effect', therefore, is to *reduce the effective cross-sectional area of a conductor, thus increasing the resistance of the conductor*. The resistance of a conductor to the drift of an alternating current, therefore, is somewhat higher than it would be to a drift of direct current, and we call this higher value its '**a.c. resistance**' to distinguish it from its resistance to direct current. 'A.C. resistance' should *not* be confused with another form of opposition to a.c. current, called 'reactance'.

At normal mains frequency (50/60 Hz), the difference between a.c. and d.c. resistance is slight, but it increases significantly at higher frequencies. In fact, at radio frequencies (300 MHz–300 GHz), the skin effect is so pronounced that there is little point in using solid conductors, so tubes (called 'waveguides') are used instead.

Resistance of conductor insulation

The resistance of the **insulating material** surrounding a conductor conforms to *exactly* the same equation as we have already met, i.e:

$$R \propto \frac{l}{A}$$

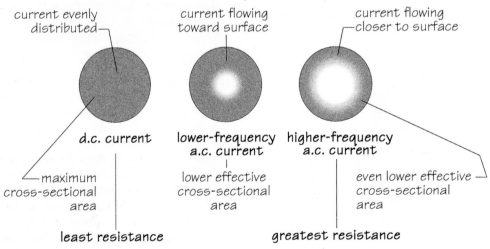

cross-section through a conductor

current evenly distributed

current flowing toward surface

current flowing closer to surface

d.c. current

lower-frequency a.c. current

higher-frequency a.c. current

maximum cross-sectional area

lower effective cross-sectional area

even lower effective cross-sectional area

least resistance

greatest resistance

Figure 5.10

As there is no such thing as a perfect insulator, a tiny current (in the order of microamperes), called a **leakage current**, passes through the insulation *from* the conductor to the surroundings (earth, or adjacent conductors).

So, in this context, the 'length' *(l)* of the insulation corresponds to its *thickness* (i.e. the path through which the leakage current passes), and the 'area' *(A)* of the insulation is the product of the *average circumference of the insulation* and the *length of the conductor* – as illustrated in Figure 5.11.

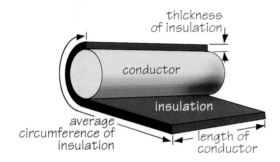

Figure 5.11

So,

$$R_{insulation} \propto \frac{l}{A} \propto \frac{\text{thickness of insulation}}{\left(\begin{array}{c}\text{average circumference of insulation}\\ \times \text{length of conductor}\end{array}\right)}$$

As the thickness of the insulation and the circumference of any given conductor are fixed values then, for any given cable, its resistance:

$$R \propto \frac{1}{\text{length of conductor}}$$

In other words, the insulation resistance is *inversely proportional to the length of the insulated conductor* – i.e. if you, say, double the length of the conductor, then its insulation resistance will be halved!

Misconceptions

Resistance 'slows down' current
No, resistance doesn't act to 'slow down' current. Its effect is to oppose the drift of current, and acts to limit the value of current .

Resistance 'blocks' current
No, resistance doesn't 'block' current, it simply limits its value!

Doubling the diameter of a conductor will halve its resistance
No! Resistance is inversely proportional to the cross-sectional area of a conductor. In other words, doubling a conductor's cross-sectional area will halve its resistance. In the case of a circular-section conductor, doubling its diameter will increase its cross-sectional area by a factor of four! So, *doubling* a conductor's diameter will reduce its resistance by a factor of *four*.

The longer a wire, the greater its insulation resistance
Resistance is directly proportional to the length of a material and inversely proportional to its cross-sectional area. In the case of an insulator surrounding a conductor, 'length' is equivalent to its thickness, and its cross-sectional area is determined by the length of the conductor multiplied by the average circumference of the insulator – in other words, the insulation's 'cross-sectional area' increases with conductor length. So, the longer the conductor, the lower the insulation resistance.

Temperature affects resistance
Yes, but indirectly. Actually, temperature affects *resistivity* which, in turn, affects resistance.

Finally . . .

Now that you have completed this chapter, are you able to achieve the objectives or learning outcomes listed at the beginning of this chapter?

Ask yourself, 'Can I . . .'

1 explain the term 'resistance'.
2 state the SI unit of measurement of resistance.
3 list the factors that affect the resistance of a material.
4 state the effect upon resistance of varying the
   a length of a material.
   b cross-sectional area of a material.
5 define the term 'resistivity'.
6 state the SI unit of measurement of resistivity.
7 solve simple problems, based on the equation:
   $$R = \rho\frac{l}{A}.$$

8 explain how the above equation relates to conductor insulation.

9 explain, in general terms, the effect of an increase in temperature upon the resistance of:

a pure metal conductors.

b insulators.

10 explain how resistance values are indicated on schematic symbols.

Online resources

The companion website to this book contains further resources relating to this chapter. The website can be accessed via the following link:

www.routledge.com/cw/waygood

Chapter 6

Practical conductors and insulators

On completion of this chapter, you should be able to

1. identify, and state the function of, each of the main components of an electric cable.
2. explain why some conductors are stranded rather than solid.
3. explain how the cross-sectional area of a stranded conductor may be determined.
4. explain why the sectional shape of larger multicore cable conductors are frequently 'wedge' shaped, rather than circular.
5. list the main properties required of a cable's insulation.
6. explain the relationship between a cable's length and the resistance of its insulation.
7. identify the function of a cable conductor by the colour of its insulation.

Introduction

An electric **cable** normally consists of one or more metal current-carrying **conductors**, each surrounded by a layer of **insulation** which, in turn, is usually covered by a tough outer **protective sheath**. In some cases, the cable may be further protected by metal tape or wire **armour**.

In this chapter, we will confine our descriptions to medium- and low-voltage cables used for the wiring of residential, commercial and industrial installations. The construction of high-voltage cables requires extensive explanations which are beyond the purpose of this chapter.

Conductors

In the electrical industry, 'wires' are more properly called '**conductors**'.

A cable may have one, two, three or more separately insulated conductors. Each insulated conductor is called a **core**, and cables with more than one core are referred to as *multicore cables* (or, in North America, *'multi-conductor'* cables).

The very best metal conductor is **silver** which, because of its cost, is limited to special applications such as relay contacts, printed circuit boards, etc. It's common to compare the conductivity of other conductors to that of silver, as indicated in Table 6.1.

Table 6.1

Conductor	Relative conductivity
Silver	100%
Copper	95%
Gold	67%
Aluminium	58%
Tungsten	30%
Iron	14%
Constantan	3.3%
Carbon	1.5%

An Introduction to Electrical Science, Waygood, ISBN 9780415810029, 2013. © Taylor & Francis

The most widely used conductor is **copper**, which has a number of advantages over its nearest rival, aluminium. Copper is second only to silver in terms of its relative conductivity (95%), and is a readily available natural resource. Copper used for general conductors is usually annealed, meaning that it is reheated and allowed to slowly cool, rendering it tough, yet soft enough to be easily drawn into single or stranded conductors. Hard-drawn copper, on the other hand, has a higher tensile strength, making it ideal for suspending from overhead electricity distribution poles.

When exposed to air, bare copper forms a conductive oxide which then acts to prevent further corrosion, allowing bare (as opposed to insulated) copper to be used for overhead lines. Copper can also be easily soldered, making it ideal for electronics applications.

The abundance of **aluminium** makes it a more economical alternative to copper, and is the second most widely used conductor. Its conductivity, relative to silver, is 58 per cent, making it a poorer conductor than copper which means it must have a larger cross-sectional area than copper in order to conduct the same amount of current.

Its light weight makes it an ideal conductor for the very long spans between overhead electricity transmission towers, although its relative low tensile strength means it must be reinforced with a steel-wire core. Like copper, an oxide forms on the surface of aluminium, preventing further corrosion but, unfortunately, this oxide is a very poor conductor which can cause problems at terminations. The oxide also makes it very difficult to solder aluminium unless specialist, abrasive fluxes are used.

A major problem with aluminium is termed 'cold flow', which causes the metal to flow away under the pressure from screw terminals and cause the termination to loosen. For this reason, special connectors and techniques are required for terminating aluminium conductors, rendering it unsuitable for residential and commercial wiring. In fact, in some countries, aluminium is banned for residential and commercial wiring following fatal fires.

The central component of any cable, then, is the **conductor** itself, which may be *solid*, or *stranded* – as illustrated in Figure 6.1.

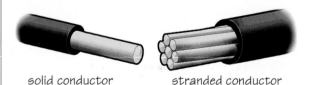

solid conductor   stranded conductor

Figure 6.1

Cables of the type used to wire residential and commercial buildings, having cross-sectional areas below 4 mm², are usually manufactured using *solid* conductors. Beyond this cross-sectional area, conductors are normally *stranded*. Stranded conductors impart greater flexibility into cables, compared with cables having solid conductors of equivalent size, making them much easier to handle.

A conductor of any given cross-sectional area can be manufactured either from a few strands of thicker wire, or from a large number of strands of thinner wire – in either case, the total area will remain the same. Generally, for any given cross-sectional area, the greater the number of strands, the greater the flexibility of the cable. This is why 'flexible cables' (commonly called 'flex' or 'cords') are used to connect portable appliances to the fixed wiring in buildings. Flexible cables enable such appliances to be regularly moved around without over-stressing and, possibly, snapping the conductors.

The strands of a stranded cable form a mathematical pattern, with the first layer being a single strand, the second layer being formed from 6 strands, the third layer from 12 strands, and so on.

Stranded cables are identified in the following fashion: '**7/0.85**'. In this example, '**7**' represents the *number of strands*, and '**0.85**' represents the **diameter** *of one strand*, expressed in millimetres.

The cross-sectional shape, or 'profile', of a conductor is not necessarily circular, whether solid or stranded. For example, in the case of larger multi-core cables, such as those used for underground services, the conductors' cross-sectional profiles are frequently 'shaped' in order to reduce the overall diameter of the complete cable compared to a cable with circular sections but of the same current rating. These shapes vary depending upon the number of cores, but are typically 'wedge' shaped. In high-voltage cables, these shapes also reduce the need to pack the internal airgaps with fillers, and to help evenly distribute the electrical stress caused by electric fields within the cables.

In the two examples illustrated in Figure 6.2, the circular-section cable has the same size outer diameter as the shaped-section cable to its right, but the cross-sectional area of its conductors is very much smaller than the shaped conductors. It also has larger voids (air gaps) which, even if packed with insulated fillers, make the cable more susceptible to moisture ingress. If these were high-voltage cables, the larger voids in the circular-section cable would also be susceptible to breakdown due to variations in dielectric stress due to the differences in the cable's insulating material

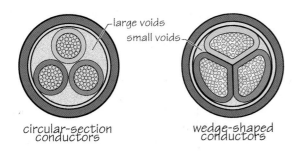

large voids
small voids

circular-section
conductors

wedge-shaped
conductors

Figure 6.2

and the material used as 'fillers' within the voids (including air).

In order to efficiently carry an electric current, the conductor must have a *low* resistance. If you refer back to the chapter on *resistance*, you will recall that resistance (symbol: *R*) is *inversely proportional to a conductor's cross-sectional area (symbol: A)*, expressed as:

$$R \propto \frac{1}{A}$$

For a given material, then, *doubling* its cross-sectional area will *halve* its resistance, and *vice versa*. This is because the area of a circular-section conductor is proportional to the *square* of its diameter and, so, *doubling the* **diameter** *will reduce its resistance by a quarter* and so on.

In the UK and Europe, the cross-sectional areas of conductors are always expressed in **square millimetres** (symbol: **mm²**). In the United States and Canada, however, a completely different method is used, based on the **American Wire Gauge (AWG)** system, which uses Imperial units of measurement: 'circular mils' for circular-section conductors, or 'square mils' for rectangular-section conductors such as busbars. A 'mil' is one-thousandth of an inch. A 'circular mil' is not really a true unit for area, as it is simply a conductor's diameter, in mils, squared, which results in a number that 'represents' a particular cross-sectional area.

For a solid conductor of circular cross-section, of diameter *d*, the cross-sectional area (*A*) is measured perpendicular to the cable's length, and is given by:

$$A = \frac{\pi d^2}{4}$$

For a circular-section stranded conductor, made up of *n* strands, each of diameter *d*, the total cross-sectional area is given by:

$$A = n\left(\frac{\pi d^2}{4}\right)$$

The cross-sectional area of a **7/0.85** cable, therefore, can be determined as follows (we must first convert millimetres into metres):

$$A = n\left(\frac{\pi d^2}{4}\right)$$

$$= 7 \times \left(\frac{\pi \times (0.85 \times 10^{-3})^2}{4}\right)$$

$$= 4.00 \times 10^{-6}\, \text{m}^2$$

$$= 4.0\ \text{mm}^2\ \text{(Answer)}$$

For cables with a larger number of strand layers, the resistance of the outer strands is actually somewhat higher than that of the inner strands. This is due to the spiralling of the outer strands, which means that the lengths (and, therefore, the resistance) of the outer strands are longer than the lengths of the inner strands. The above worked example, therefore, is only accurate for cables with relatively few strand layers.

> Note that any circuit-protective conductor (often called an 'earth' conductor) within the cable is *not* counted as a 'core'. So, a 'two-core' cable may, or may *not*, include a protective conductor.

Insulation

A cable's **insulation** (also called a **dielectric**) is the layer of material immediately surrounding the conductor(s), whose function it is to electrically isolate the conductors from each other as well as from their surroundings, in order to prevent anyone from receiving an electric shock by accidentally touching them.

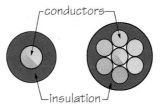

conductors

insulation

Figure 6.3

The material used as an insulator or dielectric must have each of the following *properties*:

- high resistivity
- high dielectric strength

- impervious to moisture
- strength and toughness
- flexibility
- immune to chemicals
- stability over a wide range of operating temperatures.

An insulator's dielectric strength is defined as *'the maximum electric field that can be sustained by a dielectric before breakdown occurs'*, and is expressed in volts per metre (symbol: V/m) or, in more practical terms: kilovolts per millimetre. In simple terms, it defines the maximum potential difference an insulating material can withstand before breaking down and conducting.

An insulator's dielectric strength must be appropriate for its application – e.g. a bellwire operating at 12 V clearly does not require the same degree of dielectric strength as, say, a cable operating at 400 V.

As we learnt in the chapter on *resistance*, the resistance of insulation is *inversely proportional to the length of the conductor* – i.e. the *longer* the conductor, the *lower* its insulation resistance!

For residential wiring, the minimum values of insulation resistance, measured between the live* conductors and between the live conductors and the protective ('earthing') conductor are specified in the **IET Wiring Regulations**. For nominal voltages up to and including 500 V, this value must be ≥ 1.0 MΩ, and the insulation must withstand a test voltage of 500 V(d.c.).

> *The term **'live conductor'** refers to any conductor intended to be energised in normal use – in other words, it applies to both line and to the neutral conductors, but *not* a protective (earth) conductor.

Insulation (dielectrics)

The type of insulating (or 'dielectric') material determines the safe operating temperature of a conductor (for residential wiring, typically 70°C or 90°C), as early insulation failure may result if its actual temperature is allowed to exceed its specified safe operating temperature.

Most residential installation cables are insulated using what, in earlier editions of the IET Wiring Regulations *used* to be called **'PVC'** or **'rubber'**. However, since the introduction of the 2008 edition of these Regulations, these terms are no longer used to describe insulating materials; instead, 'PVC cables' are now called **'thermoplastic-insulated'** cables, and

'rubber cables' are now termed **'thermosetting'** cables

These new terms are related to the anticipated operating temperature of the cable, and provide a simpler method of classifying insulation. This offers a solution to the problem of the ever-increasing variety of materials being used these days in the manufacture of cable insulation, by simply classifying them in terms of their *properties*, instead of their *ingredients*.

Mineral-insulated cables were first patented to a Swiss inventor, Arnold Borel, in 1896, but began commercial production, in 1936, by the British company *Pyrotenax*. In fact, within the electrical trade, these cables are generally known as *'pyro'* (short for *'pyrotenax'*) cables.

These cables consist of solid, high-conductivity, copper (or, less commonly, aluminium) conductors, surrounded by densely compacted magnesium oxide powder, and enclosed within an outer sheath of copper tubing. The magnesium oxide provides insulation between the conductors themselves, as well as between the conductors and the surrounding metallic sheath. These cables are also available with their outer metal sheath covered with a thermoplastic sheath which will protect the metal sheath against corrosive environments.

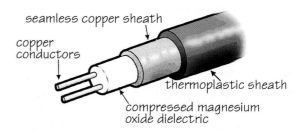

Figure 6.4

The insulating material, magnesium oxide, is non-toxic, non-hydroscopic (does not easily absorb moisture), chemically stable, has a high melting point, and high thermal and electric resistivities.

The combination of metallic sheath and the mineral insulation protects the cable from severe mechanical abuse, such as twisting, bending and compression, making it ideal for use in industrial and commercial work.

Colour coding insulation

Insulation is always colour coded so that the function of each conductor can be identified. These colours must conform to national or international codes of practice. In the UK, for example, the approved colours for identifying cable cores for a.c. power and lighting circuits are shown in Table 6.2.

Table 6.2

Function	Alphanumeric	Colour
protective (earthing) conductor		yellow & green stripe
single-phase line conductor	L	brown
line 1 conductor of a three-phase system	L_1	brown
line 2 conductor of a three-phase system	L_2	black
line 3 conductor of a three-phase system	L_3	grey
neutral	N	blue

Protective sheath

Overlaying the cable's insulation is its **protective sheath**. Its main functions are to provide limited *mechanical protection* to the insulation, to *prevent ingress of moisture* into the cable and, in the case of multicore cables, to *contain the individual cores*. A two-core thermoplastic cable with a protective sheath is illustrated in Figure 6.5.

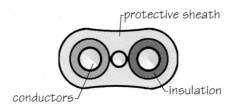

Figure 6.5

Cables used in residential installation work will normally have a thermoplastic sheath, as the likelihood of mechanical damage is low because such cables are normally reasonably protected from their surroundings by being run below floorboards, sunk into plasterwork, etc.

Armouring

Cables, such as underground distribution cables, intended for use where the possiblity of mechanical damage is high, require additional protection.

Such cables are usually **armoured**. This consists either of overlapping flat metal tapes or of circular-section metal wires (often two layers, wound in opposite directions to increase the cable's flexibility), manufactured from steel or, in some cases, aluminium.

Armouring is intended to protect the cable from the impact of hand tools such as shovels, pick axes, etc., and to prevent the possibility of penetration by rocks, etc., once the cable trench has been backfilled.

Finally . . .

Now that you have completed this chapter, are you able to achieve the objectives or learning outcomes listed at the beginning of this chapter?

Ask yourself, 'Can I . . .'

1   identify, and state the function of, each of the main components of an electric cable.
2   explain why some conductors are stranded rather than solid.
3   explain how the cross-sectional area of a stranded conductor may be determined.
4   explain why the sectional shape of larger multicore cable conductors are frequently 'wedge' shaped, rather than circular.
5   list the main properties required of a cable's insulation.
6   explain the relationship between a cable's length and the resistance of its insulation.
7   identify the function of a cable conductor by the colour of its insulation.

Chapter 7

Effect of temperature change upon resistance

On completion of this chapter, you should be able to

1 state the general effect of changes in temperature upon:
   a pure metal conductors.
   b insulators.
   c carbon (special case).
2 explain the term *temperature coefficient of resistance*.
3 state the unit of measurement of *temperature coefficient of resistance*.
4 explain the difference between *positive* and *negative* temperature coefficients of resistance.
5 solve problems on the effect of changes in temperature upon the resistance of materials.

Introduction

Variations in temperature can affect the physical properties of any material – including, amongst others, its optical, electrical and magnetic properties.

One of the physical properties of a material is its **resistivity** and, since resistance is directly proportional to resistivity, any change in resistivity will result in a change in **resistance**.

A change in temperature will *directly* affect the resistivity of a material, thereby *indirectly* affecting its resistance.

The way in which a material's resistivity and, therefore, resistance changes depends upon the nature of the material itself. Of particular interest to us, in general terms, is the behaviour of **pure metal conductors**, **alloys**, **insulators** and (a special case) **carbon**.

Pure metal conductors

In the case of **pure metal conductors**, such as copper or aluminium, an *increase* in temperature will cause an increase in its resistance. Conversely, reducing the temperature of a pure metal conductor will cause its resistance to fall. The *amount* of change in resistance for a given change in temperature, varies from metal to metal.

Alloys

Some alloys, such as **constantan** (a copper-nickel alloy), have been specially developed to maintain a fixed resistance over a wide range of temperature variations. Such alloys are used in the manufacture of measuring instruments, for example, where any change in resistance due to temperature variations can affect the accuracy of such instruments.

Insulators

In the case of an insulator, an increase in temperature causes a corresponding increase in the internal energy of its atoms which cause free electrons to be liberated to act as additional charge carriers, effectively *reducing its resistance*. If the temperature is allowed to increase, then it will eventually lead to a complete breakdown

An Introduction to Electrical Science, Waygood, ISBN 9780415810029, 2013. © Taylor & Francis

of the insulator's ability to withstand voltage. High temperatures are a major cause of insulation breakdown, which is why it is important never to interfere with the ventilation provided with electrical equipment.

In electrical installations, whenever cables are bundled together, or installed within thermal insulation (e.g. within a thermally insulated wall), they must be *de-rated* – that is, their published current ratings must be reduced in order to prevent them from overheating, which would likely result in insulation failure.

Carbon

Carbon is widely used in the manufacture of contacts used to connect the stationary and rotating conductors of electrical motors and generators. These contacts are termed '**brushes**'.

Carbon is ideal for this application because it is self-lubricating and is a relatively good conductor. It is a *special case* because, unlike most other conductors, its resistivity decreases when its temperature increases. In other words, it behaves in the *opposite* way to metal conductors.

Temperature coefficient of resistance

If the resistance of a length of **copper** wire is measured at various values of temperature from around 20°C (which is normally considered to be ambient temperature) up to around 200°C, it will be seen to *increase* in a linear fashion – as represented by the solid graph line in Figure 7.1.

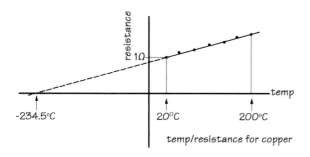

Figure 7.1

If this graph line is projected (a process called *extrapolation*) to the left, as represented by the broken graph line, it will eventually 'cut' the horizontal (temperature) axis at **−234.5°C**. In other words, at a temperature of **−234.5°C**, the resistance of the copper wire will be zero ohms.

Now, this should *not* be interpreted as meaning that, for copper, the linear relationship between resistance and temperature continues to be maintained below 20°C. What it means is that, for the range of temperatures *over which the conductor is expected to operate*, its resistance is approximately linear, and *behaves as though it would reach zero ohms at −234.5°C*.

If we now assume that the resistance at 20°C is, say, 1 Ω then, as we follow the extrapolated line back from −234.5°C to 20°C (a total of **254.5** degrees), the resistance of the copper wire will have increased by 1 Ω.

So, if a 254.5-degree rise in temperature causes the resistance of the copper wire to rise by one ohm then, for every one-degree rise in temperature, its resistance will rise by $\frac{1}{254.5}$ Ω.

> A one-degree rise in temperature has caused a corresponding $\frac{1}{254.5}$ ohm increase in the resistance of the copper wire.

This fraction is more commonly expressed as a decimal: 3.93×10^{-3}/°C or, in SI, 3.93×10^{-3}/K (**per kelvin**).

This figure has special significance, and is called the **temperature coefficient of resistance** (symbol: α_{20}, pronounced '*alpha-twenty*').

The subscript '20' simply indicates that the temperature coefficient of resistance is quoted for an ambient temperature of 20°C.

> The term **temperature coefficient of resistance** is defined as '*the incremental change in the resistance of any material as a result of a change in temperature*', and is considered to be a property of any conducting material. Its SI unit of measurement is '**per kelvin**'(**/K**) or, for everyday use, '**per degree Celsius**' (**/°C**).

The temperature coefficient of resistance of **3.93×10^{-3}/°C**, quoted above, applies to *copper*. The values for some other common conductors are listed in Table 7.1. Be aware, though, that these values *vary somewhat according to the purity of the various metals* and, therefore, may vary according to which source the figures are obtained from.

Of particular interest in Table 7.1 are the temperature coefficients of resistance for **constantan** and for **carbon**. The figure for constantan indicates that its graph line is practically *horizontal*, whereas the figure for carbon indicates that its graph line slopes in the

Table 7.1 Temperature Coefficients of Resistance at 20°

Conductors	α_{20} (/ °C)
silver	3.80×10^{-3}
copper	3.93×10^{-3}
aluminium	3.90×10^{-3}
tungsten	4.50×10^{-3}
constantan	0.008×10^{-3}
carbon	-0.50×10^{-3}

opposite direction to that for a pure metal! We say that carbon has a **negative temperature coefficient of resistance**.

Although tables of temperature coefficients of resistance usually assume an ambient temperature of 20°C, you should also be aware that *some* tables assume an ambient temperature of 0°C (or their equivalents expressed in kelvin). Temperature coefficients of resistance at 0°C are shown as α_0.

Finding resistance at various temperatures

Figure 7.2 represents the effect of temperature upon a copper conductor. Two points are considered on the graph: the resistance (R_1) of the conductor at temperature (T_1) and the resistance (R_2) of the conductor at a second temperature (T_2).

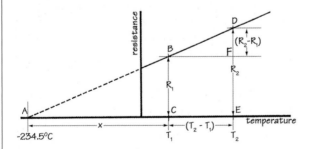

Figure 7.2

Examination of the graph will reveal two right-angled triangles, *ABC* and *BDF*. Because angles $\angle BAC$ and $\angle DBF$ are the same, these two triangles are termed **similar triangles**, and the following ratios apply:

$$\frac{BC}{DF} = \frac{AC}{BF}$$

where:

$$BC = R_1$$
$$DF = (R_2 - R_1)$$
$$AC = x$$
$$BF = (T_2 - T_1)$$

Substituting these values into our original equation:

$$\frac{R_1}{(R_2 - R_1)} = \frac{x}{(T_2 - T_1)}$$

Rearranging, to make R_2 the subject of the equation:

$$\frac{R_2 - R_1}{R_1} = \frac{(T_2 - T_1)}{x}$$

$$R_2 - R_1 = R_1 \frac{(T_2 - T_1)}{x}$$

$$R_2 = R_1 + \left(R_1 \frac{(T_2 - T_1)}{x} \right)$$

$$R_2 = R_1 \left(1 + \frac{T_2 - T_1}{x} \right)$$

Rearranging further:

$$R_2 = R_1 \left(1 + \frac{1}{x}(T_2 - T_1) \right)$$

The reciprocal of x is, of course, the **temperature coefficient of resistance** at temperature T_1 (symbol: α_{T_1}). So, the the equation becomes:

$$R_2 = R_1 \left(1 + \alpha_{T_1}(T_2 - T_1) \right) \qquad \text{—equation (1)}$$

where:

R_2 = resistance at temperature T_2

R_1 = resistance at temperature T_1.

α_{T_1} = temp coefficient of resistance, quoted at temperature T_1.

T_2 = upper temperature

T_1 = lower temperature (ambient)

This is a general equation, and applies to a temperature coefficient of resistance quoted at *any* temperature.

Most commonly, however, the coefficient is quoted at an ambient temperature of 20°C – in which case, we can replace T_1 with 20°C, the equation becomes:

$$R_2 = R_{20}\left(1 + \alpha_{20}(T_2 - 20)\right) \qquad \text{—equation (2)}$$

If, however, the coefficient is quotated (less commonly) at zero-degrees Celsius, then the equation becomes:

$$R_2 = R_0\left(1 + \alpha_0(T_2 - 0)\right)$$

Or simply:

$$R_2 = R_0\left(1 + \alpha_0 T_2\right) \qquad \text{—equation (3)}$$

Worked example 1   The resistance of a copper conductor, at 20°C, is found to be 50 Ω. Calculate its resistance at a temperature of 60°C.

Solution From the table of temperature coefficients of resistance, $\alpha_{T_1} = 3.9 \times 10^{-3}$.

either	**or**
(using equation 1):	*(using equation 2):*

$R_2 = R_1\left(1 + \alpha_{T_1}(T_2 - T_1)\right)$ $R_2 = R_{20}\left(1 + \alpha_{20}(T_2 - 20)\right)$

$\quad = 50 \times \left[1 + (3.9 \times 10^{-3})(60 - 20)\right]$ $= 50 \times \left[1 + (3.9 \times 10^{-3})(60 - 20)\right]$

$\quad = 50 \times \left[1 + (3.9 \times 10^{-3})(40)\right]$ $= 50 \times \left[1 + (3.9 \times 10^{-3})(40)\right]$

$\quad = 50 \times \left[1 + (0.156)\right]$ $= 50 \times \left[1 + (0.156)\right]$

$\quad = 50 \times \left[1.156\right]$ $= 50 \times \left[1.156\right]$

$\quad = 57.8\ \Omega$ (Answer) $= 57.8\ \Omega$ (Answer)

Variation on the general equation

An alternative way of approaching this topic results in the following variation of the **general equation**, which can be used to find the resistance at *any* temperature, *without the need to know an initial resistance at a reference temperature of either 0°C or 20°C* – providing you know the *coefficient (α_0) quoted at zero degrees Celsius.*

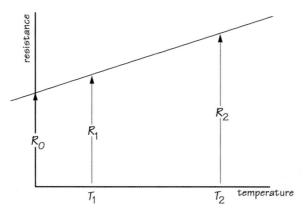

Figure 7.3

In the resistance/temperature graph shown in Figure 7.3, the resistance, R_1 is found using equation (3), above, that we have already learnt, i.e:

$$R_1 = R_0(1 + \alpha_0 T_1) \qquad \text{—equation (3)}$$

Similarly, the resistance R_2 is found from:

$$R_2 = R_0(1 + \alpha_0 T_2) \qquad \text{—equation (4)}$$

If we now divide equation (4) by equation (3),

$$\frac{R_2}{R_1} = \frac{\cancel{R_0}(1 + \alpha_0 T_2)}{\cancel{R_0}(1 + \alpha_0 T_1)}$$

As you can see, the R_0 appearing in the numerator and denominator cancel, giving us the following equation:

$$\frac{R_2}{R_1} = \frac{(1 + \alpha_0 T_2)}{(1 + \alpha_0 T_1)} \qquad \text{—equation (5)}$$

where:

R_2 = final resistance

R_1 = initial resistance

T_2 = final temperature

T_1 = initial temperature

α_0 = temp coefficient at 0°

Worked example 2   The resistance of a coil at the beginning of an experiment is found to be 25 Ω at an ambient temperature of 18°C. What will be its resistance at a temperature of 300°C? Take the coil's temperature coefficient of resistance as 0.000 24/°C at 0°C.

Solution

$$\frac{R_2}{R_1} = \frac{(1 + \alpha_0 T_2)}{(1 + \alpha_0 T_1)} \qquad \text{—equation (5)}$$

$$\frac{R_2}{25} = \frac{(1 + 0.000\ 24 \times 300)}{(1 + 0.000\ 24 \times 18)}$$

$$R_2 = 25 \times \frac{(1 + 0.000\ 24 \times 300)}{(1 + 0.000\ 24 \times 18)}$$

$$= 25 \times \frac{1.072}{1.004} = 25 \times 1.068 = 26.7\ \Omega \text{ (Answer)}$$

Worked example 3   The resistance of a coil of copper wire at the start of an experiment is measured at 173 Ω, at a temperature of 16°C. At the end of the test, the resistance was found to have risen to 215 Ω. Calculate the final temperature of the coil. Take the coil's temperature coefficient of resistance as 0.004 26/°C at 0°C.

Solution

$$\frac{R_2}{R_1} = \frac{(1+\alpha_0 T_2)}{(1+\alpha_0 T_1)} \qquad \text{—equation (5)}$$

$$\frac{215}{173} = \frac{(1+0.004\,26 \times T_2)}{(1+0.004\,26 \times 16)}$$

$$1+0.004\,26 \times T_2 = \frac{215}{173} \times (1+0.004\,26 \times 16)$$

$$1+0.004\,26 \times T_2 = \frac{215}{173} \times 1.068 = 1.327$$

$$1+0.004\,26 \times T_2 = 1.327$$

$$0.004\,26 \times T_2 = 1.327 - 1$$

$$T_2 = \frac{0.327}{0.004\,26} = 76.76°C \ (Answer)$$

Negative temperature coefficient of resistance

Earlier in this chapter, we learnt that carbon has a *negative* temperature coefficient of resistance (α_{20}), quoted as **−0.5×10⁻³/°C** in Table 7.1. This means that as its temperature *rises*, its resistance will *fall*.

This will be demonstrated in the following worked example.

Worked example 4   If the resistance of a block of carbon is 0.25 Ω at 20°C, calculate its resistance at 200°C, where α_{20} is −0.5 × 10⁻³/°C.

Solution

$$R_2 = R_{20}\left(1+\alpha_{20}(T_2 - 20)\right) \qquad \text{—equation (2)}$$

$$= 0.25 \times \left[1+(-0.5\times10^{-3})(200-20)\right]$$

$$= 0.25 \times \left[1+(-0.5\times10^{-3})(180)\right]$$

$$= 0.25 \times \left[1+(-90\times10^{-3})\right]$$

$$= 0.25 \times [1-0.009]$$

$$= 0.25 \times 0.91$$

$$= 0.228\ \Omega\ (Answer)$$

This gives carbon a characteristic, which proves very useful when it is used as 'brushes' in electric motors. As the brushes heat up due to the friction of the machine's commutator, their resistance falls.

Constantan and similar alloys

With critical measuring instrument components, it's usually essential to maintain an approximately constant resistance value over a wide range of temperature values if false readings are to be avoided. For this reason, critical circuit resistors are wound using wire manufactured from metal alloys such as 'constantan'.

'**Constantan**', also known as 'Eureka wire', is the trade-name for a copper-nickel alloy (approx. 60:40 ratio) formulated in the late 1800s by Edward Weston, an English-born American who founded the world-famous *Weston Electrical Instrument Company*, manufacturing precision measuring instruments (including their famous photographic exposure meters).

With an α_{20} of just **0.000 02 per degree Celsius**, constantan maintains an *approximately constant resistivity and, therefore, resistance over a wide range of temperatures*.

For example, if the resistance of a coil of constantan resistance wire, at 20°C, is 50Ω, let's find out what its resistance will be at 500°C.

$$R_2 = R_{20}\left(1+\alpha_{20}(T_2 - 20)\right) \qquad \text{—equation (2)}$$

$$= 50 \times \left[1+(0.000\,02)(500-20)\right]$$

$$= 50 \times \left[1+(0.000\,02)(480)\right]$$

$$= 50 \times \left[1+(0.0096)\right]$$

$$= 50 \times [1.0096]$$

$$= 50.48\ \Omega\ (Answer)$$

As you can see, this is a very small increase in resistance over quite a large temperature variation.

Summary

The **temperature coefficient of resistance** (symbol: α) is a physical property of a material which directly affects the resistivity and, indirectly, the resistance of that material.

Temperature coefficient of resistance is defined as '*the incremental change in the resistance of any material as a result of a change in its temperature*'. Its SI unit of measurement is '**per kelvin**', although it is more commonly expressed as '**per degree Celsius**'.

The temperature coefficient of resistance of most metal conductors is **positive**, meaning that an *increase* in temperature results in an *increase* in resistance. Most insulating materials (as well as the conductor, carbon), however, have a **negative** temperature coefficient of resistance, meaning that their resistances will *decrease* as the temperature *increases*.

The value of the temperature coefficient of resistance varies according to its **reference temperature**. Whereas some tables list temperature coefficients of resistance at an ambient temperature of 20°C (shown as 'α_{20}'), other tables list the coefficients at 0°C (shown as 'α_0'). *So care must be taken when reading these tables.*

To find the resistance of a material at any given temperature, the following equations may be used.

A **general equation**, which can be used for *any* reference temperature (α_{T_1}), where R_2 is the final resistance, R_1 is the initial resistance, T_2 is the final temperature, T_1 is the initial temperature, and α_{T_1} is the coefficient at T_1:

$$R_2 = R_1\left(1 + \alpha_{T_1}(T_2 - T_1)\right) \qquad \text{—equation (1)}$$

The following variation on the above equation can be used when the reference temperature is 20°C, where R_2 is the final resistance, R_{20} is the initial resistance at 20°C, T_2 is the final temperature, and α_{20} is the coefficient at 20°C:

$$R_2 = R_{20}\left(1 + \alpha_{20}(T_2 - 20)\right) \qquad \text{—equation (2)}$$

If the reference temperature is 0°C, then the following variation of the equation applies:

$$R_2 = R_0\left(1 + \alpha_0 T_2\right) \qquad \text{—equation (3)}$$

Finally, if the value of α_0 (coefficient at zero degrees Celsius) is known, then the following variation of the equation may be used, where R_2 is the final resistance, R_1 is the initial resistance, T_2 is the final temperature, and T_1 is the initial temperature:

$$\frac{R_2}{R_1} = \frac{(1 + \alpha_0 T_2)}{(1 + \alpha_0 T_1)} \qquad \text{—equation (4)}$$

Finally . . .

Now that you have completed this chapter, are you able to achieve the objectives or learning outcomes listed at the beginning of this chapter?

Ask yourself, 'Can I . . .'

1 state the general effect of changes in temperature upon:
  a pure metal conductors.
  b insulators.
  c carbon (special case).
2 explain the term *temperature coefficient of resistance.*
3 state the unit of measurement of *temperature coefficient of resistance.*
4 explain the difference between *positive* and *negative* temperature coefficients of resistance.
5 solve problems on the effect of changes in temperature upon the resistance of materials.

Online resources
The companion website to this book contains further resources relating to this chapter. The website can be accessed via the following link:
www.routledge.com/cw/waygood

Chapter 8

Ohm's Law of Constant Proportionality

On completion of this chapter, you should be able to

1 state Ohm's Law of Constant Proportionality.
2 describe the circumstances under which Ohm's Law applies.
3 describe the difference between 'linear' ('ohmic') and 'non-linear' ('non-ohmic') materials and circuit components.
4 state the relationship between potential difference, current and resistance.
5 given any two of the following quantities, determine the third:
   • potential difference
   • current
   • resistance.

Introduction

In 1827, a virtually unknown German physics teacher, **Georg Simon Ohm** (1789–1854), published the results of a series of experiments in which he had established a relationship between *the current drifting along a wire and the potential difference applied across the ends of that wire.*

These days, we consider Ohm's Law so fundamental that it is rather surprising to learn that Ohm's findings were initially greeted with derision by the scientific community, and not accepted until Ohm was nearing the end of his life! Even then, recognition first came from Britain, rather than from his native Germany, when, in 1842, he was granted membership of the Royal Society.

Figure 8.1

Yet, despite being probably the best-known 'law' in electrical engineering, it remains the most misinterpreted and least understood by electricians and technologists!

Ohm's experiments were very simple by today's standards, but must have been very difficult to perform in the early nineteenth century because, of course,

An Introduction to Electrical Science, Waygood, ISBN 9780415810029, 2013. © Taylor & Francis

there were no electrical measuring instruments as we know them today available for him to use, and nor were there any standard units of measurement for voltage or current!

He discovered that whenever he applied the same voltage to a particular conductor, the resulting current always reached the same, constant value. By experimenting further, he also discovered that, *providing the temperature and other factors did not change*, the resulting current was *directly proportional* to any variations in the voltage he applied to the conductor. So if, for example, he *doubled* the voltage, then the current would also *double*; if he *halved* the voltage, then the resulting current would also *halve*.

This can be expressed as follows:

For a given conductor, the ratio of voltage to current is a constant.

This statement became known as **Ohm's Law** or, more accurately, **Ohm's Law of Constant Proportionality**.

Expressed mathematically, we can say:

$$\frac{E}{I} = k \text{ (constant)}$$

> **Ohm's Law of Constant Proportionality** states that *'for a given conductor, the ratio of voltage to current is a constant'*.

At the time, Ohm believed that his discovery applied to *all* conductor. In other words, he believed that his 'law' was *universal*. But, as we shall learn, he was wrong!

Repeating Ohm's experiments

To perform the modern-day equivalent of Ohm's experiment, we can use the following simple circuit, comprising a variable voltage supply, a voltmeter to measure changes in the potential difference across a resistor and an ammeter to measure the resulting current.

By simply increasing the potential difference across the resistor in uniform increments, and noting the corresponding values of current, we can use our results to construct a graph.

It is standard practice, when constructing *any* graph, to plot the quantity which we intentionally vary (called the *'independent variable'*) along the horizontal axis, and the quantity which changes as a result of changes

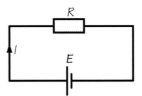

Figure 8.2

in the first quantity (the *'dependent variable'*) along the vertical axis.

So, for Ohm's experiment, it is usual to plot the voltage along the horizontal axis, and the resulting current along the vertical axis (see Figure 8.3).

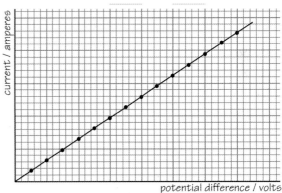

graph for a 'linear' material that obeys Ohm's Law

Figure 8.3

As you can see, the resulting graph, in this particular case, just happens to be a *straight line*, and the straight line indicates *constant proportionality* between the dependent and independent variables. Which is precisely what led Ohm to conclude that, for each of the conductors that he tested, providing the temperature (and, in fact, *all* other physical conditions) remained constant, *the ratio of voltage to current is a constant*.

For Ohm's Law to apply, then, the outcome of the experiment **must** result in a straight-line graph, indicating that the *ratio of voltage to current remains constant*.

But, unfortunately, this is *not* always the case!

For example, if we were to repeat the same experiment using, say, a tungsten-filament lamp instead of a resistor, then the result of our experiment would produce a *curved-line,* rather than a straight-line, graph (see Figure 8.4).

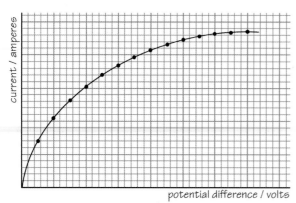

graph for a 'non-linear' material that does not obey Ohm's Law

Figure 8.4

The reason for this is that the current through the lamp causes the temperature and, therefore, the resistance of its tungsten filament to change during the course of the experiment. And one of the essential conditions of Ohm's Law is that *a changing temperature must not be allowed to affect the resistance of the load.*

A curved-line graph also means that changes in potential difference *do not* produce proportional changes in the resulting currents. That is, the ratio of voltage to current *varies continuously throughout the experiment* – confirming that tungsten does *not* obey Ohm's Law!

> **Ohm's Law** is, therefore, *by no means a universal law that applies to all loads.*

In fact, some scientists have even gone so far as to suggest that Ohm's Law should be 'demoted' from being a 'law'(!) because it is cannot be universally applied – this led one (unknown) science writer to say,

> *Ohm's law isn't a very serious law. It's the 'jaywalking' of physics. Sensible materials and devices obey it, but there are plenty of rogues out there that don't!*

Those 'sensible' materials which actually obey Ohm's Law are termed '**linear**' or '**ohmic**' materials; those 'rogue' materials which do *not* obey Ohm's Law are called '**non-linear**' or '**non-ohmic**' materials.

And there are far fewer ohmic materials than there are non-ohmic.

> **Ohm's Law** does *not* apply to all materials; it *only* applies to 'linear', or 'ohmic', materials.

Most, but not all, metal conductors are, fortunately, 'linear' or 'ohmic'. Fixed-value **resistors**, which are circuit components used to modify the natural resistance of a circuit, *must* be manufactured from 'ohmic' conductors of course, because it is important that their specified values of resistance remain constant for voltage/current variations within their power ratings.

'Non-linear' conductors and components include some metal conductors such as tungsten, as well as carbon, semiconductor devices (e.g. diodes, transistors), valves ('vacuum tubes'), electrolytes and gases. Repeating our experiment with these will produce graphs of various shapes, including curves that change direction abruptly as the applied voltage changes.

Ohm's Law and resistance

Let's remind ourselves that **Ohm's Law** states *'the ratio of voltage to current is a constant'.* Expressed mathematically,

$$\frac{E}{I} = k$$

During the course of his experiments, Ohm found that changing the conductor resulted in a different ratio of voltage to current, resulting in different value to the constant, *k*. From this, he concluded that *k* must represent some form of physical property for a particular conductor.

Furthermore, since, for a given potential difference, the current must *decrease* if the value of *k increases*, then *k* must represent *some form of opposition to the drift of current.*

Ohm is credited with naming this opposition to current '**resistance**'. In the earlier chapter on *resistance*, we learnt that its unit, the **ohm**, is defined as *'the electrical resistance between two points along a conductor, such that when a constant potential difference of one volt is applied between those points, a current of one ampere results'.* That is:

$$ohm = \frac{volt}{ampere}$$

Expressed in terms of these units' quantities, we can say that *resistance (R) is the ratio of potential difference (E) to current (I):*

$$R = \frac{E}{I}$$

This equation is often referred to as the *'Ohm's Law equation'*. This, in fact, is *incorrect,* as the equation is actually derived from the definition of the ohm and *not* from Ohm's Law!

Furthermore, this equation applies to *all* loads, *whether those loads obey Ohm's Law or not*! In other words, it applies to *both* linear (ohmic) *and* to non-linear (non-ohmic) loads.

Let's now return to those graphs that we showed, earlier in this chapter, and learn how they relate to the resistance of the loads they represent.

You will remember from your maths that the **gradient** of a line is defined as the change in vertical distance, divided by the change in horizontal distance – as illustrated in Figure 8.5.

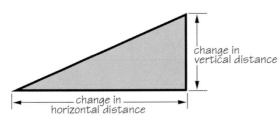

$$\text{gradient} = \frac{\text{change in vertical distance}}{\text{change in horizonal distance}}$$

Figure 8.5

For our results' graphs, this is equivalent to:

$$\text{gradient} = \frac{\text{change in current}}{\text{change in voltage}} = \frac{1}{\text{resistance}}$$

Or to put it another way, *resistance must be equal to the reciprocal of the gradient of the plotted graph line:*

$$R = \frac{1}{\text{gradient}}$$

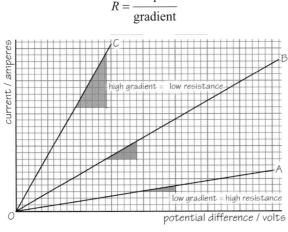

Figure 8.6

So, for a straight-line graph, the *steeper* the gradient, the *lower* the resistance it represents. For example, in Figure 8.6, load A has the highest resistance, and load C the lowest.

For a curved-line graph, *the gradient continually changes along the graph*, as illustrated in Figure 8.7, with the *steeper* gradient at point **A** representing the *lower* resistance, while the *lower* gradient at point **B** represents the *higher* resistance. For this particular curve, then, the resistance increases with higher values of potential difference.

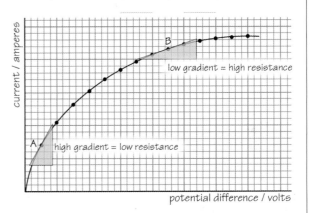

Figure 8.7

Summary

To summarise, we can say that Ohm's Law *only* applies to those loads whose ratio of potential difference to current *remains constant* over a wide range of variation in potential difference.

> If the ratio of potential difference to current is *constant* over the range of applied potential differences, then the load is **linear (ohmic)** and *obeys* Ohm's Law.

And we can say that Ohm's Law does *not* apply to those loads whose ratio of potential difference to current *changes* with variations in potential difference.

> If the ratio of potential difference to current *changes* over the range of applied potential differences, then the load is **non-linear (non-ohmic)** and *does not obey* Ohm's Law.

Hopefully, this is exactly what you would have been taught at school.

But, unfortunately, this is often quite contrary to what others may have learnt at school, or from those many textbooks which simply define Ohm's Law in terms of the equation: $R = E/I$.

Once again, this equation is derived from the *definition of the ohm*, and does **not** represent Ohm's Law!

What you have learnt, so far, in this chapter, might unfortunately contradict what you have read in other textbooks, where **Ohm's Law** is frequently expressed as follows:

> *Ohm's Law states that the current flowing though a conductor is directly proportional to the potential difference across the conductor, and inversely proportional to the resistance of that conductor.*

This statement is quoted universally but is *quite wrong!* Ohm's Law describes *a linear relationship between potential difference and current (hence its correct name* **'Ohm's Law of Constant Proportionality'***), **not** the relationship between potential difference, current, and resistance!

Whether a load obeys Ohm's Law or not, it's very important to understand that *the ratio of potential difference to current* will **always** tell you what the resistance happens to be *for that particular ratio*. So, even for a curved-line graph or, come to that, for a graph of *any* shape, you can *always* determine the resistance *at any point along that graph line* just as long as you know the voltage to current ratio *at that particular point*.

So, the following equation applies, *whether a load obeys Ohm's Law or not!* This important equation applies to *all* circuits under *all* conditions.

$$R = \frac{E}{I}$$

where:

R = resistance, in ohms

E = potential difference, in volts

I = current, in amperes

By rearranging the above equation, we can make current, potential difference, *or* resistance the subject:

$$I = \frac{E}{R} \qquad\qquad E = IR \qquad\qquad R = \frac{E}{I}$$

To help you manipulate the above equations, the triangle shown in Figure 8.8 may be useful! Place your finger over the quantity you need, and the triangle will reveal its formula.

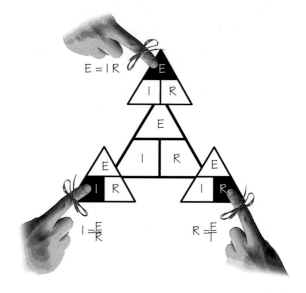

Figure 8.8

Worked examples

1   If the potential difference across a circuit is 150 V and the circuit's resistance is 50 Ω, calculate the resulting current:

e.g. $I = \dfrac{E}{R} = \dfrac{150}{50} = 3$ A (Answer)

2   If the current flowing through a conductor is 5 A, and its resistance is 25 Ω, find the potential difference across its ends:

e.g. $E = IR = 5 \times 25 = 125$ V (Answer)

3   If a circuit's potential difference is 6 V, and current is 3 A, find the resistance of the circuit:

e.g. $R = \dfrac{E}{I} = \dfrac{6}{3} = 2\ \Omega$ (Answer)

Do the circuits in each of the above worked examples obey Ohm's Law? The answer is that *we simply don't know* – there's no information supplied which tells us whether the ratio of potential difference to current remains constant at other voltages. If the ratio *does* remain constant, then they do; if the ratio *changes*, then they don't! But whether they obey Ohm's Law or not is irrelevant to the solutions, as the equations we have used apply under *any* circumstances.

4   The voltage applied to a particular circuit is 12 V, and the resulting current is 3 A. If the voltage is then increased to 36 V, the resulting current is found to be 4 A. Does the circuit obey Ohm's Law?

At the beginning of the experiment,

$$R = \frac{E}{I} = \frac{12}{3} = 4\,\Omega$$

At the end of the experiment,

$$R = \frac{E}{I} = \frac{36}{4} = 9\,\Omega$$

As the resistance has increased, the ratio of voltage to current has changed, so Ohm's Law does *not* apply.

Self-test exercise

Now complete Table 8.1 by calculating the unknown quantities. The first one (underlined) has been done for you.

Table 8.1

	I	*E*	*R*
a.	20 A	**1200 V**	60 Ω
b.	0.5 A	20 V	
c.		30 V	60 Ω
d.	1.5 A		30 Ω
e.		10 V	2 Ω
f.	4 A		6 Ω
g.	8 A	24 V	
h.		32 V	4 Ω
i.	8 mA	32 mV	
j.	5 mA		300 kΩ
k.	16 mA	8 V	
l.	1.5 mA		25 MΩ
m.	250 A	11 kV	
n.	16 A	320 V	
o.		120 V	12 kΩ

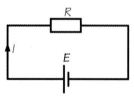

Figure 8.9

Answers

Table 8.2

	I	*E*	*R*
a.	20 A	**1200 V**	60 Ω
b.	0.5 A	20 V	**40 Ω**
c.	**0.5 A**	30 V	60 Ω
d.	1.5 A	**4.5 V**	30 Ω
e.	**5 A**	10 V	2 Ω
f.	4 A	**24 V**	6 Ω
g.	8 A	24 V	**3 Ω**
h.	**8 A**	32 V	4 Ω
i.	8 mA	32 mV	**4 kΩ**
j.	5 mA	**1500 V**	300 kΩ
k.	16 μA	8 V	**500 kΩ**
l.	1.5 mA	**37.5 kV**	25 MΩ
m.	250 A	11 kV	**44 Ω**
n.	16 A	320 V	**20 Ω**
o.	**10 mA**	120 V	12 kΩ

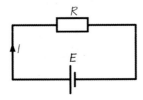

Figure 8.10

Misconceptions

Ohm's Law states that 'current is voltage divided by resistance'
No. Ohm's Law simply describes the linear relationship between potential difference and current for certain conductors.

Ohm's Law applies to *all* circuits under *all* conditions
No. Ohm's Law only applies to 'ohmic' or 'linear' circuits. These are limited to a relatively small range of metallic conductors. 'Non-ohmic' or 'non-linear' circuits or components include metals such as tungsten, and diodes, transistors, etc.

The 'Ohm's Law formula' is *I = E/R*
No. There is no 'Ohm's Law formula'! The equation referred to is derived from the definition of the ohm, *not* from Ohm's Law!

Finally . . .

Now that you have completed this chapter, are you able to achieve the objectives or learning outcomes listed at the beginning of this chapter?

Ask yourself, 'Can I . . .'

1   state Ohm's Law of Constant Proportionality.
2   describe the circumstances under which Ohm's Law applies.
3   describe the difference between 'linear' ('ohmic') and 'non-linear' ('non-ohmic') materials and circuit components.
4   state the relationship between potential difference, current and resistance.
5   given any two of the following quantities, determine the third:
   •   potential difference
   •   current
   •   resistance.

Online resources
The companion website to this book contains further resources relating to this chapter. The website can be accessed via the following link:
www.routledge.com/cw/waygood

Chapter 9

Series, parallel and series-parallel circuits

On completion of this chapter, you should be able to

1 recognise a series circuit, a parallel circuit and a series-parallel circuit.
2 recognise and interpret voltage and current 'sense' arrows.
3 explain Kirchhoff's Voltage Law.
4 explain Kirchhoff's Current Law.
5 calculate the total resistance of a series resistive circuit.
6 calculate the current flow through a series resistive circuit.
7 calculate the voltage drop appearing across each resistor in a series resistive circuit.
8 explain the potential hazard of an open circuit in a series circuit.
9 calculate the total resistance of a parallel resistive circuit.
10 calculate the current flow through each branch of a parallel resistive circuit.
11 calculate the voltage drop appearing across each resistor in a parallel resistive circuit.
12 explain the major advantages of a parallel circuit.
13 calculate the total resistance of a series-parallel resistive circuit.
14 calculate the current flow through each resistor in a series-parallel resistive circuit.
15 calculate the voltage drop appearing across each resistor in a series-parallel resistive circuit.
16 calculate the voltage drop along conductors supplying a load.

Introduction

In order for a conduction current to flow, there must be a *continuous* external conducting path, called a **circuit**, between the terminals of a source of **electromotive force** (e.g. a battery, generator, etc.) and a **load** (e.g. a lamp, etc.). This *continuous* 'electrical pathway' is termed a **closed circuit**. If there is a *break* anywhere in this path, then no current can flow, and it's termed an **open circuit**.

Circuits are categorised according to the way in which they are connected. There are *four* such categories that you should be aware of.

A typical example of each type of circuit is shown in Figures 9.1–9.4.

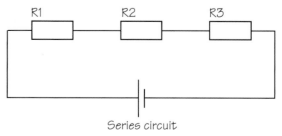

Series circuit

Figure 9.1

An Introduction to Electrical Science, Waygood, ISBN 9780415810029, 2013. © Taylor & Francis

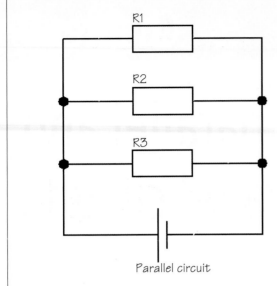

Figure 9.2

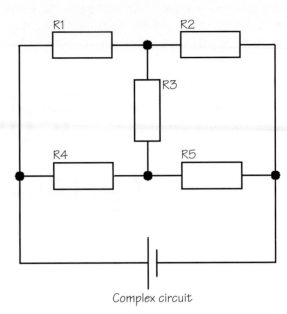

Figure 9.4

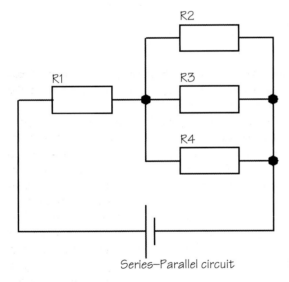

Figure 9.3

As we will see, a **series-parallel** circuit combines series and parallel elements which, as we shall learn, can be easily resolved into an equivalent series circuit and, ultimately, to a single component. There's an *infinite* number of such combinations.

The term '**complex circuit**' is actually rather misleading. It *doesn't* necessarily mean that the circuit is *complicated* (although it very often is!), but is simply the term given to a category into which we can lump *any* circuit that isn't either a series circuit, a parallel circuit or a series-parallel circuit.

Again, there is an infinite number of examples of complex circuits, but they are beyond the scope of this book, as they require special techniques (called '**network theorems**') to solve them – but you should be aware of their existence, and the fact that they *cannot* be simplified using the techniques you will learn in this chapter.

Any electrical components may be connected in the ways described above. Since most electrical components have resistance, we will now consider the effect of connecting *resistances* in series, parallel and in series-parallel.

Understanding sense arrows

Throughout this chapter, and elsewhere in this book, we will be using arrows in circuit diagrams to represent the 'directions', or 'sense', in which potential differences act and currents flow.

We call these arrows 'sense arrows', and they help us form a 'mental picture' of the 'direction' or 'sense' in which the potential differences and currents act in any given circuit at any given instant in time.

Even though we have learnt that, in metallic conductors, electric current is a flow of free electrons from a negative potential to a positive potential, it has been traditional to show '**conventional flow**' (positive to negative) in circuit diagrams.

For direct-current (d.c.) circuits, the rules for sense arrows are straightforward. As we shall learn later, they also apply to alternating-current (a.c.) circuits – but we'll worry about that later.

Voltage-source sense arrows

For potential differences, a single-headed sense arrow is used. The arrow head *always* represents the positive potential of a **voltage source** *(E)* – as illustrated in Figure 9.5.

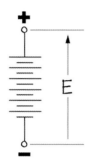

Figure 9.5

Current and voltage-drop sense arrows

For **current**, an arrow placed in the circuit always points *in the direction of conventional current flow (positive-to-negative)*.

For current to flow through individual circuit components, such as resistances, there must be a difference in potential across each of those components. As we have learnt, we call this difference in potential, a **voltage drop** *(U₁, U₂, etc.)*. The arrowhead for a sense arrow representing a voltage drop always points towards the *positive* potential – or in the *opposite direction to the current sense arrow* (see Figure 9.6).

Figure 9.7 shows **voltage source** *(E)*, **current** *(I)*, and **voltage drop** *(U)* sense arrows in a simple series resistive circuit.

You will notice that the sense of each voltage drop, U_1, U_2 and U_3, act in the opposite sense to the potential

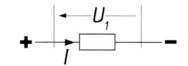

Figure 9.6

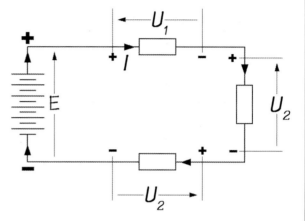

Figure 9.7

difference, *E*, across the source (battery). This agrees with Kirchhoff's Voltage Law, as described in the next section.

Once we have established our sense arrows in a circuit, all *actual* voltages and currents acting in the *same* directions as those sense arrows are assumed to be 'positive' (in the sense of the 'directions' in which they act). Any *actual* voltages and currents that then act in the *opposite* directions to those established sense arrows are then assumed to be 'negative'. For example, in Figure 9.8, sense arrows are established for a circuit in which a battery is supplying a resistor. If the resistor is removed, and replaced with, say, a battery charger which supplies a charging current back to the battery, then the 'charging' current is considered to be *negative* relative to the established sense arrow.

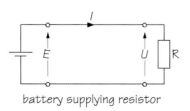

battery supplying resistor

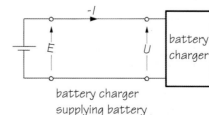

battery charger supplying battery

Figure 9.8

Sense arrows are particularly important when we come to study a.c. theory, and we will be making a great deal of use of them at that point.

Kirchhoff's Laws

Before we move on to examine series, parallel and series-parallel circuits, we need to understand *two* very important 'laws' which will help us understand the behaviour of voltages and currents in electric circuits.

These two laws are called, respectively, '**Kirchhoff's Voltage Law**' and '**Kirchhoff's Current Law**', and they are both credited to the Prussian physicist, **Gustav Kirchhoff** (1824–1887) who, astonishingly, established them while he was still a university student!

These laws describe the behaviour of *voltages* and *currents* in *all* electric circuits, and an understanding of these laws is essential to understanding series, parallel, series-parallel *or* complex circuits.

Kirchhoff's Voltage Law

Kirchoff's Voltage Law states that, *'for any closed loop, the sum of the voltage drops around that loop is equal to the applied voltage'*. A 'voltage drop', you will recall, is the potential difference that appears across a component, such as a resistor, due to the flow of current through that component ($U = I R$).

> In many textbooks, **Kirchhoff's Voltage Law** is expressed in the following form: *'In any closed loop, the algebraic sum of the voltages is zero'*. This is simply another way of stating what we have already stated, above.

Let's examine this law for various circuits, starting with a simple series circuit, as shown in Figure 9.9.

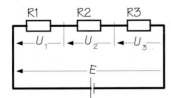

Figure 9.9

For a series circuit, there is only one 'closed loop', and this is shown in bold. If we were to individually measure the voltage drop across each resistor, U_1, U_2 and U_3, and compare them with the applied voltage, E, we would find:

$$E = U_1 + U_2 + U_3$$

To fully understand what we mean by a 'closed loop' and the relationship between the voltages around that closed loop, let's now look at another example: this time, the series-parallel circuit illustrated in Figure 9.10.

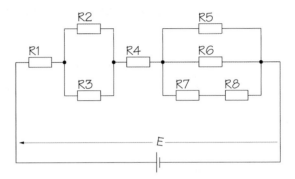

Figure 9.10

For our purposes, a 'closed loop' is *any* **route** around the circuit from the battery's positive terminal to its negative terminal.

For example, let's first examine the 'closed loop' shown in bold in Figure 9.11, which includes resistors R_1, R_2, R_4 and R_5:

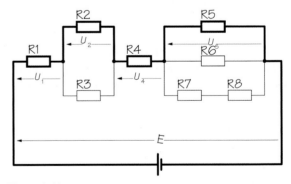

Figure 9.11

If we were to measure the voltage drop across each resistor in this particular 'closed loop', U_1, U_2, U_4 and U_5, and compare them with the applied voltage, E, we would find:

$$E = U_1 + U_2 + U_4 + U_5$$

Now, let's look at a different 'closed loop' through the same circuit, which includes resistors R_1, R_3, R_4 and R_6 (see Figure 9.12).

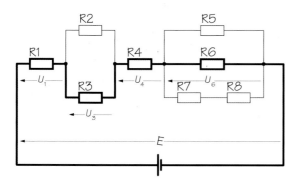

Figure 9.12

If we were to measure the voltage drop across each resistor in this particular closed loop, U_1, U_3, U_4 and U_6, and compare them with the applied voltage, E, then we would find:

$$E = U_1 + U_3 + U_4 + U_6$$

Next, let's look at yet a different 'closed loop' through the same circuit, which includes resistors R_1, R_3, R_4, R_7 and R_8 (Figure 9.13).

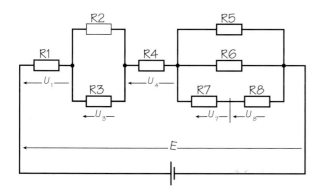

Figure 9.13

Once again, if we were to measure the voltage drop across each resistor in the closed loop, U_1, U_3, U_4, U_7 and U_8, and compare them with the applied voltage, E, we would find:

$$E = U_1 + U_3 + U_4 + U_7 + U_8$$

Finally, let's look at the 'closed loop', formed by R_1, R_3, R_4 and R_5 (Figure 9.14).

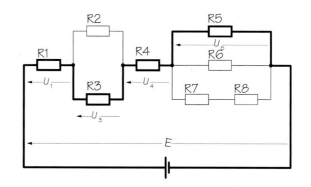

Figure 9.14

If we were to measure the voltage drop across each resistor in the closed loop, U_1, U_3, U_4 and U_5, and compare them with the applied voltage, E, we would find:

$$E = U_1 + U_3 + U_4 + U_5$$

From each of the above examples, it should be clear that it doesn't matter *which* 'closed loop' route you take around *any* electric circuit, *the sum of the voltage drops around that particular loop will* **always** *equal the applied voltage*. **It's essential that you understand this concept.**

Kirchhoff's Current Law

Kirchhoff's Current Law states that '*the sum of the individual currents approaching a junction is equal to the sum of the currents leaving that junction*'.

In many textbooks, **Kirchhoff's Current Law** is expressed in the following form: '*At any junction, the algebraic sum of the currents is zero*'. This is simply another way of stating what we have already stated above.

Let's examine another series-parallel circuit in order to understand this law.

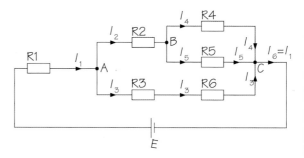

Figure 9.15

In the circuit shown in Figure 9.15, there are three 'junctions', labelled **A**, **B** and **C**.

At junction **A**, the current (I_1) *approaching* that junction is equal to the sum of the two currents, I_2 and I_3, *leaving* that junction. i.e:

$$I_1 = I_2 + I_3$$

At junction **B**, the current (I_2) *approaching* that junction is equal to the sum of the two currents, I_4 and I_5, *leaving* that junction. i.e:

$$I_2 = I_4 + I_5$$

At junction **C**, the sum of the three currents ($I_3 + I_4 + I_5$) *approaching* that junction is equal to the current, I_6, *leaving* that junction. i.e:

$$I_3 + I_4 + I_5 = I_6$$

And, of course, current I_6 is exactly the same current as I_1.

> There is a misconception amongst many students that the current approaching a junction 'splits up' at that junction. It's far more accurate to say that the current 'approaching' a junction is 'made up of' (or 'the sum of') the two currents 'leaving' that junction. In other words, the magnitude of the current approaching a junction is decided by the sum of the currents leaving that junction.

Series circuits

If a number of different circuit components are connected 'end-to-end' (or 'daisy-chained'), so that there is only *one* continuous path for the current to flow along, then these components are said to be connected in **series**. An example of a **series circuit** is a set of old-fashioned Christmas tree lights – the disadvantage with a series circuit (as you will probably have experienced!), is that if there is a break *anywhere* in the circuit, then no current can flow.

Any number of components can be connected in series. However, for convenience we will consider a series circuit with just three components (in this case, resistances), each having resistances labelled: R_1, R_2 and R_3 (see Figure 9.16).

Figure 9.16

Current in a series circuit

In a series circuit, then, because there is only *one path* for the current to flow along, the *same* current must flow through each component. We say that *'the current is common to each component'*. Wherever we place an ammeter (a current-measuring instrument) in the circuit, it will give exactly the same reading because it's measuring exactly the same current – each of the ammeters in the series circuit shown in Figure 9.17 will register *exactly* the same value.

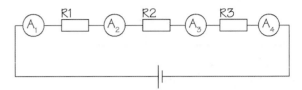

Figure 9.17

Voltage drops in a series circuit

If we placed four voltmeters (instruments for measuring voltage) across each resistance, as shown in Figure 9.18, we would find that the sum of the voltage readings across each resistance would equal the supply voltage (i.e. the potential difference across the circuit). As explained in the earlier chapter on *potential and potential difference*, these individual voltage readings are termed **voltage drops**, and are the product of the current through the individual resistance and that resistance (i.e. $U = IR$).

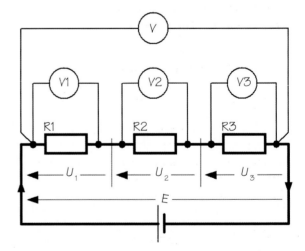

Figure 9.18

As we have seen, this relationship is credited to Kirchhoff, who realised that *the sum of the voltage drops around any closed path is equal to the supply voltage*. This may be expressed as follows:

$$E = U_1 + U_2 + U_3$$

We know, from Ohm's Law, that E or $U = IR$.

So, substituting E and V, we have:

$$IR_T = IR_1 + IR_2 + IR_3$$

Since the current is common to each resistance, we can divide throughout by I:

$$\frac{\cancel{I}R_T}{\cancel{I}} = \frac{\cancel{I}R_1}{\cancel{I}} + \frac{\cancel{I}R_2}{\cancel{I}} + \frac{\cancel{I}R_3}{\cancel{I}}$$

Simplifying this equation, we end up with:

$$R_T = R_1 + R_2 + R_3$$

where: R_T = total resistance of the circuit.

So, for a **series circuit**, the total resistance is simply *the sum of the individual resistances*:

$$R_T = R_1 + R_2 + R_3 + etc.$$

Summary

In a *series circuit*:

- the same current flows through each component.
- the sum of the individual voltage drops will be equal to the supply voltage applied to the circuit.
- the total resistance is equal to the sum of the individual resistances.

Worked example 1   A circuit comprises four components, having resistances of 2 Ω, 4 Ω, 6 Ω and 8 Ω connected in series. If the circuit is connected across a 200-V supply, calculate:

a   the total resistance of the circuit
b   the current flowing
c   the voltage drop across each component.

Solution   Always start by sketching a circuit diagram (as shown in Figure 9.19), and inserting all the information that you are given in the question (remember, in a circuit diagram, '2R' represents '2 Ω', as explained in a previous chapter).

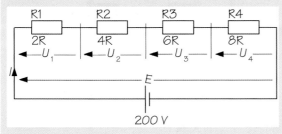

Figure 9.19

---

a   Total resistance:

$$R_T = R_1 + R_2 + R_3 + R_4$$
$$= 2 + 4 + 6 + 8$$
$$= 20\ \Omega \text{ (Answer a.)}$$

b   Circuit current:

$$I = \frac{E}{R} = \frac{200}{20} = 10\ \text{A (Answer b.)}$$

c   Voltage across each component:

$$U_1 = IR_1 = 10 \times 2 = 20\ \text{V (Answer c.1)}$$

$$U_2 = IR_2 = 10 \times 4 = 40\ \text{V (Answer c.2)}$$

$$U_3 = IR_2 = 10 \times 6 = 60\ \text{V (Answer c.3)}$$

$$U_4 = IR_1 = 10 \times 8 = 80\ \text{V (Answer c.4)}$$

To confirm the answers to part c, check that the sum of the voltage drops is equal to the supply voltage:

$$E = U_1 + U_2 + U_3 + U_4 = 20 + 40 + 60 + 80 = 200\ \text{V}$$

Potential hazard of series circuits

Many students believe that if there is a gap in a series circuit (e.g. when a lamp is removed), the voltage across that gap is zero. *Nothing could be further from the truth!*

In fact, the voltage across the gap will be *equal to the circuit's supply voltage!*

At mains level (i.e. 230 V) or higher supply voltages, an **open circuit** occurring in a series circuit, could result in *a potentially hazardous situation*. As no current flows, no voltage drops (i.e. the product of current and resistance) can occur across the healthy resistors – leaving **the full circuit voltage to appear across the break in the circuit**!

Let's look at the worked example above. Assume that R_2 fails and creates an open circuit. Since no current can now flow, then: $U_1 = 0$; $U_3 = 0$ and VU_4 = 0. Therefore:

$$\text{since } E = U_1 + U_2 + U_3 + U_4$$
$$\text{then } U_2 = E - U_1 - U_3 - U_4$$
$$U_2 = 200 - 0 - 0 - 0$$
$$= 200\ \text{V}$$

Potential hazard when working with series circuits! In the event of an **open circuit in a series circuit**, *the full supply voltage will appear across the break in the circuit – creating a potentially hazardous situation.*

Series circuits in practice

A very useful application for a series circuit is as a **voltage divider**, which is widely used in electronic circuits, and consists of a number of high-resistance resistors connected in series, as shown in Figure 9.20.

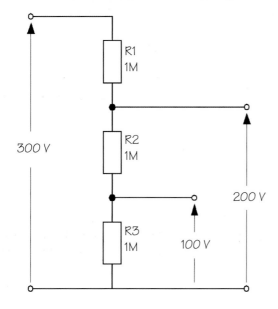

Figure 9.20

In this particular example, three identical resistors connected in series will divide the 300-V input voltage into 200-V and 100-V output voltages. But by selecting appropriate values of resistance, *any* value of output voltage can be obtained (providing it is lower than the input voltage).

Although beyond the scope of this chapter, the resistance of any load chosen for this application must be *significantly* higher in value than resistances making up the voltage divider. This is because of what is known as the 'loading effect', by which the resistance of a load alters the effective resistances of the voltage divider, causing the values of the divided voltages to change from the unloaded state. The higher the resistance of the load, compared with the resistance of the voltage divider, the less the loading effect.

Voltage dividers provide a very useful way of reducing voltages in d.c. circuits where transformers cannot work (transformers only work with a.c.).

Parallel circuits

When individual components are connected, as shown in Figure 9.21, where there is *more* than one path for current flow, then the components are said to be connected in **parallel**. Each of these individual paths is termed a **branch**.

One advantage of a **parallel circuit** is that, should a break occur in one branch, it will *not* affect the operation of the components in the other branches, as they are still connected to the supply voltage.

Any number of components can be connected in parallel. Again, for the sake of convenience, we will consider a parallel circuit with just three components, each having resistances labelled: R_1, R_2 and R_3.

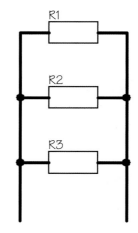

Figure 9.21

Voltage in a parallel circuit

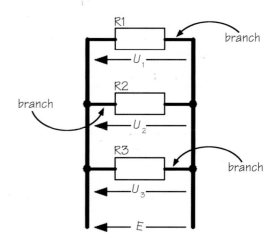

Figure 9.22

The supply voltage (E) applied across a parallel circuit *is common to each branch* – regardless of the number of branches (see Figure 9.22).

$$E = U_1 = U_2 = U_3 = etc.$$

This complies with Kirchhoff's Voltage Law, where *each branch represents an individual closed loop* (i.e. an alternative route through the circuit), so the voltage drop across a component in an individual branch will equal the supply voltage.

This is the *second major advantage* of a parallel circuit, and is the reason why most everyday circuits are connected in parallel – it ensures that *the same voltage is applied across every component*. For example, every circuit in a house is connected in parallel, ensuring that 230 V (or 120 V in North America) will appear across every component (individual lamps, socket outlets, etc.).

Current in a parallel circuit

If we placed ammeters into a parallel circuit, as shown in Figure 9.24, we would find that the sum of the current readings in each branch would equal the current drawn from the supply.

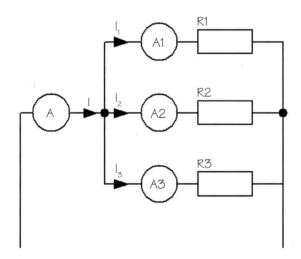

Figure 9.23

As we have learned, this relationship was discovered by Kirchhoff, who realised that *the sum of the currents approaching any junction in a circuit is equal to the sum of the currents leaving the same junction*, and can be expressed as follows:

$$I = I_1 + I_2 + I_3$$

We know, from Ohm's Law, that $I = \dfrac{E}{R}$

So, substituting for I, we have:

$$\frac{E}{R_T} = \frac{E}{R_1} + \frac{E}{R_2} + \frac{E}{R_3}$$

Since the supply voltage, E, is common to each resistance, we can divide throughout by E:

$$\frac{\cancel{E}}{\cancel{E}R_T} = \frac{\cancel{E}}{\cancel{E}R_1} + \frac{\cancel{E}}{\cancel{E}R_2} + \frac{\cancel{E}}{\cancel{E}R_3}$$

Simplifying this equation, we end up with:

$$\frac{1}{R_T} = \frac{1}{R_1} + \frac{1}{R_2} + \frac{1}{R_3}$$

where: R_T = total resistance of the circuit.

So, for a **parallel circuit**, the total resistance is given by:

$$\frac{1}{R_T} = \frac{1}{R_1} + \frac{1}{R_2} + \frac{1}{R_3}$$

Summary

In a *parallel* circuit:

- the supply voltage appears across each component.
- the sum of the individual branch currents will be equal to the supply current.
- the total resistance is found from the equation:

$$\frac{1}{R_T} = \frac{1}{R_1} + \frac{1}{R_2} + \frac{1}{R_3} + etc.$$

Worked example 2   A circuit comprises four components, each having a resistance of 2 Ω, 4Ω, 6 Ω and 8 Ω respectively. If the circuit is connected across a 96-V supply, calculate:

a   the current through each branch
b   the supply current
c   the total resistance.

Solution   Always start by sketching a circuit diagram (as shown in Figure 9.24), and inserting all the information that you are given in the question (remember, in a circuit diagram, '2R' represents '2 Ω', as explained in a previous chapter).

a   Current through each branch:
Since the supply voltage, E (= 96 V), is common to each branch

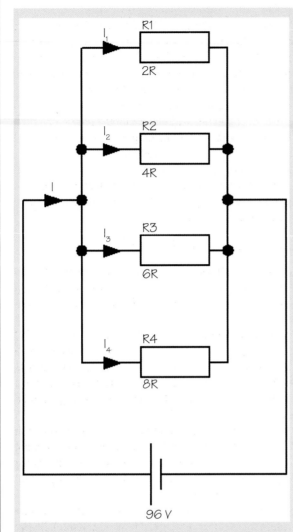

Figure 9.24

$$I_1 = \frac{E}{R_1} = \frac{96}{2} = 48 \text{ A (Answer a.1)}$$

$$I_2 = \frac{E}{R_2} = \frac{96}{4} = 24 \text{ A (Answer a.2)}$$

$$I_3 = \frac{E}{R_3} = \frac{96}{6} = 16 \text{ A (Answer a.3)}$$

$$I_4 = \frac{E}{R_4} = \frac{96}{8} = 12 \text{ A (Answer a.4)}$$

b   Supply current:
Applying Kirchhoff's Current Law

$$I = I_1 + I_2 + I_3 + I_4$$
$$= 48 + 24 + 16 + 12$$
$$= 100 \text{ A (Answer b)}$$

c   Total resistance:
This can be determined in either of *two* ways:
  either:

$$\frac{1}{R_T} = \frac{1}{R_1} + \frac{1}{R_2} + \frac{1}{R_3} + \frac{1}{R_4}$$

$$\frac{1}{R_T} = \frac{1}{2} + \frac{1}{4} + \frac{1}{6} + \frac{1}{8}$$

Using the lowest common factor:

$$\frac{1}{R_T} = \frac{12 + 6 + 4 + 3}{24}$$

$$\frac{1}{R_T} = \frac{25}{24}$$

$$R_T = \frac{24}{25}$$
$$= 0.96 \ \Omega \text{ (Answer c.)}$$

or you can use your calculator!

or:

$$R_T = \frac{E}{I} = \frac{96}{100} = 0.96 \ \Omega \text{ (Answer c.)}$$

Note: In any parallel circuit, the total resistance is *always* less than the lowest-value branch resistance.

Parallel circuits in practice

Individual electrical components (lights, socket outlets, etc.) are *all* connected in parallel with each other to ensure that they share the same supply voltage. This is essential because, as we shall learn later, for any component to operate at its rated power, it *must* be subjected to its rated voltage (which must correspond to the supply voltage). This is true whether you are looking at the wiring system of a house or the wiring system in a car.

Series-parallel circuits

A **series-parallel circuit** (some textbooks call these '*combinational circuits*') is a circuit that combines the characteristics of series and parallel circuits. Unfortunately, there is an infinite number of possible combinations but, fortunately, all such circuits can be solved by applying the same logical approach as we will now apply to the following examples.

Essentially, this approach comprises reducing any resistances connected in series to a single equivalent resistance, any resistances connected in parallel to a single equivalent resistance . . . and continuing this process until we end up with a single resistance.

Examples of series-parallel circuits

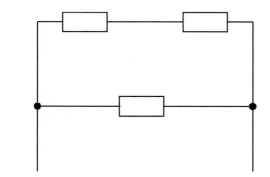

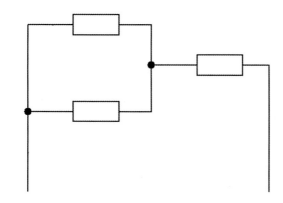

Figure 9.25

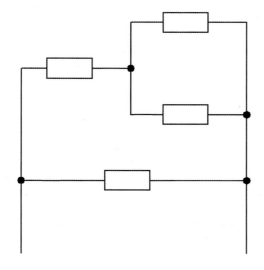

Figure 9.26

Figure 9.27

The circuits shown in Figures 9.25–9.27 are simply representative examples of series-parallel circuits. In Figure 9.25, there are two resistors in series in the upper branch, which is in parallel with a third resistor in the lower branch. In Figure 9.26, there are two resistors in parallel with each other and this combination is in series with a third resistor. In Figure 9.27 we have two resistors in parallel with each other; this combination is in series with a third (forming the upper branch), and this combination is in parallel with a fourth resistor in the lower branch.

Unfortunately, as we have already learnt, there is an *infinite* number of combinations of circuit that can be classified as series-parallel!

Solving series-parallel circuits

'Solving' a typical series-parallel circuit means that we must be able to: (a) *determine its total equivalent resistance*, and (b) *determine the voltage drop across, and the current through, each component.*

Determining the **total resistance** of a series-parallel circuit requires a logical step-by-step approach, which you cannot reasonably be expected to develop instantly! So to help you develop the skills necessary to solve series-parallel circuits, the algorithm in Figure 9.28 may prove a helpful tool.

Work through the following example, using the algorithm. Then apply it to other circuits until you are confident of the logical sequence of steps necessary for solving them. Then, phase out your use of the algorithm as you gain confidence.

Let's use this algorithm to solve the relatively simple series parallel circuit, shown in Figure 9.29. We are required to find (a) its total resistance, (b) the current through, and the voltage drop across each component.

So, starting at the top of the algorithm, we are asked '**Are there any parallel branches?**' The answer is, of course, 'yes' – we've highlighted this in Figure 9.30.

The next question is '**Are there any purely series parts to the parallel branches?**' Again, the answer is 'yes', as highlighted in Figure 9.31.

The algorithm now tells us to '**resolve purely series parts of parallel branches to a single resistance**'. So, let's go ahead and do that (we'll call the resulting resistance R_A):

$$R_A = R_1 + R_2 = 2 + 4 = 6\ \Omega$$

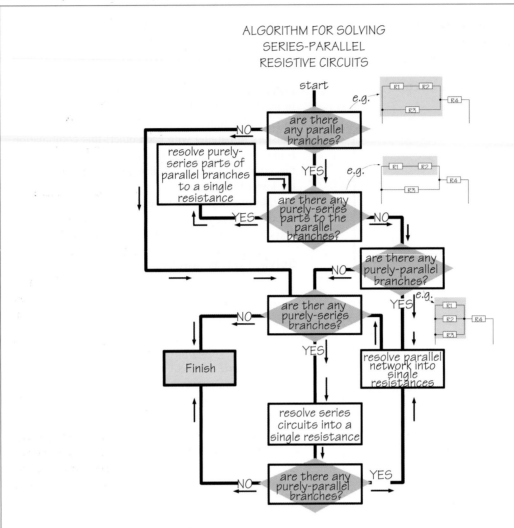

Figure 9.28

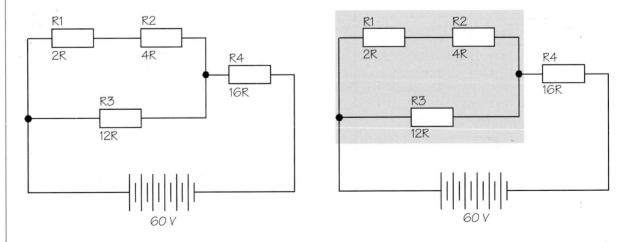

Figure 9.29   Figure 9.30

branch comprising
resistances in series

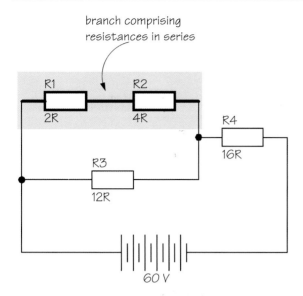

Figure 9.31

Having done this, we can redraw the circuit as shown in Figure 9.32.

two
resistances
in parallel

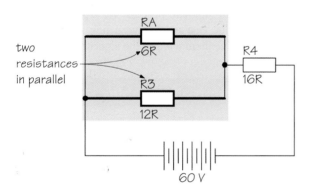

Figure 9.32

The algorithm then repeats, '**Are there any purely series parts to the parallel branches?**'. The answer is 'no', so we move on to the next question, '**Are there any parallel branches?**' and, of course, the answer is 'yes', so we are instructed to '**resolve parallel network into a single resistance**'. So, let's do that (we'll call the resulting resistance, R_B):

$$\frac{1}{R_B} = \frac{1}{R_A} + \frac{1}{R_3}$$

$$\frac{1}{R_B} = \frac{1}{6} + \frac{1}{12}$$

$$\frac{1}{R_B} = \frac{2+1}{12}$$

$$\frac{1}{R_B} = \frac{3}{12}$$

$$R_B = \frac{12}{3}$$

$$= 4\,\Omega$$

The equivalent circuit now looks like that shown in Figure 9.33.

two resistors in series

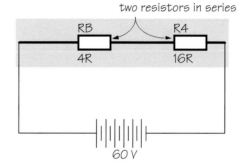

Figure 9.33

The algorithm now asks, '**Are there any purely series branches?**', to which the answer is 'yes', so we are instructed to '**resolve series circuit into a single resistance**'. Let's do that (we'll call the resulting resistance R_T):

$$R_T = R_B + R_4 = 4 + 16 = 20\,\Omega \text{ Answer (a)}$$

So, *by following our algorithm*, we have resolved (simpified) the series-parallel circuit into one equivalent resistance of 20 Ω (as shown in Figure 9.34).

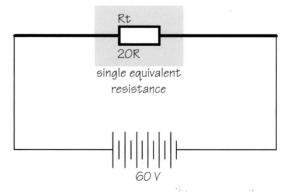

single equivalent
resistance

Figure 9.34

We have been asked to determine the *currents* flowing through, and the *voltage drops* across, each of the components. So, we'll start by redrawing the original circuit, and labelling all the voltage drops and currents (see Figure 9.35).

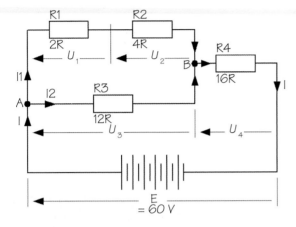

Figure 9.35

Before beginning, though, it's worth pointing out that *we cannot necessarily work out the currents and voltages in the same sequence in which they have been labelled.*

So, let's examine the circuit. We've labelled the supply voltage *I*. In accordance with Kirchhoff's Current Law, this current divides at point A in the circuit, with I_1 flowing through the upper branch of the parallel combination, and I_2 flowing through the lower branch. At point B, these two currents recombine and flow through R_4 as the supply current, *I*.

As we have already worked out the *total resistance of the circuit*, we can find the value of the supply current very easily:

$$I = \frac{E}{R_T} = \frac{60}{20} = 3 \text{ A (Answer)}$$

So, we can now quite easily find the voltage drop across R_4, which we'll label U_4:

$$U_4 = IR_4 = 3 \times 16 = 48 \text{ V (Answer)}$$

Now, we know, from Kirchhoff's Voltage Law that, in any closed loop, the sum of the individual voltage drops will equal the supply voltage. So, if we take the loop or path through R_3 (i.e. the lower branch of the parallel combination) and R_4, then:

since: $E = U_3 + U_4$

then: $U_3 = E - U_4 = 60 - 48 = 12$ V (Answer)

We can now work out the value of the current, I_2, through the lower branch of the parallel combination:

$$I_2 = \frac{U_3}{R_3} = \frac{12}{12} = 1 \text{A (Answer)}$$

If we apply Kirchhoff's Current Law to junction A, we can find the value of current, I_1, flowing through the top branch of the parallel combination:

since: $I = I_1 + I_2$

then: $I_2 = I - I_1 = 3 - 1 = 2$ A (Answer)

Finally, we are now able to determine the voltage drops, U_1 and U_2:

$$U_1 = IR_1 = 2 \times 2 = 4 \text{ V (Answer)}$$

$$U_2 = IR_2 = 2 \times 4 = 8 \text{ V (Answer)}$$

So, solving a series-parallel circuit, regardless of its complexity, involves the logical, step-by-step approach demonstrated in the above example.

Once again, you are urged not to *rely* on using the algorithm described above; instead, use it to help you *understand* the logical process for solving series-parallel circuits. Use it to solve a few such circuits, then try to solve them *without* the algorithm.

Voltage drops along cables

The lines that represent the connections (the 'wires', if you like) between components in the schematic diagrams, shown in this chapter, are considered to have no resistance. But 'real' wires, of course, *do* have resistance – albeit *low* values of resistance.

This means that, whenever a load current passes through a real wire or conductor, a voltage drop will occur along its length. So the potential difference appearing across the load must equal the supply voltage, *less* the voltage drop along the conductors supplying that load – that is:

potential difference across load = supply voltage – conductor voltage drops

Figure 9.36 should make this clear. The two voltmeters will measure the voltage drops along each of the two supply conductors.

This, of course, is really a simple series circuit, in which the resistance of each conductor is in series with the resistance of the load.

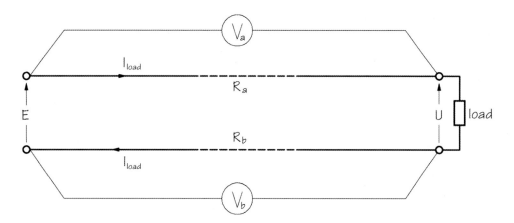

Figure 9.36

So the potential difference *(U)* appearing across the load will be the difference between the supply voltage *(E)* and the sum of the two voltage drops (voltmeter readings V_a and V_b) along the two conductors:

$$U = E - (I_{load}R_a + I_{load}R_b)$$
$$= E - I_{load}(R_a + R_b)$$

Normally, both conductors will have the same cross-sectional area and length and, so, will have identical resistances. So the above equation could then be simplified to:

$$U = E - 2(I_{load}R)$$

... where *R* is the resistance of either of the two conductors.

It is very important, of course, that the voltage drops along conductors are not excessive or it may result in a deterioration in the performance of the load being supplied. Lighting, in particular, is very susceptible to reduced voltage, and dim lighting usually indicates a low voltage. This is why the *IET Wiring Regulations* specify that the voltage drop between the supply terminals and any fixed current-using devices must not exceed **3 per cent** in the case of lighting circuits, whereas it must not exceed **5 per cent** for other circuits. For a nominal supply voltage of 230 V, this works out at 6.9 V for lighting circuits, and 11.5 V for all other circuits.

For the more common two-core cables used in residential buildings, the resistance (at 20°C) of each core, expressed in milliohms per metre, is shown in Table 9.1.

Table 9.1

Cross-sectional area	Applications	Resistance in mΩ/m
1.0 mm²	lighting circuits	18.10
1.5 mm²	lighting circuits	12.10
2.5 mm²	ring main and radial power circuits	7.41
4.0 mm²	radial power circuits	4.61

Worked example 3 The resistance of the heating element of an electric iron is 180 Ω. It is connected across a 230-V supply using a cable comprising a pair of conductors, each having a resistance of 1.2 Ω. Calculate (a) the current drawn by the electric iron, (b) the voltage across the iron's heating element, (c) the total voltage drop along the cable and (d) each core.

Solution

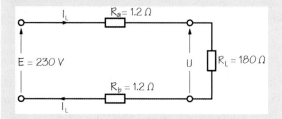

Figure 9.37

a The load current will be the supply voltage divided by the total resistance of the circuit:

$$I = \frac{E}{(R_{load} + R_a + R_b)} = \frac{230}{(180 + 1.2 + 1.2)} = \frac{230}{182.4}$$

$$= 1.26 \text{ A (Answer a.)}$$

b The voltage *(U)* appearing across the iron's heating element will be the product of the load current and the resistance of the heating element:

$$U = IR_{load} = 1.26 \times 180 = 226.8 \text{ V (Answer b.)}$$

c The voltage drop along the cable will be the difference between the supply voltage *(E)* and the voltage appearing across the load *(U)*:

$$\text{cable voltage drop} = E - U$$

$$= 230 - 226.8 = 3.2 \text{ V (Answer c.)}$$

d So, the voltage drop across each core must be:

$$R_a = R_b = \frac{3.2}{2} = 1.6 \text{ V (Answer d.)}$$

Finally . . .

Now that you have completed this chapter, are you able to achieve the objectives or learning outcomes listed at the beginning of this chapter?

Ask yourself, 'Can I . . .'

1 recognise a series circuit, a parallel circuit and a series-parallel circuit.
2 recognise and interpret voltage and current 'sense' arrows.
3 explain Kirchhoff's Voltage Law.
4 explain Kirchhoff's Current Law.
5 calculate the total resistance of a series resistive circuit.
6 calculate the current flow through a series resistive circuit.
7 calculate the voltage drop appearing across each resistor in a series resistive circuit.
8 explain the potential hazard of an open circuit in a series circuit.
9 calculate the total resistance of a parallel resistive circuit.
10 calculate the current flow through each branch of a parallel resistive circuit.
11 calculate the voltage drop appearing across each resistor in a parallel resistive circuit.
12 explain the major advantages of a parallel circuit.
13 calculate the total resistance of a series-parallel resistive circuit.
14 calculate the current flow through each resistor in a series-parallel resistive circuit.
15 calculate the voltage drop appearing across each resistor in a series-parallel resistive circuit.
16 calculate the voltage drop along conductors supplying a load.

Online resources

The companion website to this book contains further resources relating to this chapter. The website can be accessed via the following link:

www.routledge.com/cw/waygood

Chapter 10

Cells and batteries

On completion of this unit, you should be able to

1 describe the function of an electrochemical cell.
2 explain the difference between a primary cell and a secondary cell.
3 describe the three components common to all electrochemical cells.
4 explain the difference between a cell and a battery.
5 explain the terms 'capacity' and 'discharge rate', as they apply to cells and batteries.
6 briefly explain the chemical process by which a simple Voltaic cell is able to separate charge in order to provide a potential difference between its plates.
7 briefly describe the construction of a Dry Leclanché cell.
8 briefly explain the advantages of miniature cells.
9 describe each of the following, as they apply to a lead-acid cell, in terms of its plates, electrolyte and relative density:
   a fully charged state.
   b de-energising state.
   c fully de-energised state.
   d re-energising state.
10 briefly explain how a lead-acid cell's relative density is a guide to its state of charge.
11 briefly explain the disadvantages of nickel-cadmium cells, and how these are overcome using lithium-ion cells.
12 outline what is meant by the 'responsible disposal of cells and batteries'.
13 determine the potential difference and internal resistance of identical
   a cells connected in series
   b cells connected in parallel
   c cells connected in series-parallel.

Introduction

You may have heard the joke in which an American scientist is credited with inventing an 'electric rocket' which, he claims, could easily reach the moon, yet would cost the taxpayer only $5. Unfortunately, the cost of the power cord would be around $200-million!

Well, thanks to **electrochemical cells**, we *can* use many of our electrical tools, small appliances, multimedia equipment and other gadgets without having to keep them permanently plugged into our wall sockets! Although, of course, we will still require the wall socket to regularly *re-energise* those cells.

After electromagnetism, **electrochemical action** provides us with our *second most important source of electrical energy*. As well as powering our mobile phones and other electrical gadgets, **cells** and **batteries** are used to start our cars, propel submarines, provide emergency lighting, and a whole raft of other applications too numerous to mention.

In this chapter, we are going to learn about the basic behaviour of an electrochemical cell, and examine a selection of these cells and batteries together with their application in electrical engineering. Also, in this chapter, to avoid any confusion, we will consistently describe current in external circuits in terms of **electron flow**, *not* conventional flow. The use of conventional flow in this context will be unnecessarily confusing!

As we learnt in the chapter on *potential and potential difference*, an electrochemical cell uses a chemical reaction to release the energy necessary to separate charges, thus creating a potential difference across its terminals. This potential difference is then responsible for causing a drift of electric charges (free electrons)

An Introduction to Electrical Science, Waygood, ISBN 9780415810029, 2013. © Taylor & Francis

around the external circuit. These charges, of course, already exist in the conductors of the external circuit, and are not 'injected into it' by the cell! So what an electrochemical cell *does not* do, is to 'store electric charges' which it then 'pumps around' its external circuit. It is, therefore, techically incorrect to say that a cell *'stores charge'*, or that a cell *'discharges'*, or that a secondary cell can be *'recharged'*. Despite this, these terms are widely used when referring to the behaviour of cells and batteries. In this chapter, we are going to try and avoid this, by using more appropriate terms – e.g. we will say that a cell *'stores energy'*, that a cell *'de-energises'*, that a secondary cell can be *'re-energised'* and so on.

Perhaps, in school or on television, you have seen a lemon being used to demonstrate the behaviour of an electrochemical cell? By inserting two dissimilar metals, such as a galvanised nail and a copper nail, into the lemon, the resulting chemical reaction between each nail and the citric acid inside the lemon results in a small potential difference appearing across the two nails which can be measured with a voltmeter.

The same result would be obtained if, for example, we used other fruit, or even a potato, instead of the lemon. It *isn't* the lemon or the potato that is converting the energy necessary to create the resulting potential difference but, rather, *the chemical reaction that takes place within the dissimilar metals when they come into contact with an acid or alkaline.*

The 'lemon cell' comprises the same *three* components that must be present in *all* types of chemical cell: two **dissimilar electrodes** and an **electrolyte** (a dilute acid or alkaline).

The **circuit symbols** for cells and batteries are illustrated in Figure 10.1. The longer stroke always represents the positive electrode.

cell   battery

Circuit symbols for cells and batteries

Figure 10.1

Types of electrochemical cell

Electrochemical cells are either **'galvanic'** or **'electrolytic'**.

A **'galvanic cell'** creates a potential difference across its terminals, through the process of **charge separation** by the release of energy due to the chemical reaction between its electrodes and the electrolyte. Charge separation was explained in the earlier chapter on *potential difference*, and you may wish to review that chapter before continuing.

An **'electrolytic cell'**, on the other hand, uses electricity to decompose its electrolyte, a process called 'electrolysis', usually for the purpose of electroplating – such as silver or chromium plating.

Put simply, a 'galvanic cell' *produces* electrical energy, whereas an 'electrolytic cell' *uses* or *absorbs* electrical energy.

Primary and secondary cells

Galvanic cells are broadly classified as **'primary cells'** and **'secondary cells'**, where

- **primary cells** are those in which their electrochemical reaction is irreversible, which means that they *cannot be re-energised* and, so, are disposable.
- **secondary cells** are those in which their electrochemical reaction is reversible, which allows them to be *fully re-energised* and, so, are re-useable.

While **primary cells** are *'galvanic'* cells', **secondary cells** behave as *'galvanic'* cells' when they are being *de-energised*, but as *'electrolytic'* cells' when they are being *re-energised*.

Secondary cells (formerly called 'accumulators') are generally more economical than primary cells, despite their initially higher cost together with the cost of a recharging unit*, but that cost can be spread out over numerous discharge/recharge cycles before they finally fail. This makes secondary cells ideal for portable tools, video camcorders, etc.

*Strictly speaking, we should talk about a 're-energising unit' or a 'battery re-energiser', rather than a 'recharging unit' or 'battery recharger'. However, the terms 'recharging unit' or 'battery recharger' are so widely used that it would be impossible to change the terminology at this stage.

While *all* cells 'self de-energise' to some degree, when they are not in use, this is significantly lower with primary cells than it is with most secondary cells, meaning that primary cells usually have a very much longer shelf life than energised secondary cells. This makes disposable primary cells the better choice for applications where they see just occasional use – such as in torches (flashlights).

There are numerous different types of cell within each of these two general classifications, with some of the more common listed in Figure 10.2.

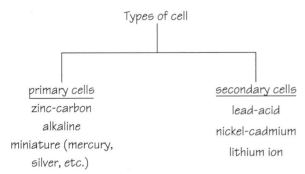

Types of cell

primary cells
zinc-carbon
alkaline
miniature (mercury, silver, etc.)

secondary cells
lead-acid
nickel-cadmium
lithium ion

Figure 10.2

We'll examine some of these cells, in various degrees of depth, later in this chapter.

Terminology

Before embarking on our investigation into the behaviour of primary and secondary cells, we need to understand some of the terminology and specifications that are common to both classes of cell.

Anode and cathode

Not surprisingly, a great deal of confusion surrounds the use of the terms '**anode**' and '**cathode**', when naming a cell's electrodes. This is due to a conflict between the ways in which these terms were applied in the past, by electrical engineers and by chemists. A conflict, incidentally, which the chemists eventually won!

Accordingly, a very brief explanation of the terms 'anode' and 'cathode', *as they relate to electrochemical*

cells (the terms are used differently for electronic devices!) won't go amiss! Even though we will be avoiding these terms in this chapter, you *will* come across them and you should be aware that, from the electrical point of view, *their definitions have changed over recent years*, rendering their use in older textbooks obsolete.

Contrary to popular belief amongst those who were brought up on older electrical textbooks, regardless of the type of cell (electrolytic *or* galvanic), by definition free electrons *always* travel through the external circuit **from the anode to the cathode**. It's as simple as that!

> Free electrons *always* travel through the **external** circuit **from a cell's anode to its cathode**.

So the terms 'anode' and 'cathode' are based on the *direction of electron flow*, and **not** on the polarity of the electrodes!

In other words, a cell's 'anode' and 'cathode' *change*, according to whether a cell is de-energising or is being energised. For a **galvanic cell** (a de-energising cell), *the negative plate is the anode* and *the positive plate is the cathode* whereas, for an **electrolytic cell** (an energising cell – i.e. with the current flowing in the opposite direction), *the positive plate becomes the anode* and *the negative plate becomes the cathode*.

> Be aware that most older textbooks invariably define the 'cathode' as being the *negative* plate and the 'anode' as the *positive* plate. *This is no longer the case!*

Because of the confusion over the use of the words 'anode' and 'cathode', we will avoid using these terms throughout the rest of this chapter. Instead, we will be referring to the electrodes as being either 'negative' or 'positive' plates.

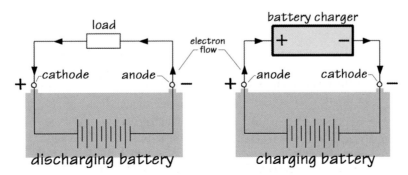

Figure 10.3

Capacity and discharge rate

The **capacity** of a cell is a measure of *its ability to deliver current to a load*. Capacity is related to the quantity of active material the cell contains, the surface area of its plates, and its temperature.

The **capacity** of any particular cell is determined experimentally by the manufacturer, by de-energising the cell at a constant current without exceeding a safe temperature, until the cell reaches its 'cut-off' voltage (i.e. the cell's minimum voltage, below which the cell is considered to be fully de-energised) and, then, multiplying that current by the time taken to reach that cut-off voltage. A cell's capacity, therefore, is the product of current and time and, so, is expressed in **ampere hours (A·h)**. The capacity of smaller cells are expressed in milliampere hours (mA·h).

A cell's capacity is dependent upon the temperature at which that cell is operating, and capacity is normally expressed at a temperature of 25°C.

From the above explanation of capacity, it would *seem* that a lead-acid car battery, for example, with a capacity of, say, 200 A·h, is *theoretically* capable of delivering *any* combination of current and time, whose product is 200 A·h, for example:

    either: 1 A for a period of 200 h
    or: 2 A for a period of 100 h
    or: 10 A for a period of 20 h
    or: 20 A for a period of 10 h
    or even: 2000 A for a period of 6 min!
    . . . and so on!

'Theoretically' maybe! But, in practice, we *have* to take care; excessively high currents cause excessively high temperatures which can actually melt a battery's electrodes or even cause a battery to explode! And 2000 A is most certainly an excessively high current! So, how do we know what the maximum 'safe' continuous de-energising current should be for a particular cell or battery?

The answer is that we must refer to another part of a battery's specification: its **discharge rate** (or, more accurately, '**de-energising rate**'), expressed in **hours**. We can use this figure, together with the corresponding capacity, to determine a cell or battery's maximum, *safe*, continuous de-energising current – that is, the maximum continuous current that can be supplied to a load without overheating the battery.

> A cell's '**discharge rate**' determines the maximum value of continuous current that cell is capable of supplying to its load without overheating.

For lead-acid cells, discharge rates are typically expressed as '**8 h**', '**10 h**' or as '**20 h**'. Often, the discharge rate is shown on specification sheets in the form: '**C/8**', '**C/10**' or '**C/20**', where '**C**' represents the capacity of the cell.

For example, the motor vehicle's lead-acid battery that we described earlier may be quoted as having a capacity/discharge rate of '**200/10**', which represents a capacity of 200 A·h at a discharge rate of 10 h, which means that it is designed to supply a *maximum* continuous de-energising current of (200 ÷ 10 =) 20 A for 10 h before the cell reaches its cut-off voltage.

An alternative way of defining a cell's 'capacity' (mainly used with larger cells or batteries) is by expressing it in **watt hours (W·h)** although, strictly speaking, 'capacity' is the *wrong* word to use in this context because a watt hour is a measure of the **energy**, not of the charge, a cell can supply.

To convert the capacity of a cell, expressed in ampere hours, into the energy it can supply in watt hours, we can use the following equation:

$$\left[\begin{array}{l} \text{energy (in watt hours)} = \text{capacity} \\ \text{(in ampere hours)} \times \text{cell voltage (in volts)} \end{array} \right]$$

For example, a 12-V lead-acid battery having a capacity of 250 A·h, will be able to supply 3000W·h or 3 kW·h (i.e. 250 A·h ×12 V).

Now that we have learnt some of the terminology involved, let's move on and study the cells themselves.

Primary cells

Simple Voltaic cell

In this section, we will start by briefly examining the operation of what is known as a '**simple Voltaic cell**', whose operation is representative of the chemical process by which *all* primary cells operate, albeit with different combinations of chemicals.

The important thing to understand, not only about this particular cell, but about all cells, is that they provide the energy required to increase the potential of a charge passing though that cell. You may wish to review the chapter on *potential and potential difference*, before proceeding.

A 'simple **Voltaic** cell' is the name we give to a cell consisting of copper and zinc electrodes, inserted into an electrolyte of dilute sulfuric acid*.

A **chemical symbol** is used to identify a particular element. For example, the symbol **H** represents **hydrogen**, **Zn** represents **zinc**, and so on. Combinations of these symbols represent compounds, with numeric subscripts indicating the presence of two or more atoms of an element. For example, the chemical symbol for (the compound) water, is H_2O, which indicates a molecule comprising *two* atoms of hydrogen and *one* atom of oxygen.

The chemical symbol for pure sulfuric acid is H_2SO_4, indicating that it is a compound comprising **hydrogen** (H_2) and **sulfate** (SO_4). When dissolved in water, sulfuric acid *dissociates* – that is, it *separates* into its constituent parts hydrogen and sulfate. During this process, the sulfate (SO_4) molecules each acquire two electrons from the hydrogen (H_2) molecules. As a result of this, the sulfate molecules become negatively charged sulfate ions (SO_4^{2-}), while the hydrogen molecules each become positively charged hydrogen ions ($2H^+$).

Remember, an **ion** is simply a charged atom or molecule – i.e. one that has become electrically unbalanced by gaining or losing one or more electrons. It's charge, positive or negative, is indicated by + or – superscript added to its chemical symbol (e.g. a negative sulphate ion, that has gained two excess electrons is therefore shown as: SO_4^{2-}), as indicated in the previous paragraph.

The ionisation of sulfuric acid (or, in fact, *any* acid or, in some cases, alkaline) in water leaves charged particles within the solution. So we can say that the solution, now termed an electrolyte, *has become a liquid conductor* and *capable of supporting current flow* which is, of course, defined as a drift of charged particles.

When the zinc electrode is inserted into the electrolyte, the zinc starts to dissolve. This process is caused by positive zinc ions (Zn^{2+}) which, attracted by the negative sulfate ions (SO_4^{2-}), detach themselves from the zinc electrode, each leaving behind two electrons, and move into the solution and combine with the negative sulfate ions to form neutral zinc sulfate ($ZnSO_4$) which plays no further part in the reaction. As more and more zinc ions detach, the zinc electrode starts to acquire a negative charge due to the surplus of electrons left behind on it.

This is quite a vigorous process and, after a short period of time, the zinc electrode has acquired a sufficiently high negative charge to prevent any further loss of positive zinc ions into the electrolyte, and any further action stops. If any more positive zinc ions now try to detach themselves, they are simply attracted back to the negatively charged electrode.

At this point, if it were possible to measure the amount of negative potential of the zinc electrode, relative to the electrolyte, it would be found to amount to **–0.76 V**.

If we now turn our attention to the copper electrode, a similar, but much milder, process occurs. That is, positive copper ions (Cu^{2+}) dissolve into the electrolyte, each leaving two electrons behind, causing the copper electrode to eventually acquire sufficient negative charge to prevent any further positive copper ions

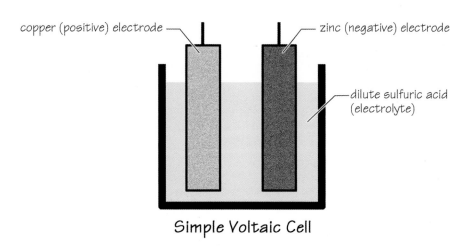

copper (positive) electrode

zinc (negative) electrode

dilute sulfuric acid (electrolyte)

Simple Voltaic Cell

Figure 10.4

from leaving. This time, if we were able to measure the potential of the copper electrode relative to the electrolyte, we would find that its potential is **+0.34 V**.

So the zinc electrode becomes the cell's **negative plate** while, *relative to the zinc electrode*, the copper electrode becomes the cell's '**positive' plate**.

At this point in our explanation you are no doubt wondering, if the copper electrode has acquired a *negative* charge, *why* its potential should be +0.34 V? The answer is that, while its potential is indeed negative, it is much *less negative* than the zinc electrode and, so, *relative to the zinc*, it is considered to be positive. If this has left you confused, by all means review 'potential', in the chapter on *potential and potential difference* before you continue on, but the key thing to remember is that *all potentials are relative to the point of reference from which they are measured.*

In fact, it is an oversimplification to say that the potentials of the two electrodes are 'measured with respect to the electrolyte'. Rather, by common agreement, their potentials are theoretically compared to that of a **hydrogen electrode**, whose reference potential has been arbitrarily assigned as being at zero volts. Even though *all* metal electrodes (as well as our hydrogen electrode) actually acquire *negative* potentials, those metals which are *less negative* than the hydrogen electrode are considered to be *positive with respect to the hydrogen electrode*. Figure 10.5 should make this clear.

The potentials of different metals, each measured with respect to this 'hydrogen electrode', are listed in what is known as the '**Electrochemical Series**', with what are described as the 'more-active' electrodes (e.g. lithium, at −3.02 V) towards the top of the series, and the 'less-active' electrodes (e.g. gold, at +1.68 V) towards the bottom of the series.

Table 10.1

Conductor	Symbol	Electrode potential/ volts
lithium	Li	−3.02
potassium	K	−2.92
barium	Ba	−2.90
sodium	Na	−2.71
aluminium	Al	−1.67
zinc	Zn	−0.76
chromium	Cr	−0.71
iron	Fe	−0.44
nickel	Ni	−0.25
tin	Sn	−0.14
lead	Pb	−0.13
hydrogen	**H**	**0.00**
bismuth	Bi	+0.20
copper	Cu	+0.34
silver	Ag	+0.80
mercury	Hg	+0.85
gold	Au	+1.68

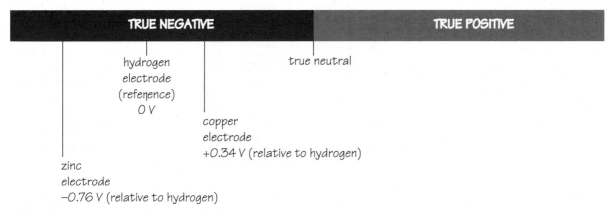

Figure 10.5

The *potential difference* between *any* pair of electrodes from this series is then obtained by simply subtracting their individual potentials – and the *further apart* they are on the series, the *greater* the resulting potential difference between them.

So, for copper and zinc, the resulting potential difference between them is:

$$+0.34 - (-0.76) = \mathbf{1.10\ V}$$

The potential difference (voltage) between the copper and zinc electrodes of a simple Voltaic cell is **1.10 V.**

So, while the electrolyte plays a key part in the creation of a potential difference between a cell's electrodes, *it's actually the materials from which the electrodes themselves are manufactured that determine the potential difference that can be achieved by any particular type of cell.*

You should also understand that this voltage represents the 'open circuit' voltage of the cell, which we call its **electromotive force (e.m.f.).** When supplying a load, the terminal voltage will always fall below the value of the cell's e.m.f., as we shall learn later.

Let's move on, now, and find out what happens when our cell's electrodes are connected to an external load.

Remember that once the two electrodes have reached their respective potentials, *no further activity will take place within the cell.* The zinc electrode has become sufficiently negative to prevent any further positive zinc ions from detaching into the electrolyte, and the copper electrode has become sufficiently negative to prevent any further positive copper ions from detaching.

However, immediately the cell is connected to an external load, things begin to happen! *Electrons start to leave the more negative zinc electrode, and travel through the load, across to the less negative ('positive') copper electrode.*

As more and more electrons leave the zinc, its negative charge starts to fall, allowing more and more positive zinc ions to detach themselves and dissolve into the electrolyte. These positive zinc ions (Zn^{2+}) now start to *repel* the positive hydrogen ions ($2H^+$) adrift within the electrolye, *driving them towards the copper electrode.*

So, a drift of positive hydrogen ions becomes established within the electrolyte, from the zinc electrode towards the copper electrode. At the same time, there is a drift of negative sulfate ions towards the zinc electrode. In other words, *the drift of positive hydrogen ions and negative sulphate ions, in opposite directions, constitutes the electric current within the electrolyte* while *the drift of free electrons constitutes the electric current through the external circuit.*

Although the current in the external circuit is by *free electron drift* and the current within the electrolyte is by *ion drift*, their magnitudes, of course, are identical – i.e. when the *external* current is, say, 1 A, the *internal* current is also 1 A. Although the charge carriers are different, it is, of course the *same* current!

As each positive hydrogen ion arrives at the copper electrode, it attracts an electron which has arrived at the copper electrode via the external circuit, forming a *neutral hydrogen atom*, and is liberated to form bubbles of hydrogen gas.

Some of this hydrogen gas simply bubbles out of the electrolyte to disperse in the atmosphere but as the action continues, more and more of these bubbles start to coat and isolate the copper electrode – a condition we call *'polarisation'*. Polarisation eventually causes the copper electrode to behave as though it was a *hydrogen electrode* (which is why we see a 'hydrogen electrode' listed amongst metal electrodes in the *Electrochemical Series*), and the potential difference between the two electrodes (zinc and hydrogen) now falls towards 0.76 V.

So, if you were wondering how it is possible to have a 'hydrogen electrode', then this is the answer.

With this reduction in the cell's electromotive force, the current in the external circuit also starts to fall. This condition is equivalent to the resistance of the cell increasing and, in fact, we describe this condition as an *'increase in the cell's internal resistance'.*

Polarisation makes this particular cell unsuitable for most applications. In fact, for practical purposes, the simple Voltaic cell is little more than a laboratory demonstration. In more practical cells (e.g. the dry cells we use to power our torches, etc.), the problem of polarisation is avoided by adding a 'depolarising' agent to the electrolyte; this material contains oxygen which combines with the hydrogen to form water, reducing the amount of hydrogen bubbles and leaving the positive electrode clear.

Unfortunately, the addition of a depolarising agent does absolutely nothing to counter the continuing destruction of the zinc electrode (and, to a lesser extent, the copper electrode) and, eventually, the electrode becomes completely eaten up and the cell dies. This is accelerated by '**local action**', which describes internal electrolytic corrosion within the electrodes themselves due to the presence of impurities within the metals.

Dry Leclanché cells

The 'dry cell' or, more accurately, the '**dry Leclanché**' cell, is also known as a '**carbon-zinc**' cell, after the materials from which its electrodes are manufactured. This is one of the most common, and inexpensive, disposable cells that we use to power our torches, electronic toys, etc.

Strictly speaking, 'dry cells' are not dry at all; rather, their electrolyte is a *paste* or *gel* (so it might be more accurate to describe it as a 'non-spillable wet cell'!), composed of ammonium chloride typically mixed with starch and flour.

Its positive 'plate' is a carbon rod located at the centre of the cell, immediately surrounded by a core of fine carbon granules mixed with manganese oxide, which act as a 'depolarising agent' to reduce the amount of hydrogen bubbles forming on the carbon electrode. The negative 'plate' is a zinc canister, into which the carbon rod and depolarising agent is inserted, with the space between filled with the electrolyte, before being completely sealed.

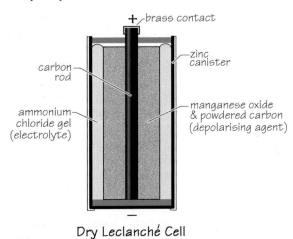

Dry Leclanché Cell

Figure 10.6

The electromotive force achieved by the dry carbon-zinc cell is approximately **1.5 V**, but this drops significantly if a large load current is drawn from the cell.

Unfortunately, as zinc ions detach into the electrolyte, the zinc container is gradually eaten away until, eventually, the cell begins to leak electrolyte. For this reason, dry cells that are approaching the end of their life should always be removed from the device they are powering to avoid the device from becoming damaged from any corrosion caused by the leaking electrolyte.

The capacity of an AA-size carbon-zinc cell is typically 400–1700 mA·h. The large battery manufacturer, *EverReady*, claims that its carbon cells have a shelf life of approximately one year while retaining 95–100 per cent of their capacity, or up to four years with a capacity of 65–80 per cent.

Other types of 'dry' disposable cells include the '**alkaline cell**'. This is similar in appearance to the carbon-zinc cell, but it uses an electrolyte of potassium hydroxide, which is an *alkaline* rather than an acid – hence its name. Alkaline cells have between three and five times the capacity of an equivalent carbonzinc cell, making them the ideal alternative for powering photographic equipment, portable audio equipment, etc.

Because the electrodes of primary cells corrode when they de-energise, *they cannot be re-energised* – unlike the next class of cells we are going to study.

Miniature cells

The **mercury** (or **mercuric oxide**) **cell** is one of several types of miniature, or 'button', cells developed as a spin-off to the U.S. space programme in order to meet the demands for miniaturisation, and they have day-to-day applications which include powering electronic watches, hearing aids, etc. They are manufactured in a wide range of diameters and thicknesses, they have an excellent shelf life, and the materials used in the cell provide it with many times the capacity of other cells of comparable size.

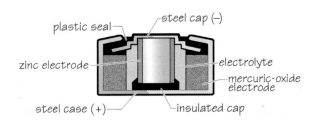

Figure 10.7

The mercury cell is so-called because its positive electrode is manufactured from a mixture of mercuric-oxide and graphite. Its negative electrode is manufactured from zinc, and its electrolyte is potassium hydroxide. The entire cell is encapsulated in a two-part stainless-steel capsule, with the main part usually (but not always) providing the positive potential, and the 'cap' part providing the negative potential to the external circuit. Its e.m.f. is 1.35 V, which remains exceptionally steady throughout its life span.

As mercury is highly toxic, a more environmentally friendly alternative is the **silver-oxide** cell which, externally, looks identical to a mercury cell, but which has a slightly higher e.m.f. of 1.5 V. This cell uses silver oxide as its positive electrode and zinc as the negative electrode, with sodium hydroxide or potassium hydroxide as an electrolyte.

In common with other types of cell, these need to be disposed of responsibly, as discussed elsewhere in this chapter.

Secondary cells

A **secondary cell** will deliver current to a load through a chemical reaction similar to that of primary cells, but its chemical reaction is *reversible*, enabling it to be restored to practically its original condition.

A secondary cell can be restored, or '**re-energised**', simply by passing a d.c. current through the cell *in the opposite direction to its de-energising current.*

When a cell stops working, it has run out of **energy**, *not* charge! The chemical reactions that take place within a cell releases **energy**, and this energy is used to separate charges in order to produce a potential difference across its terminals. So, when a secondary cell's chemical reaction stops, the cell has become 'de-energised' not 'discharged'. And we 're-energise' that cell; we don't 'recharge' it!

Lead-acid cells

Invented in the mid-nineteenth century by a Frenchman, Gaston Planté, the **lead-acid cell** is one of the most widely used types of secondary cell, and the type used in practically all motor vehicles. More compact lead-acid cells are also used widely for other applications, such as emergency lighting, uninterruptible power supplies (UPS), etc.

A lead-acid cell has an open-circuit voltage of around 2.1 V when fully energised, falling to a little less than 2 V when fully de-energised.

Construction

Prior to initial charging, the active material used on *both* electrodes is manufactured from a mixture of lead sulfate and lead oxide. During the initial charging, however, this material is converted to a hardened paste of brown-coloured **lead dioxide** (also known as 'lead peroxide') which forms the positive plate, and to a grey-coloured **porous metallic lead** (also known as 'spongy lead') which forms the negative plate.

These so-called 'active materials' are relatively weak, so they need mechanical support which is provided by an **electrode grid** manufactured from a hard lead alloy. In addition to providing support for the active materials, the grid also acts to conduct the current between the active materials and the external load.

This combination of active material and supporting grid is called a '**plate**'.

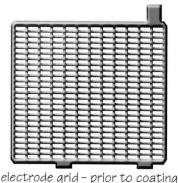

electrode grid – prior to coating

Figure 10.8

The positive and negative plates are kept apart from each other by means of **separators**, which are a little larger in area than the plates themselves. These are sheets of insulation, typically manufactured from polyethylene, which are microporous in order to enable ions to pass between adjacent plates of opposite polarity.

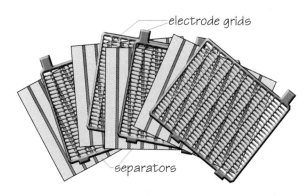

electrode grids

separators

Figure 10.9

Combinations of plates and separators make up the individual cells.

A number of individual cells are inserted into a polypropylene container, connected together in series, and a lid fitted. Lead bars are used to interconnect plates of like polarity, and to provide posts by which the battery can be connected to its load. In the case of a nominal 12-V car battery, six cells are used.

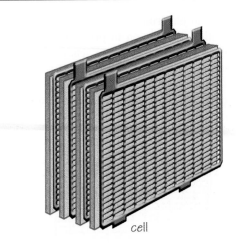

cell

Figure 10.10

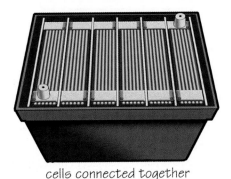

cells connected together
in series

Figure 10.11

Behaviour of lead-acid cell during discharge/charging cycle

The action of a lead-acid cell during its de-energising/re-energising cycle involves chemical reactions between the lead compounds on its electrodes, and changes to the relative density (specific gravity) of the electrolyte due to the formation of water.

As well as being reversible, the chemical reaction of a lead-acid is rather more complicated than that of a primary cell (with the behaviour at the positive electrode being particularly complicated!). Fortunately, its overall behaviour can be neatly summed up in the form of what is known as a '**reversible chemical equation**', as follows:

$$PbO_2 + Pb + 2H_2SO_4 \rightleftharpoons 2PbSO_4 + 2H_2O$$

This equation might *look* complicated but, in fact, it is very simple. The double-headed equals-sign simply means that the process is *reversible*; the upper part (pointing to the right) indicating the 'completely de-energised' state, and the lower part (pointing to the left) indicating the 'fully energised' state.

So the left-hand side of the equation represents the situation *before* the cell de-energises. The chemical symbol, PbO_2, represents lead dioxide (the positive plate), Pb represents grey metallic lead (the negative plate), and $2H_2SO_4$ represents the electrolyte, dilute sulfuric acid. These are the chemicals present within the cell *when it is fully energised*.

The right-hand side of the equation represents the situation *after* the cell has become fully de-energised. The chemical symbol, $2PbSO_4$, represents lead sulfate (a white-coloured compound) which has replaced the lead dioxide and metallic lead on the electrodes, while the chemical symbol, $2H_2O$, indicates the formation of water which dilutes the electrolyte. These are the chemicals present within the cell *after it has become de-energised*.

So let's move on and consider each of the four stages of the cell's condition: (a) *fully energised*, (b) *de-energising*, (c) *fully de-energised* and (d) *energising*.

Fully energised cell

Table 10.2

positive plate:	Brown lead dioxide
negative plate:	Grey porous metallic lead.
electrolyte:	Maximum acid content. Maximum relative density.

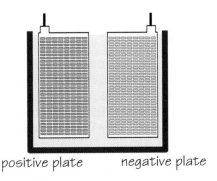

positive plate   negative plate

Figure 10.12

Cell de-energising

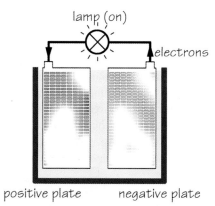

Figure 10.13

Table 10.3

positive plate:	Brown lead dioxide being converted to (white) lead sulfate.
negative plate:	Grey porous metallic lead being converted to (white) lead sulfate.
electrolyte:	Formation of water, diluting the acidity. Relative density falling.

Fully de-energised cell

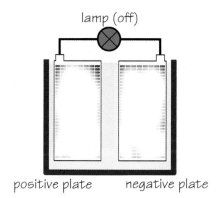

Figure 10.14

Table 10.4

positive plate:	Electrode now mainly white lead sulfate.
negative plate:	Electrode now mainly white lead sulfate.
electrolyte:	Maximum water content/ minimum acid content. Minimum relative density.

Re-energising cell

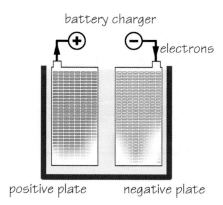

Figure 10.15

Table 10.5

positive plate:	White lead sulfate being converted back to brown lead dioxide. Water converted to hydrogen and oxygen gases.
negative plate:	White lead sulfate being converted back to grey porous metallic lead. Water converted to hydrogen and oxygen gases.
electrolyte:	Acid content increasing. Relative density increasing.

Relative density

As you can see from Tables 10.2–10.5, the **relative density** of the electrolyte is at its maximum when the cell is fully energised, and at its minimum when the cell is completely de-energised.

So, the simplest way to determine the capacity of a lead-acid cell or battery is to measure the electrolyte's **relative density**. The term 'relative density', has long replaced the older term 'specific gravity' which, in this context, is a measure of *the mass of electrolyte compared to the mass of an equal volume of pure water*.

Pure sulfuric acid has a relative density of 1.84 – which means that it has 1.84 times the mass of an equal volume of pure water. When sulfuric acid is prepared as an electrolyte, by diluting it with water, its relative density is typically about **1.27**. During de-energising, the electrolyte becomes even more diluted, due to the formation of water during the chemical reaction, and its relative density gradually falls to about **1.17** when the cell is completely de-energised.

The electrolyte's relative density (RD) is a guide to the cell's condition.

*Completely energised cell's RD = **1.27** at 15°C*
*Completely de-energised cell's RD = **1.17** at 15°C*

Relative density varies with temperature, and the figures above are quoted for a temperature of 15°C. Battery rooms are normally equipped with a temperature-compensation chart which allows the measured relative density to be adjusted for the ambient temperature.

The relative density of an electrolyte is measured using a **hydrometer** (illustrated in Figure 10.16). Enough electrolyte is drawn into the hydrometer, using the rubber bulb, to enable the graduated float to rise. The relative density is then read off the scale engraved on the float, at the level of the electrolyte.

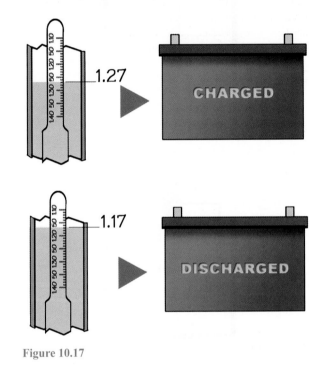

Figure 10.17

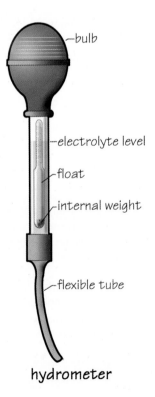

bulb

electrolyte level

float

internal weight

flexible tube

hydrometer

Figure 10.16

Always beware of the corrosive properties of sulfuric acid!

Figure 10.18

Energising lead-acid cells and batteries

Lead-acid cells are re-energised, using purpose-build **battery chargers**. Since it is necessary to pass the energising current in the *opposite* direction to the cell's de-energising current, the battery charger must be connected with great care to ensure that its positive terminal is connected to the battery's positive electrode, and its negative terminal is connected to the battery's negative electrode.

Professional battery chargers, of the type you are likely to encounter in workshops or electrical substations, allow several ways in which to re-energise lead-acid cells or batteries, these include: **constant-current**, **constant-voltage** and **trickle**.

- **Constant-current re-energising**. A constant re-energising current of around 7 per cent of the figure representing the cell's capacity is a safe rule of thumb. For example, if the cell has a rated capacity of, say, 250 A·h, then we should use a current of 7 per cent of 250, or 17.5 A. The cell's relative density should then be monitored, hourly, and when it ceases to increase further, the re-energising can cease.
- **Constant-voltage re-energising**. This method will cause a high initial value of re-energising current. As the cell's potential difference, which opposes the battery charger's voltage, will gradually increase during re-energising, the current will gradually decrease. With badly sulfated cells (see below), this method may result in excessively high temperatures (in excess of 50°C), and is best avoided.
- **Trickle re-energising**. This method is also called 'float re-energising', and is carried out by connecting the cell across a source which provides just enough re-energising current to make up for any loss through de-energising. This method is widely used in electricity substations, for example, to ensure that the batteries remain at their full capacity in order to meet the occasional requirements for tripping circuit breakers, etc.

Care and maintenance of lead-acid cells and batteries

Should you ever be tasked with maintaining lead-acid cells or batteries, you must be constantly aware of the **corrosive properties of electrolyte**, and of the **potentially explosive nature of the gas** vented from batteries during their charging process.

Accordingly, you must *always* wear eye protection, and the rubber aprons and gloves that will be provided by your employer and, of course, *never* smoke near batteries or, in particular, inside a battery room. Battery rooms are normally provided with an emergency eye-wash station, and you should always make yourself aware of where this is and how to use it.

It's important that lead-acid cells and batteries are regularly cleaned, and their terminals and connections lightly coated with petroleum jelly in order to prevent corrosion. Cleaning can be carried out by washing the outside of the batteries using a solution of baking soda, followed by clean water.

The electrolyte level should *never* be allowed to fall below the level of the plates and, whenever necessary, topped up using distilled water. Allowing the electrolyte level to remain below the plates will cause any lead sulfate to dry out and harden – a process called 'sulfation' – which is irreversible, and will reduce the effective active area of the plates, reducing the cell's capacity.

If you ever have to prepare electrolyte, you must remember to *always add acid to water,* **never** *the other way around!* Failure to do so could result in an explosion as the reaction can be extremely vigourous, producing large quantities of heat.

Performance ratings for lead-acid vehicle batteries

As we have already learnt, the capacity of a battery is normally expressed in **ampere hours** or in **watt hours**. However, the performance of a vehicle battery now includes two additional specifications, termed its '**cold cranking capacity**' and its '**minimum reserve capacity**'.

> **Cold-cranking current** is defined as *'that current which can be continuously delivered by a cell, at –18°C, for 30 s, after which each cell must deliver a terminal voltage of 1.2 V or higher'.*

For example, a 12-V car battery having a cold-cranking current of, say, '350 A', means that is must be able to deliver 350 A for 30 s, at –18°C, without the battery's voltage falling below 7.2 V (i.e. $6 \times 1.2 = 7.2$ V).

> **Minimum reserve capacity** is defined as *'the time, in minutes, that a cell will support a full accessory load (about 25 A), at 25°C, while sustaining a terminal voltage of 1.75 V or better'.*

You can think of the minimum reserve capacity of a car battery as being a sort of 'capacity insurance', indicating how long you have to find a service station after, say, your vehicle's alternator or fan belt fails. For example, a minimum reserve capacity of, say, '150' means that you have 150 minutes! Less, of course, if you switch off most of the car's accessories.

Nickel-cadmium (Ni-Cd) cells

Nickel-cadmium (**Ni-Cd**) cells (widely referred to as 'ni-cads') are re-energisable cells that use nickel oxide hydroxide and metallic cadmium foil as their positive and negative electrodes, and potassium hydroxide as an electrolyte. The two plates, kept apart by a separator which is impregnated with the electrolyte, are rolled up rather in the same way as a 'Swiss roll'.

Fully energised Ni-Cd cells achieve an electromotive force of 1.25–1.35 V. One of the characteristics of a Ni-Cd cell is that its voltage remains relatively constant during de-energising until, eventually, it suddenly falls to around 0.9 V with little or no warning.

One of the major drawbacks with Ni-Cd cells is their so-called '**memory effect**', which results in them failing to achieve their rated capacity. This occurs when the cell is repeatedly re-energised after having been only partially de-energised, and causes the cell to appear as though it 'remembers' the lower capacity.

Yet another drawback with Ni-Cd cells is an effect called '**voltage depression**', which is usually the result of repeated over-energising. Voltage depression describes a condition in which the cell's voltage falls far more rapidly than normal, giving the impression that it is not maintaining its capacity.

Nickel-cadmium cells have been widely used for powering a wide range of portable electric handtools, such as electric drills, etc., as well as for a great many consumer electronics applications, including camcorders, laptop computers, mobile telephones, etc.

Lithium-ion (Li-ion) cells

The main drawbacks of nickel-cadmium cells, i.e. their 'memory effect' and 'voltage depression', have been overcome thanks to the introduction, during the early 1990s, of **lithium-ion** (**Li-ion**) cells and batteries.

As well as eliminating the problems of memory effect and voltage depression, lithium-ion cells are very much lighter and, physically, they are significantly smaller than other re-energisable cells of similar capacity. An interesting comparison is that a lithium-ion battery can supply around 150 W·h of energy per kilogram mass of battery, compared with just 25 W·h of energy per kilogram for a lead-acid battery! That's an enormous difference!

Although currently more expensive than a corresponding nickel-cadmium battery, lithium-ion batteries now have become the standard for professional and consumer electronics equipment such as camcorders, mp3 players, laptop computers, mobile telephones, etc., and they are now starting to appear as light-weight battery packs for power tools.

A lithium-ion cell's electrodes can be manufactured from various materials. Typically, the positive electrode consists of a thin aluminium foil, coated with lithium and cobalt metal oxides, and the negative electrode consists of a thin copper foil, coated in graphite. The electrodes are separated with porous polyethylene film. The electrolyte is lithium hexafluorophosphate ($LiPF_6$), dissolved in an organic solvent.

The electrodes, their separator and the electrolyte is wound ('Swiss roll' style) into a sealed container.

While re-energising, lithium ions are transferred through the electrolyte from the positive plate to the negative plate; while de-energising, lithium ions are transferred in the opposite direction. During this process, little change takes place to either the positive or the negative plate.

The average e.m.f. for a lithium-ion cell is 3.6 V (it varies from 4.2 V down to about 3 V); which is equivalent to *three* nickel-cadmium cells, thus reducing the number of cells required for similar applications.

Lithium-ion cells require their own, dedicated, two-stage charging unit. Initially, the charger limits the energising current, until the voltage across the electrodes reaches 4.2 V, at which point it then changes to a constant voltage operation as the current reduces towards zero. Under normal operating conditions, these cells are capable of between 300 and 500 energise/de-energise cycles.

Responsible disposal of cells and batteries

As you are aware, cells and batteries contain hazardous and toxic materials, so it is important that we dispose of them responsibly in order to reduce any risk to the environment or to human health. This means *not* disposing of them with other general municipal waste that finds its way into landfill sites but, instead, making use of the **battery collection points** that are now available at sales points.

In order to encourage users to dispose of cells and batteries responsibly, under the European Battery Directive 2006/66/EC, manufacturers are now required to bear the cost of collecting, treating and recycling industrial, automotive and portable batteries, together with the costs of any publicity campaigns intended to inform the public of these arrangements. Countries not covered by this European directive have their own corresponding requirements which amount to the same thing.

This includes the requirement to provide facilities at sales points, where cells and batteries can be safely discarded for collection and safe disposal under the Directive.

Manufacturers are also required to provide visible, legible and indelible markings on their batteries and/or packaging which, among other requirements, displays the **WEEE (Waste Electronic and Electrical Equipment Directive 2008***)* pictogram, together with the relevant chemical symbols, as in Figure 10.19.

Pb/Hg/Cd

Figure 10.19

This pictogram also appears on any electrical product that contains embedded battery packs which are intended to last for the life of that product (e.g. electric toothbrushes), and indicates that the product itself must also be disposed of separately from municipal waste.

Batteries

As we have already learned, a **battery** is simply *a number of cells connected together*. In most cases, cells are connected either in **series** or in **parallel**, although **series-parallel** connections, of course, are also possible.

It is *not* recommended that we connect different types of cell together, or to mix partially de-energised cells with fully energised cells, so in this section, we will assume that each cell is of *the same type* and has *the same level of energy*.

Cells in series

Cells are connected in **series** when the positive terminal of one cell is connected to the negative terminal of the next, as illustrated in Figure 10.20.

$$E = E_1 + E_2 + E_3$$

Figure 10.20

We connect cells in series when we want to increase the available e.m.f. The total e.m.f. of cells connected in series is the sum of their individual e.m.f.s.

$$E_{total} = E_1 + E_2 + E_3 + etc.$$

Of course, *we can only achieve e.m.f.s that are multiples of that of an individual cell* – we *cannot*, for example, achieve an e.m.f. of, say, 10 V, using 1.5-V dry cells! We could only achieve 10 V by using a completely *different type* of cell, such as five 2-V lead-acid cells.

The total internal resistance of cells connected in series is the sum of their individual internal resistances, just like resistors in series:

$$R_{total} = R_1 + R_2 + R_3 + etc.$$

The total discharge current available from cells connected in series must exceed the discharge current delivered by an individual cell.

Cells in parallel

Cells are connected in parallel when all their positive terminals are connected together and all their negative terminals are connected together, as illustrated in Figure 10.21.

We connect cells in parallel when we want to maintain the same e.m.f. as a single cell, but wish to increase the available discharge current.

$$E_{total} = E_1 = E_2 = E_3 = etc.$$

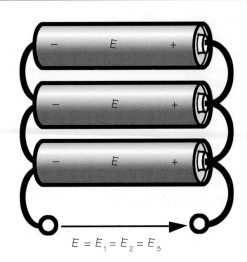

$$E = E_1 = E_2 = E_3$$

Figure 10.21

The maximum discharge current available from cells connected in parallel is the sum of the currents deliverable by each cell.

The total internal resistance is determined in exactly the same way as we determine the total resistance of resistors in parallel:

$$\frac{1}{R_{total}} = \frac{1}{R_1} + \frac{1}{R_2} + \frac{1}{R_3} + etc.$$

The alternative, and quicker, method is to simply divide the internal resistance of one cell by the total number of cells:

$$\text{total internal resistance} = \frac{\text{internal resistance of one cell}}{\text{number of cells}}$$

Cells in series-parallel

If we want to increase *both* the terminal e.m.f. *and* the amount of discharge current deliverably, then we could connect (identical) cells in **series-parallel**, as illustrated in Figure 10.22.

In the example in Figure 10.22, the total e.m.f. will be three times the e.m.f. of an individal cell, and the complete battery will be capable of delivering a discharge current that is three times that of any individual branch cell (which, in turn, is that of any individual cell).

The total internal resistance of this arrangement can be determined in exactly the same way as we learnt to determine the total resistance of any series-parallel resistive circuit. In this example, the total internal resistance will be one-third of the internal resistance of one branch.

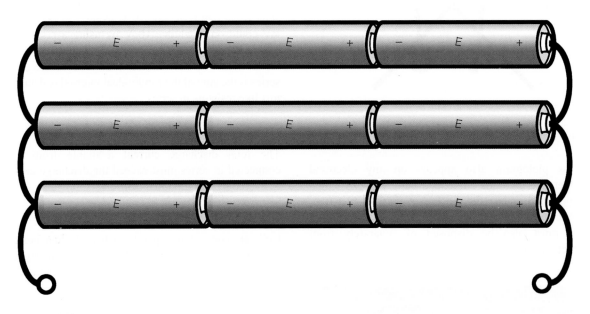

Figure 10.22

Finally . . .

Now that you have completed this chapter, are you able to achieve the objectives or learning outcomes listed at the beginning of this chapter?

Ask yourself, 'Can I . . .'

1 describe the function of an electrochemical cell.
2 explain the difference between a primary cell and a secondary cell.
3 describe the three components common to all electrochemical cells.
4 explain the difference between a cell and a battery.
5 explain the terms 'capacity' and 'discharge rate', as they apply to cells and batteries.
6 briefly explain the chemical process by which a simple Voltaic cell is able to separate charge in order to provide a potential difference between its plates.
7 briefly describe the construction of a dry Leclanché cell.
8 briefly explain the advantages of miniature cells.
9 describe each of the following, as they apply to a lead-acid cell, in terms of its plates, electrolyte and relative density:
  a fully charged state.
  b de-energising state.
  c fully de-energised state.
  d re-energising state.
10 briefly explain how a lead-acid cell's relative density is a guide to its state of charge.
11 briefly explain the disadvantages of nickel-cadmium cells, and how these are overcome using lithium-ion cells.
12 outline what is meant by the 'responsible disposal of cells and batteries'.
13 determine the potential difference and internal resistance of identical
  a cells connected in series
  b cells connected in parallel
  c cells connected in series-parallel.

Online resources

The companion website to this book contains further resources relating to this chapter. The website can be accessed via the following link:
www.routledge.com/cw/waygood

Chapter 11

Internal resistance of voltage sources

On completion of this chapter, you should be able to

1 explain what is meant by the *internal resistance* of a voltage source.
2 sketch the *'equivalent circuit'* of a voltage source.
3 explain the difference between a voltage source's
  a electromotive force
  b internal voltage drop
  c terminal voltage.
4 explain the relationship between a voltage source's
  a electromotive force
  b internal voltage drop
  c terminal voltage.
5 solve simple problems on the effects of a voltage source's internal resistance.

Internal resistance explained

The load current supplied by *any* voltage source, whether it be a cell or battery, a generator, or a transformer, as well as passing through the load must, of course, *also pass through the voltage source itself.*

This is because the voltage source is part of the complete circuit around which the load current flows. In other words, the 'inside' of the voltage source is *in series* with its external load.

So what exactly do we mean by the 'inside' of a voltage source?

Well, in the case of a generator, for example, the load current must also pass through the windings (coils) in which the voltage is generated by the rotating machine.

Because windings are wound from copper wire, this 'internal current' is, of course, a *drift of free electrons –* just as it is through the external (load) circuit.

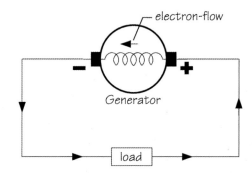

Figure 11.1

In the case of a cell or battery, the load current passes through the electrolyte in which its electrodes are immersed. However, this current is *a drift of ions* (charged atoms), rather than of free electrons, and may flow in the *opposite* direction (or even different ions flowing in *both* directions!) to the electrons in its external circuit (although this direction is unimportant to what follows).

Whether the 'internal current' is a drift of free electrons *or* a drift of ions *isn't really important*. It's simpler just to think of it as being 'current'! What *is* important is that, internally, this current is *opposed* by what we call the 'internal resistance' *(R_i)* of the voltage source.

An Introduction to Electrical Science, Waygood, ISBN 9780415810029, 2013. © Taylor & Francis

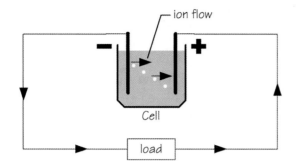

Figure 11.2

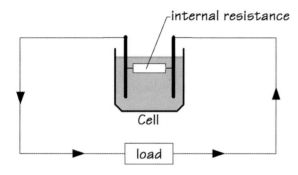

Figure 11.3

In the case of a **generator**, the internal resistance is *the resistance of the windings* within which the voltage is generated. In the case of a **cell** or **battery**, the internal resistance is made up of the combined effect of *two* components: the first is due to the resistivity of the internal conducting parts (the electrodes and any metallic connections), and the second, called 'ionic resistance', is due to the electrochemical opposition within the electrolyte. The first increases as the cell's electrodes dissolve (in the case of non-rechargeable cells) because their effective cross-sectional area reduces; the second increases as the cell de-energises.

The internal resistance of a fully energised, healthy cell is typically expressed in milliohms but, as the cell deteriorates, its internal resistance may rise markedly. A cell's internal resistance will also vary according to its state of charge.

The internal resistance of a generator's windings is also low, but will increase somewhat when the windings are hot (see the chapter on the *effect of temperature on resistance*), whenever the machine is running for prolonged periods.

The nature of a voltage source's internal resistance is unimportant; what *is* important is that it *exists* and it affects the behaviour of the voltage source as follows.

The effect of the voltage source's internal resistance is to *cause an internal voltage drop (IR) to occur within the voltage source itself.* The direction of this voltage drop is such that *it **always** acts to oppose the electromotive force of the voltage source, and to reduce the value of terminal voltage applied to the load.*

The *larger* the load current, the *larger* the corresponding internal voltage drop, and the *lower* the terminal voltage applied to the load.

We have all experienced this effect whenever we have inadvertently switched on our car's headlights *before* starting the engine. The very large load current drawn by the starter motor causes a correspondingly large internal voltage drop to occur within the battery, resulting in a significant drop in its terminal voltage and a corresponding reduction in the brightness of the headlights (because lamps operate at their rated power *only* when subjected to their rated voltage – in this case, 12 V).

Of course, if there is no load connected to the voltage source, then no load current will flow, and there will be *no internal voltage drop.* The voltage appearing across the voltage source's terminals under this condition is called its **open-circuit voltage** or **no-load voltage** – which exactly equals the **electromotive force** (e.m.f.) of that source.

> A voltage source's **open-circuit**, or **no-load**, **voltage** corresponds to its **electromotive force**.

Representing internal resistance in a schematic diagram

In the series of three **schematic diagrams** that follow, for simplicity, we will use a chemical **cell** to represent a voltage source, but we could also have used a generator or *any* other voltage source.

In Figure 11.4, we show where the internal resistance (R_i) *really* is located – between the plates, or 'electrodes', of the cell. The internal resistance is, of course, in *series* with the load, because the load current passes through both. The grey area represents the cell's container.

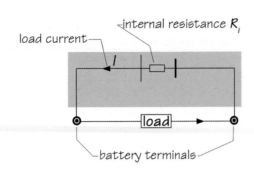

Figure 11.4

Of course, we don't *have* to show the internal resistance between the plates of the cell (even though that's where it *actually* is!). So, in Figure 11.5, we have moved it from between the plates, so that it sits *alongside* the cell – electrically, this is quite correct, even though it no longer represents its true *physical* location. The circuit symbol for the cell now represents an '**ideal' cell** (one having absolutely *no* internal resistance) producing its full electromotive force *(E)*, while the internal voltage drop *(IR$_i$)* now appears across the 'separate' internal resistance – acting in the *opposite* direction to the e.m.f. But remember, this is only a graphic representation (we call it an '**equivalent circuit**') – the internal voltage drop still *actually* occurs between the plates of the cell, not outside of them.

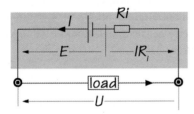

Figure 11.5

In fact, we can move the internal resistance to just about *anywhere around the circuit*, just as long as it remains *in series with the load*. So, in Figure 11.6, we have moved it even further around so that it is now adjacent to the load – again, this is electrically correct even though it is nowhere near its actual physical location.

We've moved the internal resistance to this particular position because it helps clarify the relationships between the cell's **electromotive force** *(E)*, its **internal voltage drop** *(IR$_i$)*, and the **terminal voltage** *(U)* which actually appears across the load, that is:

$$U = E - (IR_i)$$

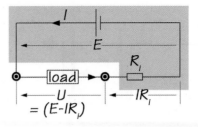

Figure 11.6

In other words, the voltage source's terminal voltage *(U), when supplying a load,* is its electromotive force *(E), less* its internal voltage drop *(IR$_i$)*.

> The *shaded* part of Figure 11.6 is called the **equivalent circuit** of the voltage source, and consists of an '**ideal' voltage source** (i.e. one having no internal resistance), in series with an **internal resistance**.

Alternatively, we could calculate the terminal voltage by simply multiplying the load current by the resistance of the load:

$$U = I\,R_{load}$$

The **load current** *(I)* itself is determined by simply dividing the cell's **electromotive force** *(E)* by the total resistance of the circuit – i.e. by the sum of the **load resistance** *(R$_L$)* and **internal resistance** *(R$_i$)*:

$$I = \frac{E}{\left(R_L + R_i \right)}$$

To simplify matters, we will ignore the resistance of any wires connecting the voltage source to the load but, for long conductor runs, the conductor's voltage drop *would*, of course, need to be taken into account to determine the load current and terminal voltage.

What should now be obvious is that the *larger* the load current *(I)*, the *larger* the internal voltage drop *(IR$_i$)*, and the *smaller* the resulting terminal voltage *(U)* appearing across the load!

The *only* time that the cell's terminal voltage will equal its electromotive force is when the load is disconnected, and no load current is flowing to cause an internal voltage drop – that is:

$$U = E - (IR_i)$$
$$U = E - (0 \times R_i)$$
$$U = E$$

... and this is the reason why the cell's **electromotive force** is also termed its '**open-circuit voltage**' or '**no-load voltage**'.

> By rearranging the equation, $U = E - IR_i$, we could easily determine the internal resistance of a cell, that is:
>
> $$R_i = \frac{E - U}{I}$$

Now would be a very good time to define exactly what is meant by the term 'electromotive force'. From Kirchhoff's Voltage Law, we can define **electromotive force** as follows:

> In accordance with Kirchhoff's Voltage Law, an **electromotive force** is equal to the sum of the voltage drops around any closed loop – *including the internal voltage drop within the source itself.*

We can express this definition, mathematically, as:

$$E = I_{load} \left(R_{load} + R_{internal} \right)$$

It is possible to directly measure the e.m.f. of a cell or battery, using a voltmeter. The very high resistance of this type of instrument draws a tiny current, ensuring that the resulting internal voltage drop is insignificant.

> **Worked example 1** A battery has an electromotive force of 9 V and an internal resistance of 600 mΩ. What will be its terminal voltage when connected to (a) a 20-Ω load, (b) a 10-Ω load and (c) no load?
>
> **Solution** First, sketch a fully labelled circuit diagram, indicating the values of the components (as shown in Figure 11.7).

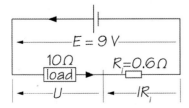

Figure 11.7

a With a load resistance of 20 Ω, the resulting load current will be:

$$I = \frac{E}{\left(R_L + R_i \right)} = \frac{9}{\left(20 + 0.6 \right)} = \frac{9}{20.6} = 0.437 \text{ A}$$

Next, we can determine the terminal voltage *(U):*

$$U = E - \left(IR_i \right) = 9 - \left(0.437 \times 0.6 \right)$$
$$= 9 - 0.262 = 8.738 \text{ V (Answer a.)}$$

b With a load resistance of 10 Ω, the load current will be:

$$I = \frac{E}{\left(R_L + R_i \right)} = \frac{9}{\left(10 + 0.6 \right)} = \frac{9}{10.6} = 0.849 \text{ A}$$

Next, we can determine the terminal voltage *(U):*

$$U = E - \left(IR_i \right) = 9 - \left(0.849 \times 0.6 \right)$$
$$= 9 - 0.509 = 8.491 \text{ V (Answer b.)}$$

c With *no load*, the load current will be zero, so we can determine the terminal voltage *(U):*

$$U = E - \left(IR_i \right) = 9 - \left(0 \times 0.6 \right)$$
$$= 9 - 0 = 9.0 \text{ V (Answer c.)}$$

In the above worked example, we say that the 10-Ω load is '**heavier**' than the 20-Ω load. In this context, 'heavier' *always* refers to the magnitude of the resulting *load current.* That is, *the larger the load current* (or the lower the load resistance), *the 'heavier' the load.*

> **Important!** A '**heavy**' load is one that draws a 'heavy' (large) current. So, a low-resistance load is a 'heavy' load, and a high-resistance load is a 'light' load.

We can also describe a load in terms of its **power**. A high-power load represents a 'heavy' load, and a low-power load represents a 'light' load. For example, a 200-W (watt) lamp represents a 'heavier' load than a 100-W lamp.

As you can see from the previous worked example, the *heavier* load (i.e. in this case, the 10-Ω load) results in the *lower* terminal voltage. In fact, a *really* heavy load may cause a *significant* fall in terminal voltage. For example, suppose we connect a 1-Ω load to the previous example.

With a load resistance of just 1 Ω, the load current will be:

$$I = \frac{E}{(R_L + R_i)} = \frac{9}{(1 + 0.6)} = \frac{9}{1.6} = 5.63 \text{ A}$$

And the resulting terminal voltage *(U)* will be just . . .

$$U = E - (IR_i) = 9 - (5.63 \times 0.6) = 9 - 3.38 = 5.62 \text{ V!}$$

As you can see, this is now a *substantial* drop in the battery's terminal voltage compared to its electromotive force!

The *main* problem with a battery, however, is that its internal resistance can increase markedly when the battery deteriorates and its electrodes become consumed by the chemical reaction. If a car battery fails due to a higher-than-normal internal resistance, a voltmeter will still indicate a 'normal' terminal voltage of 12 V. A voltmeter actually draws a tiny current (microamperes) and, so, the battery's internal voltage drop will also be tiny. So, in order to properly test any battery, it's necessary to simulate a typical load resistance – only then will a voltage test indicate the battery's true state. All battery voltage testers incorporate a load in their internal circuitry for this very purpose.

> **Worked example 2**   Let's assume that the cell in Worked Example 1 has deteriorated somewhat, and its internal resistance has risen to, say, 1.75 Ω. Let's find out what effect this will have on the terminal voltage for (a) a 20-Ω load and (b) a 10-Ω load.
>
> **Solution**
>
> a   With a load resistance of 20 Ω, the resulting load current will be:
>
> $$I = \frac{E}{(R_L + R_i)} = \frac{9}{(20 + 1.75)} = \frac{9}{21.75} = 0.41 \text{ A}$$
>
> Next, we can determine the terminal voltage *(U)*:
>
> $$U = E - (IR_i) = 9 - (0.41 \times 1.75)$$
> $$= 9 - 0.72 = 8.28 \text{V (Answer a.)}$$
>
> b   With a load resistance of 10 Ω, the load current will be:
>
> $$I = \frac{E}{(R_L + R_i)} = \frac{9}{(10 + 1.75)} = \frac{9}{11.75} = 0.76 \text{ A}$$

Next, we can determine the terminal voltage *(U)*:

$$U = E - (IR_i) = 9 - (0.76 \times 1.75)$$
$$= 9 - 1.33 = 7.67 \text{ V (Answer b.)}$$

Practical values of internal resistance

Cells

It's useful to have some idea of the typical values of internal resistance of various common **cells**:

- A non re-energiseable **alkaline** cell, for example, has an internal resistance within the range of 150–300 mΩ (depending on its physical size, e.g. AAA, AA, etc.) when new. These disposable cells are widely used in torches, smoke detectors, etc.
- A fully charged **lead-acid** cell has an internal resistance of around 50 mΩ. This very low value of internal resistance makes it an ideal cell for the manufacture of car batteries which must supply very heavy starting currents.
- A new, fully charged **nickel-cadmium** (**NiCad**) cell has an internal resistance of around 75 mΩ. NiCads are widely used in portable tools, etc., but are being replaced by lithium-ion types.
- A new, fully charged **lithium-ion** cell has an internal resistance of around 320 mΩ, which increases to around 340 mΩ after around a thousand charging cycles. These cells are now widely used for powering laptop computers, camcorders, mobile telephones, etc.

Remember that these figures relate to *cells*, not batteries. Batteries are groups of cells connected together in series, parallel or series-parallel. So the internal resistance of batteries must be determined from the number of cells and the ways in which they are connected. For example, a car battery consists of six lead-acid cells, connected in series, giving a total of approximately (6×50 =) 350 mΩ.

Generators

It's *not* possible to give equivalent figures for the internal resistance of **generators**, as there are simply too many factors involved: the type of winding configuration, the length of windings' conductor, its cross-sectional area, the machine's operating temperature and so on. Each generator needs to be assessed individually.

We will be examining the effect of a generator's internal resistance when we study generators in a later chapter.

Summary

- All voltage sources have an *internal resistance (R$_i$)*.
- A voltage source's internal resistance is *in series* with any connected **load resistance (R$_L$)**.
- The **equivalent circuit** of a voltage source comprises an **'ideal' voltage source**, producing an electromotive force *(E)*, connected *in series* with its internal resistance.
- With a load connected, the **load current** will be the 'ideal' voltage source's electromotive force *(E)* divided by the sum of the internal resistance *(R$_i$)* and the load resistance *(R$_L$)*, that is:

$$I = \frac{E}{R_i + R_L}$$

- The load current causes an **internal voltage drop (IR$_i$)** to occur across the voltage source's internal resistance.
- The voltage appearing across the load, is called the **terminal voltage (U)**, and is the *difference* between the voltage source's **electromotive force (E)** and its **internal voltage drop (IR$_i$)**, that is:

$$U = E - (IR_i)$$

- For any given load resistance, *the greater the internal resistance, the lower the voltage source's terminal voltage.*

- For any given internal resistance, *the lower the load resistance, the lower the voltage source's terminal voltage.*
- When there is no load connected to the voltage source, the terminal voltage will equal its electromotive force, because there is no internal voltage drop.

Finally . . .

Now that you have completed this chapter, are you able to achieve the objectives or learning outcomes listed at the beginning of this chapter?

Ask yourself, 'Can I . . .'

1 explain what is meant by the *internal resistance* of a voltage source.
2 sketch the *'equivalent circuit'* of a voltage source.
3 explain the difference between a voltage source's
   a electromotive force
   b internal voltage drop
   c terminal voltage.
4 explain the relationship between a voltage source's
   a electromotive force
   b internal voltage drop
   c terminal voltage.
5 solve simple problems on the effects of a voltage source's internal resistance.

Online resources

The companion website to this book contains further resources relating to this chapter. The website can be accessed via the following link:

www.routledge.com/cw/waygood

Chapter 12

Energy, work, heat and power

On completion of this chapter, you should be able to

1 define each of the following terms:
   a energy
   b work
   c heat
   d power.
2 specify the SI unit of measurement for each of the following:
   a energy
   b work
   c heat
   d power.
3 state the fundamental equation for the work done by an electric circuit.
4 state the fundamental equation for the power of an electric circuit.
5 derive alternative equations for the work done by, and the power of, an electric circuit.
6 define the term 'efficiency'.
7 solve problems on energy, work, power and efficiency.
8 explain how electricity supply companies bill their residential consumers.
9 read an analogue energy meter.
10 describe the relationship between electrical energy and heat.
11 solve problems on work and heat.

Note: unfortunately, quantities used in thermo-dynamics share the same symbols as completely different quantities used in electrical engineering. So, in this unit, you should be aware that

- in **electricity**: U = potential difference; Q = electric charge
- in **thermodynamics**: U = internal energy; Q = heat

Introduction

The study of electricity is really the study of **energy** and of **energy conversion**. Unfortunately, the word 'energy', together with the related terms, 'work', 'heat' and 'power', have 'everyday meanings' and are frequently used interchangeably by the layman. But, if you are working in the electricity industry, then it is important for you to understand the *scientific* meanings of these terms.

Energy, work and heat

Energy

For a machine to do work, it must have a source of **energy**. This source of energy may be petrol, diesel oil, coal, gas, etc. As humans we also require a source of energy, which is obtained from the food which we eat.

Energy is defined as '*the ability to do work*'.

An Introduction to Electrical Science, Waygood, ISBN 9780415810029, 2013. © Taylor & Francis

Although this is the widely accepted definition for energy, as expressed in most textbooks, it does rather assume that you know what is meant by *work* – so let's look at what we mean by 'work'!

Work

Work (symbol: W) is usually defined as '*force multiplied by the distance moved in the direction of that force*'. Unfortunately, while this makes perfect sense in mechanical engineering, it is rather difficult to relate this to electricity! A far more understandable definition is one used in the study of **thermodynamics** (heat science):

> **Work** is defined as '*the conversion of energy from one form into another*'.

The **Law of the Conservation of Energy** may be expressed as: '*energy cannot be created nor destroyed; it can only be converted from one form into another*'. 'Work' is the name given to the *process* of energy *conversion* which takes place whenever one form of energy is changed into another.

For example, a generator converts kinetic energy (the energy of motion) into electrical energy – so the machine is doing *work*. An electric motor converts electrical energy back into kinetic energy – so it, too, is doing *work*.

So any machine that *changes one form of energy into another* is doing **work**.

> Unfortunately, the definitions of **energy** and **work** are rather like the 'chicken and the egg' situation, in that to understand *one* requires the understanding of the *other!*

If you are confused over the difference between energy and work, you might like to think of 'energy' as being the equivalent of your 'savings', and 'work' as the equivalent of you 'spending'! You cannot 'spend' unless you have 'savings', and 'spending' converts those 'savings' into 'goods'!

In the same way, you cannot do **work** ('spend') unless you have **energy** ('savings'), and doing work ('spending') converts that energy ('savings') into **another form** of energy ('goods').

Heat

You may have been taught at school that heat is a 'form of energy' or that it is 'thermal energy'.

Well for many years, now, **heat** (symbol: Q) hasn't been thought of in quite that way. Instead, we say that everything whose temperature is above absolute zero temperature (zero kelvin) contains **internal energy** – *not* 'thermal energy' or 'heat energy'.

Internal energy is the sum total of all the energies associated with the atomic structure of a body, including the vibration of those atomic particles due to their temperature.

So if heat *isn't* a form of energy, then what is it? Well, these days, heat is considered to be a *process* – specifically, **the process of energy transfer between bodies due to any temperature difference between them**. And heat transfer *always* takes place from the warmer body to the cooler body.

> **Heat** is defined as *the transfer of energy from a warmer body to a cooler body*.

Relationship between work and heat

When we switch on an electric kettle, the internal energy of the water (as well, of course, as of the kettle itself!) starts to increase. So the supplied electrical energy is being *converted* into internal energy.

As *energy conversion* is taking place, **work** is being done.

As the water's internal energy continues to increase, it causes the temperature of that water to rise above room temperature. As a result, some of that energy will now be *transferred* (lost) from the hot water into the cooler surroundings.

Because this energy is being lost due to a *temperature difference*, we describe it as **heat** transfer.

We will learn more about the relationship between work and heat later in this chapter.

> Scientists describe both work and heat as '*energy in transit*' – either *in transit from one form to another* (work) or *in transit from a higher temperature to a lower temperature* (heat). In fact, energy can *only* be manipulated either by 'work' or by 'heat'.

Power

Suppose two vehicles, say, a *Range Rover* and a *Ford Focus*, are each supplied with, say, exactly one litre of fuel. Both cars would do **work** as they convert the chemical energy supplied by that litre of fuel into kinetic energy, as they move along a road. However, the *Range Rover* would use up its supply of energy (i.e. its fuel) much faster than the *Focus* and, therefore, not travel as far before running out of fuel. This is because the *Range Rover* has a much more **powerful** engine than the *Focus,* and expends energy at a much higher *rate*.

> Power is *the rate of doing work or of heat transfer*.

So **power** (symbol: *P*) is simply a *rate* (i.e. energy divided by time). It is technically incorrect to talk about 'electrical power', 'mechanical power', etc. – these terms are quite meaningless.

You can compare this with the way in which we use the word 'speed' which, like power, is also a *rate* (e.g. metres per second) – there are no 'types' of speed! Speed is simply a rate! In the same way, there are no different 'types' of power; power is also simply a rate!

When engineers *say* 'power loss', what they actually *mean* is the 'rate of energy loss'.

Units of measurement

Energy, work and heat: the 'joule'

So **work** describes the process of changing *energy* from one form to another, while **heat** describes the process of *energy* transfer due to temperature difference – therefore, **work**, **heat** and **energy** *all* share the same SI unit of measurement, the **joule** (symbol: **J**), named after the English scientist, James Prescott Joule.

> Another unit of measurement of energy or work that you will come across is the **kilowatt hour** (symbol: **kW·h**), also known as the '**unit**' (after '**Board of Trade Unit**' – the Board of Trade, at one time, governed how much energy companies could charge their consumers for energy). Because the joule is actually a *very* small unit of measurement, power utility companies charge their customers for the energy they use in kilowatt hours. A kilowatt hour is simply a 'big' version of the joule. How much bigger? Well, as we shall learn later, it's equivalent to 3.6 MJ.

Power: the watt

As **power** is the *rate of doing work* (or of *heat transfer*), it follows that its unit of measurement is the **joule per second**. However, in common with many other SI derived units, this unit is given a special name: the **watt** (symbol: **W**), named in honour of the Scottish engineer, James Watt. So a 'watt' is just another way of saying, 'joule per second'.

> You should be aware that the imperial unit of measurement of power is the **horsepower** (symbol: **HP**). Although the horsepower has been phased out in the United Kingdom since the adoption of SI, you may come across older electric motors whose nameplates specify their rated power output in horsepower, rather than in watts. To convert horsepower to watts:
>
> **1 HP = 746 W**
>
> You have probably noticed that car engines are now rated in kilowatts, rather than in horsepower. Some books refer to 'horsepower' being used to measure 'mechanical power' but, as we have learnt, there is really no such thing as 'mechanical power', 'electrical power', etc!
>
> There is absolutely no reason why you shouldn't measure the power of an electric fire in horsepower – in fact, a 'three-bar 3-kW electric fire' could just as correctly be called a 'three-bar 4-horsepower electric fire'!

Energy, work, heat and power in electric circuits

In an earlier chapter, we learnt that the volt is defined as *'the potential difference between two points such that the energy used in conveying a charge of one coulomb from one point to the other is one joule'*. This can be expressed as:

$$E = \frac{W}{Q}$$

where:

E = voltage, in volts

W = work, in joules

Q = electric charge, in coulombs

Rearranging this equation, to make W the subject:

$$W = EQ \qquad \text{—equation (1)}$$

but we have also learnt that $Q = It$, so, substituting for Q in equation (1), we have:

$$W = EIt$$

where:

$$W = \text{work, in joules}$$

$$E = \text{voltage, in volts}$$

$$I = \text{current, in amperes}$$

$$t = \text{time, in seconds}$$

The above equation is the fundamental equation for the work done, or energy expended, by an electric circuit.

As **power** is defined as *the rate of doing work*, we can express this as:

$$P = \frac{W}{t} \qquad \text{—equation (2)}$$

. . . substituting the fundamental equation for work into equation (2), we have:

$$P = \frac{EIt}{t}$$

which is simplified to:

$$P = EI$$

where:

$$P = \text{power, in watts}$$

$$E = \text{voltage, in volts}$$

$$I = \text{current, in amperes}$$

As resistance is the *ratio of voltage to current*, we can derive alternative equations for both work and power:

For Work:

$$W = EIt \qquad \text{— (eq.3)}$$

But, from Ohm's Law,

$$E = IR$$

Substituting for E, in (eq.3):

$$W = (IR)It$$

For Power:

$$P = EI \qquad \text{— (eq.4)}$$

But, from Ohm's Law,

$$E = IR$$

Substituting for E, in (eq.4):

$$P = (IR)I$$

$$\boxed{W = I^2Rt} \qquad \boxed{P = I^2R}$$

Again, from Ohm's Law,

$$I = \frac{E}{R}$$

Substituting for I, in eq.3:

$$W = E\left(\frac{E}{R}\right)t$$

$$\boxed{W = \frac{E^2}{R}t}$$

Again, from Ohm's Law,

$$I = \frac{E}{R}$$

Substituting for I, in eq.4:

$$P = E\left(\frac{E}{R}\right)$$

$$\boxed{P = \frac{E^2}{R}}$$

Summary of equations for work and power

For **work**:

$$W = EIt \quad W = I^2Rt \quad W = \frac{E^2}{R}t \quad W = Pt$$

. . . and for power:

$$P = EI \quad P = I^2R \quad P = \frac{E^2}{R} \quad P = \frac{W}{t}$$

Worked example 1   Calculate the work done by a resistor when connected across a 230-V supply, if it draws a current of 10 A for 1 min.

Solution   **Important**. Don't forget to convert minutes to seconds.

$$W = EIt$$

$$= 230 \times 10 \times (1 \times 60)$$

$$= 138\,000 \text{ J (Answer)}$$

Worked example 2   Calculate the current drawn by a 100-W lamp from a 230-V supply.

Solution

Since $P = EI$

then $I = \dfrac{P}{E} = \dfrac{100}{230} = 0.435$ A (Answer)

Worked example 3   Calculate the rate at which energy is lost when a current of 13 A flows along a conductor of resistance 0.5 Ω.

Solution   **Note**: in this question, we are asked to calculate the *rate* at which energy is lost – this is the same as 'calculate the *power* loss'.

$$P = I^2R = 13^2 \times 0.5 = 84.5 \text{ W (Answer)}$$

Worked example 4   The voltage drop across a resistance of 200 Ω is found to be 12 V. Calculate (a) the power developed by the resistor, and (b) the work done by the resistor in 30 s.

Solution

a $P = \dfrac{E^2}{R} = \dfrac{12^2}{200} = 0.72$ W (Answer a.)

b $W = Pt = 0.72 \times 30 = 21.6$ J (Answer b.)

Be aware that *some* electrical textbooks may use the symbol '*W*' for '**power**'. This has been done (ill advisably!) because '**W**' is the symbol for the 'watt'. This can lead to confusion, as '*W*' (in italics) actually represents '**work**', *not* 'power'!!!

So this is a confusing and extremely slipshod method of writing any equation, because you should *never ever* write an equation in terms of the *units* used. After all, you would *never* (or shouldn't!) write, **V** = **Ω×A**, so *why* would you do it for power?

In equations, *always* use the symbol for a 'quantity' (e.g. **P** for '**power**') never the symbol for that quantity's' unit of measurement' (e.g. **W** for '**watt**'). That is:

P = 300 W *never* **W = 300 W**

Misconceptions about energy transfer

Many students believe that *electrons* deliver energy from the supply to the load, after which they return to the supply, depleted of their energy, in order to 'collect' more energy which they deliver the next time they move around the circuit! This 'conveyor belt' model of energy transfer is quite wrong, as a brief review of the behaviour of free electrons will reveal!

You will recall from the earlier chapter on the *electron theory of electricity* that the free electrons in a metal conductor are in a constant state of vigorous, chaotic, motion and, whenever a potential difference is applied across that conductor, the electrons drift, *very* slowly, along the conductor (in the order of millimetres per hour!). In the case of alternating current, the electrons don't even move through the circuit at all; they merely vibrate backwards and forwards as the alternating supply voltage continuously changes its magnitude and direction. From this, *it should be clear that the idea of free electrons acting like an 'energy conveyor belt', is a complete misconception.*

So, how *does* energy transfer between a circuit's supply and its load? The simple answer is that, although we know how it *doesn't* supply energy, no one truly knows how it *does*, although there are various theories.

For most of these theories, the 'flow' of energy from the supply to the load in an electrical circuit can only be satisfactorily explained using post-graduate mathematics.

Perhaps the most well-known of these theories is one that describes the interaction of a circuit's magnetic and electric fields to create what is known as the 'Poynting field', named in honour of the British physicist, John Poynting (1852–1914), which causes energy to 'flow' in a direction perpendicular to both these fields, and causes that energy to be *transferred through empty space around – not through – the conductors*.

Unfortunately, there is very little point in discussing this theory very much further, as the mathematics involved is extremely abstract, and the process simply cannot be described in terms that a non-physicist can easily understand. However, Figure 12.1, which has been extremely simplified, attempts to convey the general idea.

In Figure 12.1, a pair of conductors connects a battery to a load, and (conventional) current flows in a clockwise direction around the circuit, as shown. The conductors are surrounded along their entire length by a magnetic field, and an electric field stretches between the upper (positive) conductor and the lower (negative) conductor. The white arrows are perpendicular to both the magnetic *and* the electric fields, and represent the Poynting field which acts to transfer energy *entirely through the empty space between the supply and the load.*

Energy transfer from supply to load probably takes place through the empty space around a circuit's conductors —*not* through the conductors themselves.

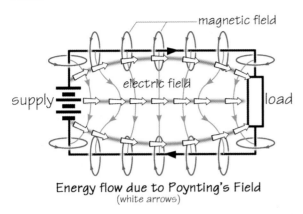

Figure 12.1

Power ratings

Electrical appliances, such as incandescent lamps, are required to have the **rated power** at which they are designed to operate displayed. The rated power is *always* shown together with the rated voltage, because an appliance *will only develop its rated power at its rated voltage.*

This information is termed **nameplate data**. In the case of lamps, for example, the nameplate data is printed either on the glass envelope or on the bayonet/screw mount, in the form: '**100 W/230 V**', etc.

If an electrical appliance, such as a lamp, is operated *below* its rated voltage, then *the power developed can be substantially reduced from its nameplate power rating.*

This demonstrates the importance of an electricity network company's responsibility for *maintaining the supply voltage within its legal limits.* In the U.K. this is 230 V (+10%/–6%) – in other words, the supply voltage must not rise beyond 253 V or fall below 216.2 V.

However, with some types of electrical device, its power rating indicates the *'maximum power'* at which *that device is designed to operate*, and *not* its normal 'operating power'.

For example, unlike lamps, **resistors** can operate quite normally well below their power rating, but if they are allowed to *exceed* their power rating, *then they will likely overheat and burn out.* So a resistor can operate at *any* combination of voltage and current, *providing* its rated power is not exceeded.

Power in series, parallel and series-parallel circuits

In the circuit in Figure 12.2, two resistors of 10 Ω and 5 Ω are connected across a supply of 30 V. Let's

calculate the power of the complete circuit, and the power of each resistor.

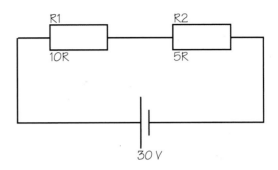

Figure 12.2

total resistance: $R_T = R_1 + R_2 = 10 + 5 = 15\,\Omega$

circuit current: $I = \dfrac{E}{R_T} = \dfrac{30}{15} = 2\,A$

circuit power: $P = EI = 30 \times 2 = 60\,W$

power of R_1: $P_1 = I^2 R_1 = 2^2 \times 10 = 40\,W$

power if R_2: $P_2 = I^2 R_2 = 2^2 \times 5 = 20\,W$

Let's look at the relationship between the power developed by the individual resistors compared to the power of the complete circuit. It's plain to see that, for a series circuit, *the total power is the sum of the power of each resistor.*

Now, let's look at a parallel circuit, in Figure 12.3, comprising two resistors, each of 10 Ω, connected across a supply of 30 V. Let's calculate the power of the complete circuit, and the power of each resistor.

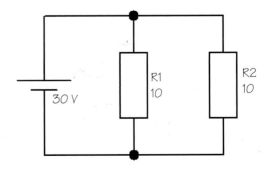

Figure 12.3

$$\frac{1}{R_T} = \frac{1}{R_1} + \frac{1}{R_2}$$

$$\frac{1}{R_T} = \frac{1}{10} + \frac{1}{10}$$

total resistance: $\frac{1}{R_T} = \frac{2}{10}$

$$R_T = \frac{10}{2} = 5\ \Omega$$

power of the circuit: $P = \dfrac{E^2}{R_T} = \dfrac{30^2}{5} = 180$ W

power of R_1: $P_1 = \dfrac{E^2}{R_1} = \dfrac{30^2}{10} = 90$ W

power of R_2: $P_2 = \dfrac{E^2}{R_2} = \dfrac{30^2}{10} = 90$ W

Once again, *the total power of the circuit is equal to the sum of the power of each resistor.*

If we were to repeat this for a *series-parallel* circuit or, in fact, for *any type of circuit*, we would obtain the *same* result!

So, we can state the general rule that:

> The total power of *any* type of circuit is the sum of the powers developed by that circuit's individual components.

Efficiency

Not *all* of the energy supplied to a load goes into doing useful work. For example, some of the energy supplied to an electric motor is lost overcoming *friction*, *windage* (air resistance), *vibration* and *heat transfer* into the surrounding atmosphere. So the useful work available from a motor is *always* somewhat less than the electrical energy supplied to the motor! Rotating machines, such as motors (**M**), are always less efficient that stationary machines, such as transformers, due to the friction of their bearings, windage losses, etc.

In practice, we talk about '**input power**', '**output power**' and '**power losses**'. But, as we have learnt, 'power' is simply a 'rate' so it is, strictly speaking, incorrect to talk about power in these ways, so a word of explanation of what we *really* mean, here, is necessary.

Figure 12.4

- By '**input power**', what we *really* mean is *the rate at which (electrical) energy is being supplied to the load.*
- By '**output power**', what we *really* mean is *the rate at which useful energy (in the case of a motor, kinetic energy) is utilised by the load.*
- By '**power loss**', what we *really* mean is *the rate at which energy is lost* overcoming friction, windage, heat transfer, etc.

Motors are *always* rated according to their **output power**, as this is what is available from the motor to drive their load and, so, it is their output power that matters, *not* their input power. The output power of a machine is often called its 'brake power' – in fact, in North America, the term 'brake horsepower' is often used to describe the output power of a machine. At present, the *output* power of an American motor is *always* expressed in horsepower, whereas its input power is always expressed in watts.

The ratio: *output power divided by input power* is termed **efficiency** (symbol: η —the Greek letter '*eta*').

$$\eta = \frac{P_{output}}{P_{input}}$$

Efficiency is simply a *ratio* and, so, it does *not* have any units of measurement. Instead, it is expressed either as a **per unit** value (e.g. '**0.85**') or, more commonly, as a **percentage** value (e.g. '**85%**'). In the above equation, efficiency is a 'per-unit' value. To change a per-unit value to a percentage value, we must multiply by 100:

$$percentage = (per\ unit \times 100)\%$$

There is simply no such thing as 'lossless' energy transfer, so efficiency can *never* reach unity (or 100%). Even the most efficient electrical machine, the transformer, can only achieve a full-load efficiency of around 0.95–0.98 p.u. (or 95%–98%).

Worked example 5   What is the efficiency, expressed both as a per-unit value and as a percentage value, of an electric motor whose output power is 3900 W when its input power is measured as 5000 W?

Solution

$$\eta = \frac{P_{output}}{P_{input}} = \frac{3900}{5000} = 0.78 \text{ p.u.(Answer)}$$

$$\text{or } 0.78 \times 100 = 78\% \text{ (Answer)}$$

Worked example 6   A 230-V electric motor, drawing a current of 12 A, is 75% efficient. Calculate its input and output powers.

Solution

Input Power, $P_{input} = EI = 230 \times 12 = 2760 \text{ W}$

$$\text{Since } \eta = \frac{P_{output}}{P_{input}}$$

$$P_{output} = \eta P_{input} = \frac{75}{100} \times 2760 = 2070 \text{ W (Answer)}$$

Worked example 7   A 400-V electric machine, operating at an efficiency of 80%, has an output power of 3 kW. Calculate the current supplied to the machine.

Solution   In this problem, we have to first find the machine's input power, from which we can then find the current supplied.

$$\text{Since } \eta = \frac{P_{output}}{P_{input}}, \text{t}$$

$$\text{then } P_{input} = \frac{P_{output}}{\eta} = \frac{3000}{0.80} = 3750 \text{ W}$$

$$\text{Since } P_{input} = EI,$$

$$\text{then } I = \frac{P_{input}}{E} = \frac{3750}{400} = 9.38 \text{ A (Answer)}$$

Residential electricity utility bills

During the 1860s, the then Chancellor of the Exchequer, William Gladstone, asked Michael Faraday, "But, what is the use of electricity?" Faraday replied that he didn't really know, but continued "however there is every possibility that you will soon be able to tax it!" Little did Faraday realise that his humorous response would eventually become a reality!

When an electricity supply company bills its residential consumers, it is charging them for the *energy consumed*, or the *work done by that energy*.

The supply company is not interested in the *rate* – i.e. the power – at which a *residential* consumer uses that energy, although it *does* care when it comes to industrial consumers, as we shall learn later.

As we have learnt, in SI the unit of measurement for work or energy is the joule. The joule represents a very small quantity of energy, so more practical units are the megajoule or gigajoule. However, for some reason, electricity supply companies have never adopted these units but use, instead, the **kilowatt hour** (**kW·h**), which is equivalent to 3.6 MJ.

> Some students describe the kilowatt hour as being *'the amount of power consumed over a period of one hour'*. This is misleading, as you do not 'consume' power! It is far more accurate to describe it as *'amount of energy consumed, over a period of one hour, at a rate of one kilowatt'*.

In the UK, the kilowatt hour is also known as a '**unit**', which is short for '**Board of Trade unit**'. The Board of Trade no longer exists, but was at one time the government organisation charged with determining the prices that electricity authorities could charge their customers.

The kilowatt hour is derived as follows:

$$\text{power} = \frac{\text{work}}{\text{time}}$$

$$\text{so, work} = \text{power} \times \text{time}$$

Now, if we measure power in kilowatts (kW), and time in hours (h), then:

$$\text{work} = \text{kilowatts} \times \text{hours} = \text{kilowatt hour}$$

So, to calculate the work done, or energy consumed, by a load in kilowatt hours, the following special equation can be used:

> work (in kilowatt hours) = power (in kilowatts) × time (in hours)

From this, we can say that *a kilowatt hour is the work done by a load, at a rate of one kilowatt, over a period of one hour*.

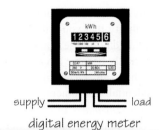

supply ═══════ load

digital energy meter

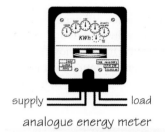

supply ═══════ load

analogue energy meter

Figure 12.5

Worked example 8 A 60-W lamp is left switched on, continuously, for a period of six hours every day. How much energy is expended over a period of one week, and how much does it cost over that period, if the electricity supply company charges 17.5 p per kilowatt hour?

Solution

$$\text{work (in kilowatt hours)}$$
$$= \text{power (in kilowatts)} \times \text{time (in hours)}$$
$$= \left(\frac{60}{1000}\right) \times (6 \times 7) = 2.52 \text{ kW} \cdot \text{h (Answer a)}$$
$$\text{cost} = 2.52 \times 17.5 = 44.1 \text{pence (Answer b)}$$

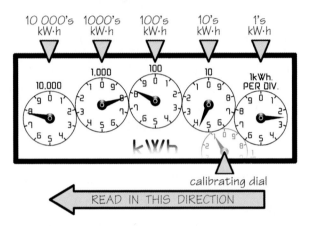

Figure 12.6

Measuring energy consumption

The energy used by residential consumers is measured using an instrument called an **energy meter**.

The energy consumed during a billing period is the difference between the current meter reading and the previous meter reading.

Energy meters are the property of the electricity supply company, and it is illegal to tamper with them. They are installed at the service entry point of a property. In modern residences, they are usually installed in an external cabinet, where they can be accessed and read without the meter reader having to enter the property. Most older properties, however, have their energy meters indoors, often in relatively inaccessible places – which means that meter readers have to enter the property in order to take a reading. Not only is this inconvenient, but it requires the householders to be at home whenever the meter reader calls.

Modern energy meters provide a digital display which practically anyone can read. However, a great many older properties still have energy meters

with *analogue* displays, which are relatively difficult to read. As it is not unreasonable to expect anyone working in the electricity industry to be able to read an analogue energy meter, we will now learn how this is done.

Typically, analogue energy meters have six dials, one (usually coloured red) of which is only used for calibration purposes and, so, can be ignored when reading the meter. Each dial is directly gear-driven by the one to its right and, so, *alternate dials rotate in opposite directions*. The remaining five dials are read, in sequence, *from right to left*, and each dial represents kilowatt hours expressed in: **units**, **tens**, **hundreds**, **thousands** and **tens of thousands** – as illustrated in Figure 12.6.

As explained, an energy meter's dials are read *from right to left*, and the figures are written down in the same sequence. When a pointer is between two numbers, the *lower* number is read, as shown in the example in Figure 12.7.

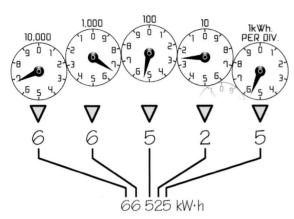

Figure 12.7

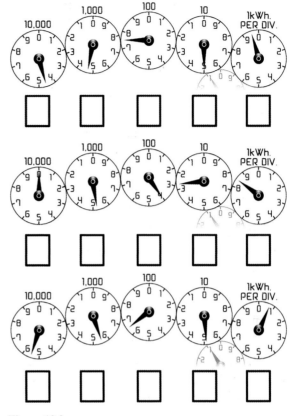

Figure 12.9

Not all readings are as obvious as the example shown in Figure 12.7.

Consider the example in Figure 12.8.

Front right to left:

- **dial 1**: reads **0**.
- **dial 2**: reads **5**, because dial 1 has just started a new revolution.
- **dial 3**: reads **9**.
- **dial 4**: reads **6**, because dial 3 has not yet completed a full revolution
- **dial 5**: reads **2**.

So the energy meter reads: **26 950 kWh**.

Self-test exercise

For each of the following, enter the meter readings in the spaces provided in Figure 12.9.

Answers are given at the end of this chapter.

Electricity and heat

Internal energy

Everything above absolute zero temperature (zero kelvin) possesses **internal energy** (symbol: U). The term 'internal energy' describes the sum-total of *all* the various energies associated with molecules (in particular, the *motion* or *vibration* of those molecules)

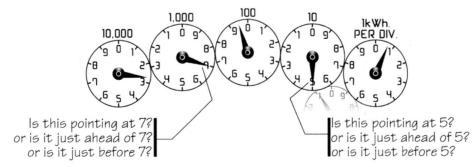

Figure 12.8

in a body, and is closely linked to temperature. Molecules are in a constant state of vibration; the higher the temperature, the greater the vibration, and the greater the internal energy.

An increase in the temperature of a body, then, will cause its internal energy to rise (and *vice versa*). However, the 'state' (e.g. solid, liquid or gas) of the body is also important. For example, ice and water can co-exist at 0°C, but water will *always* have a higher internal energy than ice at that particular temperature. Similarly, water and steam can co-exist at 100 °C, but steam will *always* have a higher internal energy than water at that temperature.

Change in internal energy

We cannot measure the absolute ('total') internal energy of a body. We can only measure the *change* in its internal energy. The change in a body's internal energy depends upon its **mass** (symbol: m), the change in its **temperature** (symbol: T) and upon a constant called the **specific heat capacity** (symbol: c) of the body.

Specific heat capacity is a constant that depends on the material involved, and is defined as *the energy required to raise the temperature of that material by one degree*. The SI unit of measurement for specific heat capacity is the **joule per kilogram kelvin** (symbol: **J/kg·K**) or, in 'everyday' units, the **joule per kilogram degree Celsius** (symbol: **J/kg·°C**).

The relationship described above is given by the equation:

$$\Delta U = m\,c\,(T_{final} - T_{initial})$$

where:

$$\Delta U = \text{change in internal energy}$$
$$m = \text{mass}$$
$$c = \text{specific heat capacity}$$
$$T = \text{temperature}$$

> **Important**! In the above equation, the Greek letter, Δ ('delta'), placed in front of the 'U', is simply the mathematical 'shorthand' meaning *'change in'*. So, 'ΔU' *(spoken as 'delta-U')* simply means *'change in internal energy'*.

Let's now look at an example of the relationship between work, heat and internal energy.

Whenever we switch on an electric kettle, what happens?

Well, an electric current flows through the kettle's heating element, causing electrical energy to be *converted* into internal energy. So **work** *(W)* is being done on the water (as well as on the kettle itself). This additional internal energy causes the existing internal energy of the water to rise, and this is accompanied by a rise in its temperature. As the temperature rises above the ambient temperature, energy is lost to the surrounding atmosphere through **heat** transfer *(Q)*.

In other words, the increase in the internal energy *(ΔU)* of the water and the kettle must be *the difference between the **work done** (W) by the electricity on the water and the **heat transfer** (Q) away from the water*. We can express this as follows:

$$W - Q = \Delta U$$

If we now substitute for U, we have:

$$W - Q = m\,c\,(T_{final} - T_{initial})$$

where:

$$W = \text{work done}$$
$$Q = \text{heat transfer}$$
$$m = \text{mass}$$
$$c = \text{specific heat capacity}$$
$$T = \text{temperature}$$

This equation can be illustrated as shown in Figure 12.10.

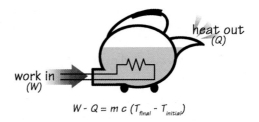

$$W - Q = m\,c\,(T_{final} - T_{initial})$$

Figure 12.10

The following worked examples demonstrate how the above equations are used.

> **Worked example 9**   An electric kettle contains 1.5 kg of water at 20°C. If the kettle draws a current of 10 A from a 230-V supply, calculate the time it would take to raise the temperature of the water to boiling point (100°C), if the average heat loss during this period is 50 kJ. The specific heat

capacity of water is 4190 J/kg°C (ignore the work done on the kettle itself).

Solution   It's *always* a good idea to start by making a sketch, showing all the information given in the problem (Figure 12.11).

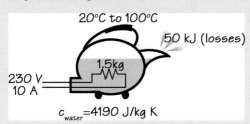

Figure 12.11

$$W - Q = mc(T_{final} - T_{initial})$$

But we know that the work done $W = E\,I\,t$, so:

$$(EIt) - 50\,000 = 1.5 \times 4190 \times (100 - 20)$$
$$(230 \times 10 \times t) - 50\,000 = 502\,800$$
$$2300t = 502\,800 + 50\,000$$
$$t = \frac{552800}{2300} \simeq 240 \text{ s}$$

As there are 60 s in one minute,

$$t = \frac{240}{60} = 4\,\text{min (Answer)}$$

Worked example 10   Repeat the previous worked example but this time take into account the metal from which the kettle itself is manufactured. Assuming the kettle itself has a mass of 0.5 kg and has a specific heat capacity of 500 J/kg·°C.

Solution   This time, we have to add together the change in internal energy of *both the water* **and** *the kettle*.

$$W - Q = \Big[m\,c\,(T_{final} - T_{initial})\Big]_{WATER}$$
$$+ \Big[m\,c\,(T_{final} - T_{initial})\Big]_{KETTLE}$$
$$(EIt) - 50\,000 = [1.5 \times 4190 \times (100 - 20)]$$
$$+ [0.5 \times 500 \times (100 - 20)]$$
$$(230 \times 10 \times t) - 50\,000 = [6285 \times 80] + [250 \times 80]$$
$$2300 \times t = 502\,800 + 20\,000 + 50\,000$$
$$t = \frac{572\,800}{2300} = 249 \text{ s}$$

As there are 60 s in one minute,

$$t = \frac{249}{60} = 4\,\text{min 9 s (Answer)}$$

Self-test exercise

By using the appropriate equations for work and power, complete Table 12.1 as it applies to the circuit. The first two have been done for you.

Self-test exercise –answers

Answers are shown in bold at the end of this chapter.

Misconceptions

Electrical power is different from other types of power
Strictly speaking, no. There are no 'types' of power! Power is simply a *rate* – the rate of doing work. So there is no such thing as 'electrical power', 'mechanical power', etc. You could, however, argue that '*electrical* power' is the 'rate of transfer of *electrical* energy'.

Horsepower measures mechanical power; watts measures electrical power
The horsepower is simply an imperial unit of measurement for power, whereas the watt is the SI unit of measurement for exactly the same thing. We could, if we wished, describe the power of an electric fire in horsepower, and a motor car's engine output in watts – in just the same way as we can use inches or millimetres to measure distance!

Isn't the output of an electric motor measured in horsepower?
Before the adoption of SI, the 'horsepower' was used to describe the output power of electric motors – these days, we use the watt. But, in North America, the horsepower is *still* used to measure the output power of motors.

Table 12.1

	Work (joules)	Power (watts)	Voltage (volts)	Current (amperes)	Resistance (ohms)	Time (seconds)
a.	**30 000**	1000	200	**5**	**40**	30
b.	20 000	500	**50**	10	5	40
c.			100		20	60
d.				5	100	50
e.	5000	100	20			
f.		1000		4		20
g.			25		25	60
h.	3000			2	1000	
i.		50		2		120
dj.	2500	5000	200			

The terms power, work and energy can be used interchangeably

In everyday language, this is usually the case. But in science and engineering, the difference between the two is very important. Power is the rate at which work is done, or energy is expended. If you like, power is equivalent to 'kilometres per hour', whereas 'work or energy' is equivalent to 'kilometres travelled'.

Energy is delivered from the supply to the load by the electrons moving through the circuit.

No, this is impossible as, in d.c. circuits, no single electrons move fast enough to travel between the supply and the load and, in a.c. circuits, the electrons simply vibrate backwards and forwards.

Heat is 'thermal energy'

Heat is the transfer of energy from a warmer body to a cooler body. So 'heat' describes the *transfer* of energy – not energy itself.

Isn't heat measured in calories? Or BTUs?

The 'calorie' is simply a unit of measurement for energy, used in the – now archaic –

'centimetre-gram-second-ampere' (cgsA) system of measurement, that predates the SI system. And the BTU (British Thermal Unit) is an imperial system unit of measurement. Energy, work and heat are *all* measured in joules in SI.

Heat and work are unrelated

Heat and work are very much related. The difference between the work done *on* a mass, and the heat transfer *from* that mass represents the change in energy of that mass.

Finally . . .

Now that you have completed this chapter, are you able to achieve the objectives or learning outcomes listed at the beginning of this chapter?

Ask yourself, 'Can I . . .'

1 define each of the following terms:
   a energy
   b work
   c heat
   d power.
2 specify the SI unit of measurement for each of the following:

a energy
b work
c heat
d power.
3 state the fundamental equation for the work done by an electric circuit.
4 state the fundamental equation for the power of an electric circuit.
5 derive alternative equations for the work done by, and the power of, an electric circuit.
6 define the term 'efficiency'.
7 solve problems on energy, work, power and efficiency.

8 explain how electricity supply companies bill their residential consumers.
9 read an analogue energy meter.
10 describe the relationship between electrical energy and heat.
11 solve problems on work and heat.

Self-test exercise – answers

44 749 kWh   5428 kWh   55 650 kWh

Table 12.2

	Work (joules)	Power (watts)	Voltage (volts)	Current (amperes)	Resistance (ohms)	Time (seconds)
a.	**30 000**	1000	200	**5**	**40**	30
b.	20 000	500	50	10	5	40
c.	30 000	500	100	5	20	60
d.	125 000	2500	**500**	5	100	50
e.	5000	100	20	**5**	**4**	**50**
f.	**20 000**	1000	*250*	4	**62.5**	20
g.	**1500**	**25**	25	**1**	25	60
h.	3000	**4000**	**2000**	2	1000	**0.75**
i.	**6000**	50	**25**	2	**12.5**	120
j.	2500	5000	200	**25**	**8**	**5**

Online resources
The companion website to this book contains further resources relating to this chapter. The website can be accessed via the following link:
www.routledge.com/cw/waygood

Chapter 13

Magnetism

On completion of this chapter, you should be able to

1 state the fundamental law of magnetism.
2 explain the term 'magnetic field'.
3 explain the term 'magnetic flux'.
4 explain the term 'flux density'.
5 state the direction allocated to magnetic flux.
6 sketch the pattern of the magnetic field, surrounding:
   a a bar magnet
   b a horse shoe magnet.
7 use the 'Domain Theory' to explain
   a magnetising an unmagnetised ferromagnetic material
   b why a north or south pole cannot exist in isolation
   c saturation.
8 list four methods of making a magnet.
9 explain the difference between permanent and temporary magnets.

Important! Throughout this chapter, the capitalised words 'North' and 'South' (including 'Magnetic North') refer to those **locations** on the Earth. The non-capitalised words 'north' and 'south' refer to **magnetic polarities**.

It's important to understand that the word '**poles**' have *two* meanings. In the 'geographic' sense, the 'poles' are those *locations* that correspond to the Earth's axis of rotation, and are termed the 'Magnetic North Pole' and the 'Magnetic South Pole'. In the 'magnetic' sense, 'poles' refer to *polarities of magnets*.

Introduction

According to legend, around 900 BC, a Greek goatherd named Magnus discovered that the iron nails in his shoes were attracted to the stones in a field where he grazed his goats. The stones in Magnus's field were naturally magnetised ferrites that we know, today, as '**magnetite**' – named after 'Magnus's stones'.

As fascinating as this legend is, it is far more likely that the origin of the word 'magnetite' is derived from the region of Greece known as 'Magnesia'.

Magnesia is located in south-eastern Greece, and features a peninsula which partly encompasses a huge and spectacular bay on the Aegean coast, and whose modern capital is the city of Volos. Magnesia is famous as the home of a number of Greek mythical heroes – including *Jason* and *Achilles*. Jason was the heroic leader of the *Argonauts* and their quest for the 'golden fleece', while Achilles was the hero of the Trojan War and a central character in Homer's *'Iliad'* whose only weakness gave us the modern expression *'Achilles' heel'*.

But more relevant to this chapter, in ancient times Magnesia was a major source of large deposits of a black mineral called ferrous-ferric oxide, but more commonly known as **magnetite** – the most naturally magnetic of all the earth's minerals.

Magnetite is commonly found deposited on the Earth's surface, often in the form of octahedron (eight-faced) nodules, as illustrated in Figure 13.1. It seems possible that the composition of magnetite *may* have allowed it to become magnetised as the result of the magnetic fields that would have accompanied countless lightning strikes occurring over billions of years.

Figure 13.1

An Introduction to Electrical Science, Waygood, ISBN 9780415810029, 2013. © Taylor & Francis

Around the fifth century BC, the Greeks discovered that magnetite had the strange property of being able to attract and pick up small pieces of iron. Today, we call this mysterious property '**magnetism**', a word derived from a Greek word, meaning *'Magnesia stone'*.

Different sources credit both the ancient Greeks and the ancient Chinese with discovering that if a splinter of magnetite is freely suspended horizontally, so that it can rotate, it will *always* come to rest pointing in an approximately North–South direction. The first person to write about this property was an eleventh-century Chinese scientist, Shen Kua (1031–1095), who explained that it helped his countrymen improve the accuracy of their navigation, with ocean-going junks travelling from Cylon to Africa. And there is certainly a great deal of evidence to suggest that, by the twelfth century, the Chinese were using 'lodestones' (pieces of magnetite) for the purpose of navigation . . . as well as for fortune telling!

It also seems certain that the Vikings used lodestones to improve their, already excellent, navigational skills.

Interestingly, Chinese lodestones were designed to be 'South pointing', in accordance with oriental mythology.

Preserved examples of ancient Chinese lodestones include thin, slightly concave slivers of magnetite, shaped like fishes, and intended to float in small bowls of water enabling them to rotate. The peculiar, 'spoon-shaped' lodestone, illustrated in Figure 13.2, was designed to balance on its bowl, so that it could easily rotate under the influence of the earth's magnetic field, with its handle indicating South.

Figure 13.3

If an ordinary piece of steel is stroked with a piece of magnetite, it will become **magnetised** and, just like the lodestone, when freely suspended, will always come to rest with one of its ends pointing roughly towards the Earth's geographic North Pole (located in the Arctic), and with its opposite end pointing approximately towards the Earth's geographic South Pole (located in the Antarctic).

Until strong, artifical magnets could be created using electromagnetism, lodestones were the *only* source of magnetism and, as late as the seventeenth century, these weak natural magnets were still being used to 'regenerate' the failing strength of compass needles used by Royal Navy warships, by stroking them in the appropriate direction with a lodestone kept on board for that very reason. A metal needle, magnetised in this way, formed the basis of what, eventually, became the **magnetic compass** (Figure 13.4) we are familiar with today.

Figure 13.2

The word '**lodestone**' (as illustrated in Figure13.3) is actually a Middle-English (the English spoken during the four centuries following the Battle of Hastings) word, meaning *'leading stone'*, and was used to describe the first primitive magnetic compasses.

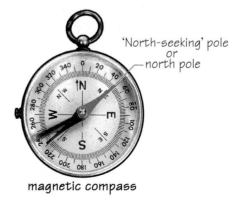

'North-seeking' pole
or
north pole

magnetic compass

Figure 13.4

Because of its useful navigational properties, one end of a magnet or compass needle came to be called its '**North-seeking pole**', while the other end was called its '**South-seeking pole**'. Over time, the use of the word '**seeking**' has fallen into disuse, and we now simply say: '**north pole**' and '**south pole**' which, by convention, are also used to name their *magnetic polarities*.

> The poles of a magnet, or of a compass needle, are named after the directions in which they point.

But, in fact, a compass needle doesn't actually point to the Earth's geographic or True North pole at all, but to a nearby *location* which we call 'Magnetic North'. It's very important to understand that 'Magnetic North' is the name we give to that *location*, and is *not the magnetic polarity at that location!*

The reason that the compass acts in this way is because the Earth itself behaves as though it contains a gigantic bar magnet buried deep within its surface – with its axis slightly displaced from the Earth's axis of rotation.

To distinguish between these different *locations*, we call them 'True North' and 'Magnetic North', and 'True South' and 'Magnetic South', respectively. Again, it's important to understand that 'Magnetic North' and 'Magnetic South' are *locations*, and *not* the magnetic polarities of those locations.

Fundamental law of magnetism

The fundamental law of magnetism states that '**like poles repel, and unlike poles attract**'.

In accordance with this law then, if a compass needle's north (-seeking) pole is attracted towards the Earth's Magnetic North, then the **magnetic polarity** of that location must be **south**.

So the Earth behaves as though it has a 'gigantic bar magnet' buried deep within its surface, with its south magnetic pole located in the Northern hemisphere, and its north magnetic pole located in the Southern hemisphere – as illustrated in Figure 13.5.

This 'giant bar magnet' is surrounded by a **magnetic field**, and a needle of a compass placed *anywhere* within that field will always align itself to lie along that field, and point towards the south magnetic pole buried deep below Magnetic North.

As we have already learnt, the axis of this 'giant bar magnet' does not coincide with the Earth's axis of rotation but, instead, 'wobbles' around it – rather in the same way that a spinning top wobbles when it starts to lose momentum. As a result of this, Magnetic North is

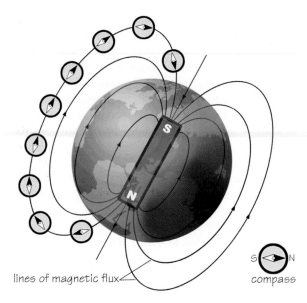

lines of magnetic flux

compass

Figure 13.5

always changing its position relative to True North as, of course, is Magnetic South with True South.

The angle between True and Magnetic North (Figure 13.6) we call the **'angle of declination'** – an angle that navigators, when using compasses, must *always* compensate for when plotting their routes. But, as if to make life difficult for navigators, this angle is continuously changing! So if you were to follow your compass north today, you would arrive at a different place than you would have arrived at, say, 10–15 years' ago! Accordingly, navigation maps are regularly updated to incorporate changes in the angle of declination.

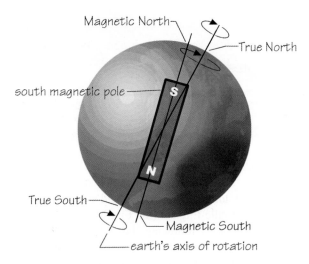

Magnetic North

True North

south magnetic pole

True South

Magnetic South

earth's axis of rotation

Figure 13.6

Furthermore, depending on where you are on the Earth, the angle of declination will be different; at some locations the declination will be minimal whereas, at other locations, the declination will be large.

Currently, Magnetic North is located in Canada's Arctic Ocean and is moving north-west towards Siberia at the rate of around 40–50 km per year. As well as moving in this general direction, Magnetic North also 'wanders around' its assumed route, so that on any given day, it can be some distance from where it has been predicted *to* be!

Magnetic fields

Like gravity, magnetism is an invisible force, for which there has been *no satisfactory explanation*. However, while we can neither see, nor explain, the nature of this force, we *can* observe and quantify its **effects**.

Unlike gravity, however, which is a force of *attraction* between masses, magnetism represents a force of *repulsion* as well as attraction, and obeys the fundamental law, which states:

> **Like poles repel; unlike poles attract**.

Knowing this, if you ever need to identify the poles of an unmarked magnet you can do so using a **compass**. A compass needle's north pole (usually coloured black or red) will be *attracted* by and, therefore, identify the magnet's south pole. And, of course, the compass needle will be *repelled* by a magnet's north pole (Figure 13.7).

The *region surrounding a magnet*, in which its effects can be observed, is termed a **magnetic field**.

> A **magnetic field** is the area surrounding a magnet in which its effects may be observed.

There is actually a way in which we can apparently 'see' the shape of a magnetic field – by conducting a very simple, but classic, experiment using a magnet, a sheet of card, and some iron filings.

As, no doubt, every schoolboy knows, if iron filings are sprinkled onto a sheet of card which has been placed over a magnet, and the card is then gently tapped so that it vibrates, the filings will be seen to arrange themselves in distinctive patterns which represent the magnetic field. The filings trace out *what appear to be* 'lines of force', called **magnetic flux** (Figures 13.8 and 13.9).

What you are actually seeing, as a result of this experiment, are *not* really lines of force – it just *looks* that way, because the 'tip' of one iron filing is attracted to the 'tail' of another, causing the filings to form closed 'chains'!

However, representing a magnetic field by lines is a very useful model, first presented by the scientist Sir Michael Faraday (1791–1867).

In Faraday's 'model' of a magnetic field, the lines of magnetic flux are allocated **direction**, determined by the *direction in which a compass needle would point* when placed within the field. As a compass needle always points *along the lines of magnetic flux*, towards a magnet's *south* pole, this direction is always **from a magnet's north pole towards its south pole**.

> We say lines of magnetic flux 'leave' the north pole of a magnet, and 'enter' the magnet at its south pole.

We can show this by moving a plotting compass (a small compass, used for experiments) around a magnet, as illustrated in Figure 13.10.

From the above images, as well as from the directions in which the plotting compasses point, it can clearly be seen that the lines of magnetic flux 'leave' and 'enter' a magnet in the areas close to the *ends* of that magnet – indicating that the *poles are concentrated at the ends of a magnet*. This can be further demonstrated by plunging a magnet into a box of small steel tacks or

a magnet's south pole
attracts a
compass's north pole

a magnet's north pole
attracts a
compass's south pole

Figure 13.7

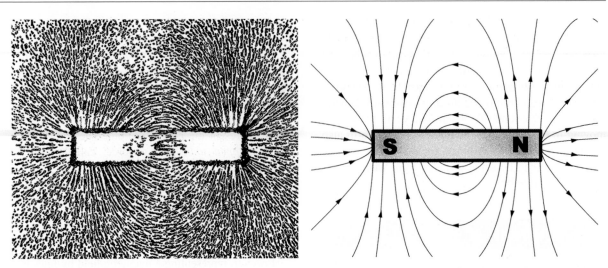

Figure 13.8 magnetic flux surrounding a bar magnet, traced with iron filings (left), and how we represent them in a drawing (right)

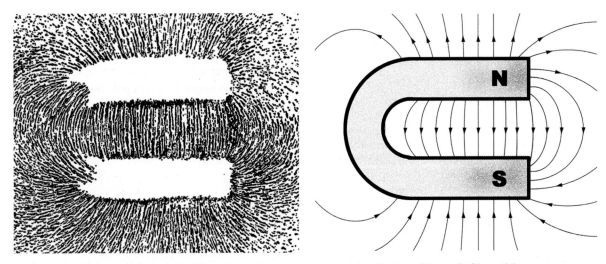

Figure 13.9 magnetic flux surrounding a horseshoe magnet, traced with iron filings (left), and how we represent them in a drawing (right)

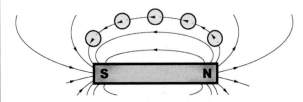

Figure 13.10

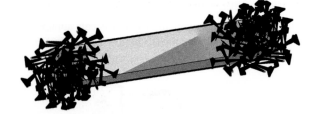

Figure 13.11

iron filings; when the magnet is withdrawn, you find that the majority of the tacks or filings are clustered around the ends of the magnet with few, if any, towards its centre.

As already mentioned, Faraday is credited with first representing a magnetic field using lines of magnetic flux, and for establishing the following 'properties' of these lines. But, once again, we must continually remind ourselves, that these lines of magnetic flux *don't actually exist*; they are merely a very useful model to help us to form a 'mental picture' of the behaviour of an otherwise invisible magnetic field. However, like all models, it does have its limitations.

So, according to Faraday's 'lines of magnetic flux' model:

- lines are *elastic*, enabling them to stretch or contract
- parallel lines, acting in the *same* direction, *repel* each other
- parallel lines, acting in the *opposite* direction, *cancel* each other
- lines *never cross each other*
- lines *always take the path of least opposition*
- there is no known *'insulator'* to the lines
- the *density* of the lines is an indication of the *strength* of the magnet.

The first four of these properties are illustrated in Figures 13.12–13.15.

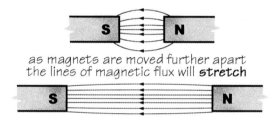

Figure 13.12

When we say that lines of magnetic flux *'take the path of least opposition'*, we need to understand that ferrous metals provide a *very* much easier medium in which to support the formation of flux than air – in fact, these metals are *thousands* of times better at supporting the formation of magnetic flux than air is. As a result, a piece of ferrous metal placed within a magnetic field

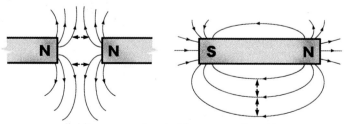

parallel lines, running in the same direction
repel each other

Figure 13.13

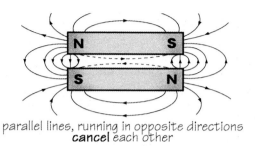

parallel lines, running in opposite directions
cancel each other

Figure 13.14

lines of magnetic flux
cannot cross

Figure 13.15

will always *significantly* distort that field – as in the example in Figure 13.16.

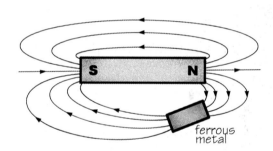

Figure 13.16

This emphasises how important it is *not* to have any ferrous metals anywhere near a compass when a directional reading is being taken, as you are very likely to obtain a false reading due to resulting distortion to the Earth's field.

Magnetic mines explode when the metal hull of a ship distorts the Earth's magnetic field as it passes in the vicinity of the mine; they are *not* set off because their magnetic sensor is attracted towards the hull.

Although there appears to be no method of insulating magnetic flux, we *can* protect sensitive instruments from flux by **shielding** them. 'Shielding' takes advantage of the sort of distortion described above, by using a ferrous-metal container to *divert* the magnetic flux *around* the sensitive instrument – as illustrated in Figure 13.17.

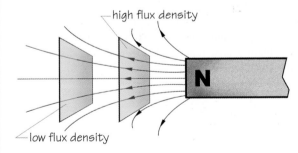

ferrous shield

Figure 13.17

The *intensity* of the lines of flux is an indication of the strength of a magnet. We call this **flux density** (symbol: B), which is defined as *'the flux per unit area'*, and we will discuss this in detail in the chapter on *magnetic circuits*. Figure 13.18 shows how the flux density varies, according to where it is measured (note that although the flux is shown in two dimensions, it is actually in three dimensions).

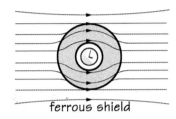

high flux density

low flux density

Figure 13.18

We have already referred to the fundamental law of magnetism, 'like poles repel while unlike poles attract', so let's look at how the 'properties' of the lines

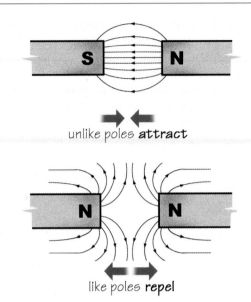

unlike poles **attract**

like poles **repel**

Figure 13.19

of magnetic flux, in our model of the magnetic field, contribute to this law.

In the case of 'unlike poles', it is the contraction of the 'elastic' lines of flux that act to pull the opposite poles together. In the case of the 'like poles', it is the repulsion between parallel lines of flux that act to push similar poles apart.

The Domain Theory of Magnetism

Whenever a bar magnet is broken in half, two new magnets are created, no matter how many times this is repeated. This is because *no pole can exist in isolation*. All magnets exist as *dipoles* – that is, they always have *pairs* of poles.

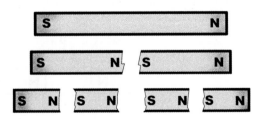

Effect of breaking a magnet in half

Figure 13.20

When contemplating this behaviour, the nineteenth-century German scientist Wilhelm Eduard Weber (pronounced *'Vay-ber'*) theorised that if this action could be repeated often enough, then we would ultimately arrive at a molecule, or even an atom, which would behave just like a tiny bar magnet. This 'molecular magnet' was eventually termed a '**domain**' and this, in turn, led to what has became known as the **Domain Theory of Magnetism**.

According to the **Domain Theory of Magnetism**, every electron not only orbits the nucleus of its atom, but also *spins about its own axis*. This causes each electron to generate its own magnetic field (rather like the Earth does) – causing it to behave like an incredibly small magnet. The polarity of the electron's magnetic field depends upon the direction of its spin – as illustrated in Figures 13.21 and 13.22.

Once again, we must continue to remind ourselves that this is only a 'model', and it is most unlikely to be what is *really* happening inside an atom!

In most atoms, pairs of electrons spin in *opposite* directions – effectively cancelling out each other's magnetic fields. Such atoms, therefore, exhibit no overall magnetic field (Figure 13.23, of course, is not to scale!).

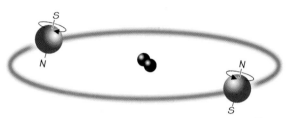

Figure 13.23

In magnetised materials, however, *more electrons spin in one direction than in the other.* For example, in Figure 13.24 (showing only one shell) four electrons spin in one direction, while only two spin in the opposite direction.

Such atoms or, more likely, molecules therefore exhibit a natural magnetic field, and are called **domains**. Magnetite is largely made up of molecules which are domains.

Figure 13.21

Figure 13.22

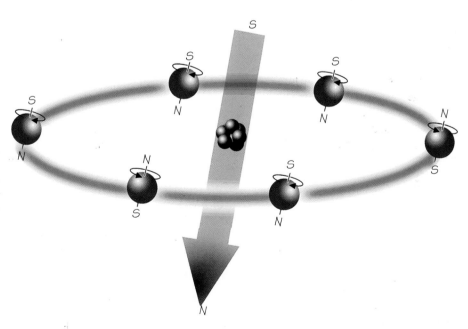

Figure 13.24

A **domain** is a molecule or an atom which exhibits a magnetic field.

In the following illustrations, for the sake of simplicity, we will represent a domain as a miniature magnet, as shown in Figure 13.25.

DOMAIN

Figure 13.25

In **unmagnetised** ferromagnetic materials, the domains arrange themselves roughly in closed 'chains', with the north pole of each domain attracted to the south pole of another domain – as illustrated in the (very much!) simplified Figure 13.26. Because these chains are closed, their poles tend to cancel out the effect of each other, and no overall magnetic effect is noticeable externally. (Note that although we have represented these domains in two dimensions, they of course actually exist in three dimensions.). Of course, the actual distribution of domains would not be as neat and as well organised as they have been illustrated here!

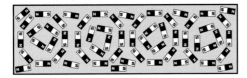

Figure 13.26 domains form 'chains' in unmagnetised ferromagnetic materials

In **magnetised** ferromagnetic materials, however, the domains are arranged in 'rows' – leaving columns of unattached north poles at one end of the material, and columns of unattached south poles at the opposite end. This explains why the poles of a magnet are concentrated at each end.

The Domain Theory also explains why, if a magnet is broken in several places, *new* magnets are created, and why *poles cannot exist in isolation*.

Making magnets

The Domain Theory helps us understand what probably happens when we make magnets, and why some materials become **permanent magnets** while others become only **temporary magnets**.

In order to make an unmagnetised ferromagnetic material into a magnet, its domains must be realigned to form rows. To do this, the domains must be forced out of their natural 'chain' formations and, then, allowed to come to rest forming 'rows'. This can be brought about by using an **external magnetic field** (in some cases, simply the Earth's natural magnetic field), to influence the final direction of the domains.

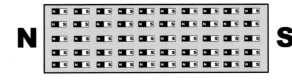

Figure 13.27 domains forming 'rows' in magnetised ferromagnetic materials

Magnets can be made in several ways, *all* involving the transfer of energy while under the influence of an external magnetic field, including by

- **stroking** the material with another magnet.
- **hammering** the material, *while aligned in a North–South direction.*
- **heating** the material, and allowing it to cool *while aligned in a North–South direction.*
- **electromagnetic induction**, in which a direct current is passed through an insulated coil that has been wound around the material.

By **stroking** a ferromagnetic material with another magnet, the domains' poles are attracted by the (opposite) pole of the magnet, and 'dragged' from their natural chain formations and realigned into rows.

By **hammering** or **heating**, energy is imparted into the ferromagnetic material which results in the domain chains becoming excited and breaking up. If then

   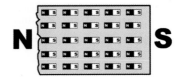

Figure 13.28 broken magnets result in two new magnets, each with its own north and south poles

left aligned in a North–South direction, the Earth's natural magnetic field will cause the domains to form rows as their energy levels return to normal. This, incidentally, explains why tools, such as screwdrivers, often become partially magnetised after rattling around in a toolbox.

Each of the methods described above are fairly *inefficient* ways of creating magnets, and the resulting magnet will be relatively weak.

Until 1820, when the Danish physicist Hans Christian Øerstedt (1777–1851) discovered the relationship between electricity and magnetism, all magnets were difficult to manufacture and they produced weak magnetic fields. But Øerstedt's discovery led to the most efficient, and practical, method of making a powerful magnet: by **electromagnetic induction**. As we will learn in the next chapter, a direct current flowing through a coil creates a very strong magnetic field. So, a current-carrying coil wound around a length of ferromagnetic material will very effectively force domain chains into rows.

Magnetic saturation

Once all the domains that *can* be aligned into rows *have* been aligned, the magnet is said to be **saturated**, and its magnetic field has reached its maximum intensity, and *it is not possible to strengthen the magnetic field beyond this point*. We will examine this more closely in the next chapter.

Induced magnetism

When a magnet is brought into contact with a non-magnetised piece of ferrous metal, it causes that metal itself to become magnetised. We call this effect '**induced magnetism**'.

If the ferrous metal is **iron**, then it will lose its induced magnetism immediately the magnet is removed. If it is **steel**, then it will *retain* its induced magnetism after the magnet has been removed – as explained in the next section.

The part of the ferrous metal closest to the magnet's pole acquires the *opposite* polarity to that pole, while the far end of the metal acquires the same polarity – essentially elongating the original magnet. This phenomenon can be explained in terms of the domain theory, as the domains in the ferrous metal become aligned under the external influence of the permanent magnet (Figure 13.29).

This can be repeated numerous times, as is the case in the example shown in Figure 13.30.

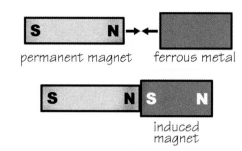

Figure 13.29

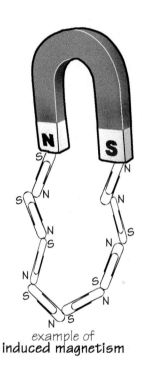

example of
induced magnetism

Figure 13.30

Ferromagnetic materials

We will explain the term 'ferromagnetism' in the next chapter. But, in the meantime, it is acceptable to think of ferromagnetic materials as being **iron**, or alloys that contain iron, such as **steel**.

For materials such as **iron**, the chains of domains are held together very weakly, and their domains realign into rows very easily under the influence of an external magnetic field. But once that external magnetic field is removed, the domains then very easily return to their chain formations. Materials such as iron, therefore, lose most of their magnetism immediately any external magnetic field is removed, and are termed **temporary magnets**.

Temporary magnets are:

- easy to magnetise
- lose their magnetism immediately an external magnetic field is removed.

For materials such as **steel**, the chains of domains hold very strongly together, and a great deal of external energy must be applied to realign them into rows. Once realigned, however, it then takes a great deal of energy to break up this new alignment. Such materials, therefore, retain most of their magnetism once any external magnetic field is removed, and are termed '**permanent magnets**'.

Interestingly, some of the strongest permanent magnets are manufactured from alloys of iron together with materials which, themselves, are magnetically very weak, including aluminium and copper. For example, '**alnico**' (a trade name for a metal alloy used to manufacture magnets) is an alloy of iron, aluminium, cobolt, copper and nickel. One of the very strongest magnets (over twenty times more powerful than alnico) is manufactured from an alloy of cobolt and platinum, and contains no iron at all.

Permanent magnets are

- difficult to magnetise
- retain their magnetism after the external magnetic field has been removed
- difficult to demagnetise.

Both permanent and temporary magnets have useful applications in electrical engineering.

- **Permanent magnetic materials** are used in the manufacture of analogue electrical measuring instruments, loudspeakers, security sensors, etc.
- **Temporary magnetic materials** are used in the manufacture of transformers, relays, contactors, etc.

Misconceptions

A compass needle indicates the direction of True North
A compass needle lies along the Earth's natural lines of magnetic flux, which link Magnetic North and Magnetic South, *not* True North and True South.

The term 'Magnetic North' refers to the magnetic polarity of that location
No. The terms 'Magnetic North' and 'Magnetic South' are used to distinguish their *locations* from True North and True South. They are *not* the magnetic polarities of those locations!

The end of a compass needle which points to Magnetic North is really a south pole
No. Because the magnetic polarity of Magnetic North is a south pole, it attracts the north pole of a compass needle.

Lines of magnetic flux exist
No they don't! They are simply a 'model' to help us visualise the pattern of a magnetic field.

Finally . . .

Now that you have completed this chapter, are you able to achieve the objectives or learning outcomes listed at the beginning of this chapter?

Ask yourself, 'Can I . . .'

1 state the fundamental law of magnetism.
2 explain the term 'magnetic field'.
3 explain the term 'magnetic flux'.
4 explain the term 'flux density'.
5 state the direction allocated to magnetic flux.
6 sketch the pattern of the magnetic field, surrounding:
   a a bar magnet
   b a horse shoe magnet.
7 use the 'Domain Theory' to explain
   a magnetising an unmagnetised ferromagnetic material
   b why a north or south pole cannot exist in isolation
   c saturation.
8 list four methods of making a magnet.
9 explain the difference between permanent and temporary magnets.

Online resources

The companion website to this book contains further resources relating to this chapter. The website can be accessed via the following link:

www.routledge.com/cw/waygood

Chapter 14

Electromagnetism

On completion of this chapter, you should be able to

1 explain evidence how an electric current is accompanied by a magnetic field.
2 determine the shape and direction of the magnetic field around a current-carrying straight conductor.
3 determine the shape and direction of the magnetic field around a current-carrying single loop conductor.
4 determine the shape and direction of the magnetic field around a current-carrying coil.
5 explain the term 'flux density', specifying its SI unit of measurement.
6 calculate the flux density of simple magnetic fields.
7 sketch the magnetic field resulting from a current-carrying conductor placed in a permanent magnetic field, and use this to explain why the conductor is subject to a force.
8 apply Fleming's Left Hand Rule to determine the direction of force on a current-carrying conductor placed in a magnetic field.
9 calculate the value of the force acting on a current-carrying conductor placed in a magnetic field.
10 describe the principle of operation of a simple d.c. motor.

Important: *conventional* current direction (positive to negative) is used throughout this unit.

Introduction

Until the early nineteenth century, **electricity** and **magnetism** were generally considered to be *separate* phenomena. Even those scientists who instinctively believed in a connection between the two had no experimental evidence with which to support their views.

However, one evening in 1820, while preparing an experiment for his students, the Danish physicist, Hans Christian Øersted (1777–1851), noticed that whenever he passed an electric current through a wire, the needle of a nearby compass deflected from its normal North–South direction.

Øersted was already one of those scientists who believed that there must be a connection between electricity and magnetism, so his discovery may not have been quite as accidental as it has since been made out to be. Nevertheless, this observation spurred him into performing further experiments which convinced him that an electric current must create a magnetic field – indicating that *there was indeed a direct relationship between electricity and magnetism*.

Øersted's observation triggered a great deal of research into this relationship, not only by Øersted himself, but by others such as Ampère, all of whom realised that the magnetism produced by an electric current could produce *forces* which, in turn, might be harnessed to produce useful *motion*.

Furthermore, if the relationship between magnetism and electricity should prove to be *reversible*, then it might then also be possible to convert motion into electricity!

We call this relationship between electricity and magnetism '**electromagnetism**'. More specifically, electromagnetism is that branch of science concerned with those forces that result between electrically charged particles in motion.

Electric currents and magnetic fields

Øersted's experiments showed that a magnetic field *surrounds a current-carrying conductor*, and that the *direction of that field* depends upon the *direction of that current*.

An Introduction to Electrical Science, Waygood, ISBN 9780415810029, 2013. © Taylor & Francis

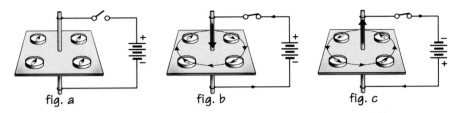

Figure 14.1

One of his experiments is very easy to reproduce, using simple equipment, as illustrated in Figure 14.1.

A conductor is passed vertically through a horizontal card, on which a number of plotting compasses have been placed around the conductor.

In Figure 14.1a, the switch supplying the vertical conductor is open, so there is no current, and all the compasses are seen to be pointing in their usual North–South direction.

In Figure 14.1b, the switch is closed and current moves through the conductor. In this case, the direction of *conventional*-current drift (i.e. positive to negative) is downward. When viewed from above, it will be seen the individual compass needles have now assumed a *clockwise* direction around the conductor.

Finally, in Figure 14.1c, the battery connection is reversed, so that the conventional current drift through the conductor is now upwards. This time, it will be seen that the individual compass needles have reversed direction, and have now assumed a counter-clockwise direction around the conductor.

From this simple experiment, we can conclude that:

---

1   An electric current is always surrounded by a magnetic field.
2   The 'direction' of the magnetic field depends upon the direction of the current.

When we describe a magnetic field as 'surrounding a current', it applies whether or not there is a conductor present. For example, a magnetic field will also surround an **arc**, which is a current passing through air or a vacuum.

Remember, as we learned in the chapter on *magnetism*, a magnetic field consists of **magnetic flux** the direction of which, by common agreement, is determined by the direction in which a compass needle will point, when placed within that magnetic field – i.e from north to south.

Electromagnetic fields

Before proceeding further, we need to learn how the *direction* of an electric current is represented, graphically, in this and in all other texts.

Imagine releasing an arrow in the same direction as the current. With the arrow moving *away* from you, you will see its *flight feathers* represented by a cross (×) but, if the arrow is coming *towards* you, then you will see its *point*, represented by a dot (•). This is called the '**cross/dot convention**'.

So, in a cross-sectional view of a conductor, it is conventional to show a cross (×) to indicate that the drift of current is *away* from you (i.e. *into* the page), and to show a 'dot' (•) to indicate that the drift of current is *towards* you (i.e. *out* of the page).

Some textbooks represent current flow away from you as a plus (+) sign, rather than as a cross, but we will avoid this as it can be confused for polarity.

This convention applies to *both* conventional flow *and* to electron flow – so it is *very* important to know

a **cross** indicates
current flowing **away** from you

a **dot** indicates
current flowing **towards** you

Figure 14.2

which current direction is being used whenever electromagnetism is being discussed. Most textbooks assume **conventional flow** (i.e. positive to negative) and, for this reason, we will also use conventional flow throughout this and all other chapters, *except when otherwise specified*.

> For electron flow, *all* the rules that you will be learning in this chapter are simply *reversed*.

Magnetic field surrounding a straight conductor

Now, let's return to our experiment. By using small, transparent 'plotting compasses', it is possible to plot *the shape and direction* of the magnetic field surrounding a current, as follows (the black part of the compass needle represents its 'North-seeking' pole), by tracing out that field.

As you can see in Figure 14.3, when viewed from either end of the conductor, the magnetic field assumes the shape of a series of concentric circles which surrounds the conductor (or, more accurately, the *current*). As you will also see, the field is more concentrated immediately surrounding the conductor, than it is further away from the conductor.

So what is going on? Well, *the field actually starts expanding from within the conductor itself*, as illustrated in Figure 14.4.

When the current is initially switched on, it starts to rapidly increase from zero to a maximum value, limited by the resistance of the circuit.

Initially, a circular magnetic field, of infinitesimal radius, forms at the centre of the conductor. As the current increases, another tiny circular magnetic field is

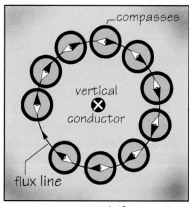

current drift
away from reader

compasses indicate flux forms
clockwise concentric circles
around current

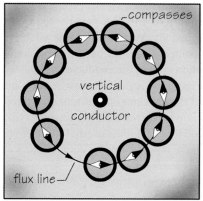

current drift
towards reader

compasses indicate flux forms
counterclockwise concentric circles
around current

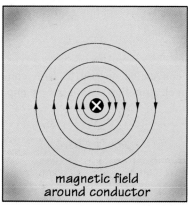

magnetic field
around conductor

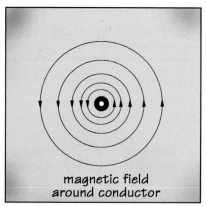

magnetic field
around conductor

Figure 14.3

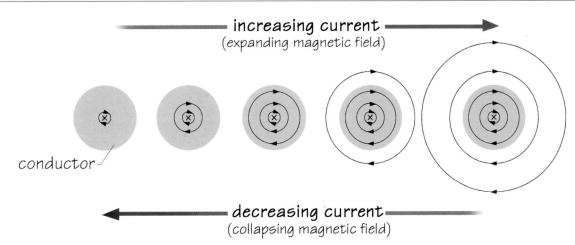

increasing current
(expanding magnetic field)

conductor

decreasing current
(collapsing magnetic field)

Figure 14.4

formed at the centre, which forces the existing magnetic field to expand outwards, rather like the ripples that radiate away from the point where a pebble is dropped into a pool. As the current increases further, yet more tiny magnetic fields are formed at the centre, forcing the others to expand even further outward. This process repeats itself, until the expanding circular magnetic fields start to surround the conductor externally, and it continues until the current has reached its maximum value, at which point the field becomes constant.

If you refer back to Faraday's model of a magnetic field, described in the chapter on *magnetism*, you will recall that the field's lines of force behave somewhat like rubber bands, and the tension within the circular flux lines causes them to be closer together towards the centre of the conductor than they are further away.

When the current is switched off, this process reverses itself, as the tension within each circular magnetic field causes them all to collapses inwards.

> It's the action of these expanding and collapsing concentric magnetic field rings, 'cutting' the conductor internally, which is partially responsible for the '**skin effect**' described in the chapter on *resistance*. As we shall learn later, whenever a magnetic field passes through a conductor, it induces a voltage into the conductor which opposes the current at the centre of the conductor, forcing it to travel closer to the surface.

Although the magnetic field surrounding a current-carrying conductor doesn't have a north or a south

pole, we can still allocate *direction* to the concentric circles. This is, of course, *the direction a compass needle would point if placed within the field* – from north to south.

Although these concentric circles of flux are shown in *two* dimensions, they are, of course, actually *three-dimensional* and enclose the current along its entire path throughout a circuit. It would, therefore, be more accurate to think of them as *concentric tubes*, rather than as concentric circles.

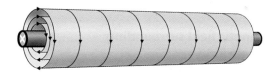

flux surrounds current in concentric tubes

Figure 14.5

However, concentric tubes are rather difficult to represent on a diagram, so it is standard practice to show the flux as concentric circles.

As we have seen from our explanation of Øersted's experiment, the *direction* of the magnetic field of a current is **clockwise** for current drifting *away* from you, and **counterclockwise** for current *drifting* towards you. To help you remember this, there are alternative rules we can use.

The first is called the '**corkscrew rule**'. With this rule, the direction of the field is the same direction in which a corkscrew is turned, when pointing in the same direction as the current.

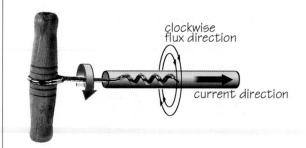

Figure 14.6

An alternative rule for determining the direction of flux around a current is the '**right-hand grip rule**'. If you grip a conductor, with your thumb pointing in the direction of the current, then the 'curl' of your fingers will represent the direction of the field.

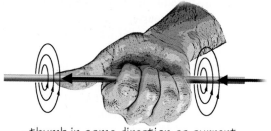

- thumb in same direction as current
- flux direction same as curl of fingers

Figure 14.7

Figure 14.8

Figure 14.9

Magnetic field surrounding a single loop

If a length of current-carrying conductor is twisted to form a **single loop**, then the resulting magnetic flux lines would appear to leave the loop from one of its faces, and re-enter the loop via its opposite face. In the example, in Figure 14.8, the loop behaves as though its *front* face was a north pole (the face from which the flux leaves) and its opposite face would behave as a south pole (the face into which the flux re-enters).

A quick way of determining which face of the loop is north and which is south is to imagine the letter's **N** and **S**, pivoted so that they are free to 'spin' in the same direction as the arrow heads placed as shown in Figure 14.9.

The letter 'N' (for 'north') will spin counterclockwise, whereas the letter 'S' (for 'south') will spin clockwise. So, in the left-hand illustration, in Figure 14.9, the current is flowing around the loop in the same direction as the letter 'N' would spin and, so, the front face of

the loop is a *north* pole. In the right-hand illustration, in Figure 14.9, the current is flowing around the loop in the same direction as the letter 'S' would spin and, so, the front face of the loop is a *south* pole.

If an insulated conductor is wound to form *several loops*, then it will form a **coil**. And when a current passes through a coil, the magnetic field surrounding each individual loop acts to *reinforce* those around adjacent loops to form a single, strong, magnetic field – as shown in the coil section in Figure 14.10.

As you can see, the shape of the magnetic field surrounding a current-carrying coil closely resembles the field surrounding a bar magnet, with its poles concentrated towards opposite ends of the coil. In the example in Figure 14.10, the magnetic flux 'emerges' from the left-hand end – indicating that end is its north pole – and 'enters' the right-hand end – indicating that end is its south pole.

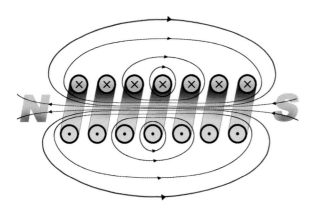

Figure 14.10

The polarity of each end of a coil may be determined using a variation of the '**right-hand grip rule**' which we learnt about earlier. In this case, simply grasp the coil with your *right* hand (Figure 14.11), such that the curl of the fingers represents the direction of current around the coil, and the thumb will then point towards the **north pole** end of the coil. And the opposite end, of course, will be its **south pole**.

Magnetic flux density

Magnetic fields are represented by lines of **magnetic flux** (symbol: Φ, pronounced 'phi'). As we learnt in the chapter on *magnetism*, these lines are *imaginary* and simply represent a model for showing the shape or pattern of a magnetic field. Although they are imaginary, we can still quantify them, and the SI unit of measurement we use to do this is the **weber** (symbol: **Wb**) – pronounced *'vay-ber'* – after the German physicist, Wilhelm Eduard Weber (1804–1891).

In simple terms, we can think of each, individual, magnetic flux line within a magnetic field as representing one weber (we won't worry about its actual definition at this stage, but will return to it later).

The 'closeness' of a field's lines of magnetic flux (or number of webers) indicates the *intensity* of that field at any given point. This is termed the **magnetic flux density** (symbol: **B**) of the field at that particular point. Just as with a permanent magnet, the flux density is always greatest in the areas nearest the poles of the coil – as can be seen in the example in Figure 14.12.

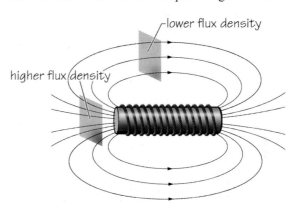

Figure 14.12

From Figure 14.12 it can be seen that more lines of flux pass through the shaded rectangle near the poles, than through the rectangle of identical cross-sectional area placed midway along the coil. So we can define **flux density** as the *'flux per unit area'*, expressed as follows:

$$B = \frac{\phi}{A}$$

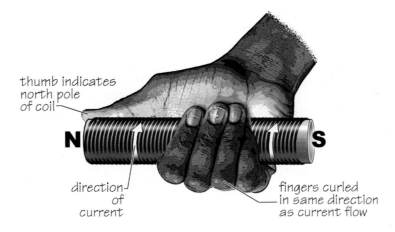

Figure 14.11

where:

B = flux density, in teslas (T)

ϕ = flux, in webers (Wb)

A = area, in square metres (m^2)

The unit of measurement for flux density is the **tesla** (symbol: **T**), named in honour of the Serbian-American engineer, Nikola Tesla (1856–1943), and is equivalent to a 'weber per square metre'.

Worked example 1   The pole face of a magnet measures 30 mm by 30 mm. The flux leaving the pole is 0.25 mWb. Calculate the flux density at the pole face.

Solution   **Note!** You must change millimetres to metres, and milliwebers to webers.

$$\text{area of poleface} = (30 \times 10^{-3}) \times (30 \times 10^{-3})$$
$$= 900 \times 10^{-6} \ m^2$$
$$B = \frac{\phi}{A} = \frac{0.25 \times 10^{-3}}{900 \times 10^{-6-3}} \approx 0.28 \ T \ \text{(Answer)}$$

Worked example 2   A current-carrying coil, of inside diameter 50 mm, generates 1500 mWb of magnetic flux. Calculate the flux density within the coil.

Solution   **Note!** The millimetres must be changed to metres, and milliwebers to webers.

$$B = \frac{\phi}{A_{coil}} = \frac{\phi}{pr_{coil}^2}$$
$$= \frac{1500 \times 10^{-3}}{p\left(\dfrac{50}{2} \times 10^{-3}\right)^2} = \frac{1500 \times 10^{-3}}{p \times 25^2 \times 10^{-6-3}}$$
$$= \frac{1500}{1963 \times 10^{-3}} \approx 764 \ T \ \text{(Answer)}$$

Electromagnets

The flux density of the magnetic field produced by a coil can be *significantly* (*hundreds* or, even *thousands*, of times!) increased by winding the coil around a **ferromagnetic** (iron alloy) core, rather than around a hollow tube.

The reason for this was explained by the Scottish engineer and academic Sir James Ewing (1855–1935) as follows: when the core is magnetised, the individual magnetic field of each of its millions of domains align with and, therefore, *reinforce*, the relatively weak magnetic field due to the current alone, thus creating a very strong combined field.

Ferromagnetic materials, such as silicon steel, are used as **cores** for electromagnets, and for the **magnetic circuits** of electrical machines. We will learn what is meant by a 'magnetic circuit' in the next chapter.

Electromagnets are very much stronger than the strongest permanent magnet, and are used whenever we want to use electric current to create movement, such as in circuit breakers, motor starters, relays, etc.

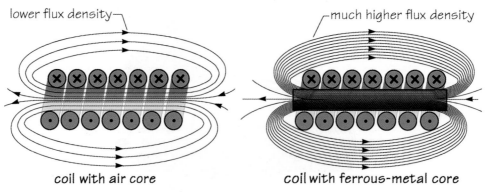

coil with air core   coil with ferrous-metal core

Figure 14.13

Force on conductors

In this chapter, we've learnt that whenever a current drifts through a conductor, a magnetic field surrounds that current. It follows, therefore, that if currents move through each of two, parallel conductors, then the resulting magnetic fields will react with each other, and produce a force between the two conductors. This force may be one of **attraction** or of **repulsion**, *depending upon the relative directions of the electric currents in the conductors*.

If we apply Faraday's properties of lines of magnetic flux, which we learnt from the chapter on *magnetism* (specifically, 'parallel flux lines acting in the *same* direction *repel* other', while 'parallel flux lines acting in *opposite* directions *cancel*') then, as illustrated in Figure 14.14, if the currents are drifting *in opposite directions*, the resulting force between them will be one of *repulsion*; whereas if the currents are drifting *in the same direction as each other*, then the resulting force will be one of *attraction*.

You will recall that the SI unit of electric current, the **ampere**, is defined in terms of *the force between two straight, parallel, current-carrying conductors*. The above description explains the reason for this force.

And these forces can be considerable. The exceptionally high fault currents that result from short-circuits in power systems can actually cause severe distortion to any parallel conductors, such as busbars, carrying such currents.

The forces between magnetic fields are *very* important in electrical engineering because, as we shall see, the operation of **electric motors** is entirely dependent upon these forces.

Now let's turn our attention to the behaviour of a current-carrying conductor, placed within a permanent magnetic field – such as that illustrated in Figure 14.15. Again, the two fields will react with each other, causing a **force** to act upon the conductor.

Since (according to Faraday's rules on their behaviour) lines of magnetic flux never cross, they will either crowd together to *reinforce* each other whenever they act in the *same direction* or, when they act in *opposite directions*, they will *weaken* each other. The resulting **force** will then attempt to push the conductor out of the permanent field.

This is called '**motor action**' because, as we shall see, this is the basis of how electric motors, as well as other devices, such as dynamic loudspeakers, work.

The above principle is also used to extinguish electric **arcs** in certain types of circuit breaker. By placing strong magnets (or electromagnets) either side of the circuit breaker's main contacts, the arc formed when the contacts break will be forced sideways, causing it to lengthen. Because a lengthened arc is unstable, it will extinguish relatively easily.

The **magnitude** of the resulting force on the conductor depends upon the flux density of the permanent magnetic field, the current in the conductor, and the length of conductor within the field, as expressed below:

$$F = BIl$$

where:

F = force, in newtons (N)

B = flux-density, in teslas (T)

l = length of conductor in field, in metres (m)

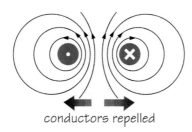
conductors repelled

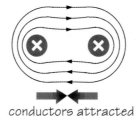
conductors attracted

Figure 14.14

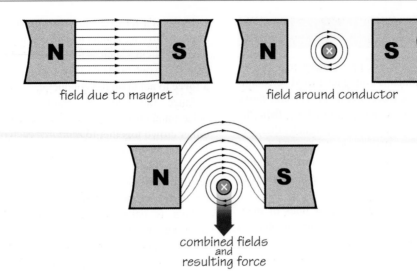

field due to magnet   field around conductor

combined fields
and
resulting force

Figure 14.15

Worked example 3 A current of 5 A drifts through a conductor placed between the poles of a magnet. If 50 mm of the conductor is within the magnetic field, and the flux density of the field is 500 mT, calculate the resulting force on the conductor.

Solution

$$F = BIl = (500 \times 10^{-3}) \times 5 \times (50 \times 10^{-3})$$
$$= 125 \times 10^{-3}\,\text{N} \quad \text{or} \quad 125\,\text{mN (Answer)}$$

Fleming's Left-Hand Rule

The **direction** of the force on a conductor may be determined from first principles, by drawing the individual magnetic fields, separately – as illustrated earlier – or, instead, by using **Fleming's Left-Hand Rule*** for *conventional* current. This rule was created by the British engineer and academic, Sir John Ambrose Fleming (1849–1945), to help his students determine the direction of rotation of an electric motor's armature.

*Fleming, in fact, devised *two* such rules. His 'Left-Hand Rule' (for conventional flow) applies to motors, whereas his 'Right-Hand Rule' (for conventional flow) applies, as we shall learn, to generators. To avoid confusing the two, you might want to think that, in the UK, '*motor* cars drive on the *left*' (motors = left-hand rule).

To apply Fleming's Left-Hand Rule, extend the first finger (index finger), the second finger, and thumb of the left hand at right angles to each other, as illustrated in Figure 14.16:

- the **f**irst finger indicates the direction of the permanent **f**ield
- the se**c**ond finger indicates the direction of the **c**urrent
- the thu**m**b will then indicate the direction of the resulting **m**otion.

Basic motor action

The effect of the force acting on a current-carrying conductor in a permanent magnetic field is made

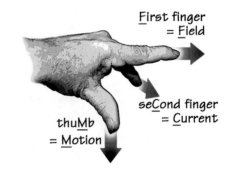

First finger = Field

seCond finger = Current

thuMb = Motion

Figure 14.16

use of in a simple **electric motor** and **moving coil instruments** (i.e. ammeters and voltmeters). However, it is *not* our purpose, in this particular chapter, to study these devices in any depth but, rather, to demonstrate a practical application for the force on a conductor.

Figure 14.17 is very much simplified, with many of its important features not illustrated for the purpose of clarity.

If a current is passed, counterclockwise, through the horizontal loop of wire, pivoted between the poles of a permanent magnet, as shown in Figure 14.17, opposing forces will act on each side of the loop, causing it to rotate.

If you apply Fleming's Left-Hand Rule to the left side of the loop (current drifting towards you), you will find that the force acts to push that conductor *upwards*. If you apply Fleming's Left Hand Rule to the right side of the loop (current drifting away from you), you will see that the resulting force acts to push that conductor *downwards*. The result of these two opposing forces, is to create a torque (rotational force) that will then act to rotate the coil clockwise.

Another way of looking at this is to examine the field pattern by combining the permanent field with the fields surrounding each side of the loop, which should make it perfectly clear in which direction the coil will rotate (see Figure 14.18).

Commutation

Now let's turn our attention to a very practical problem: How do we

- supply current to and from a continuously rotating coil?
- maintain the current flow around the armature in the same direction relative to the field, to ensure that the torque continues to act in the same direction?

The second requirement is necessary, of course, in order for the motor to rotate continuously in the same direction!

The answer is by using a *rotatary reversing switch*, called a **split-ring commutator** – as shown in Figure 14.19.

In the *simplified* Figure 14.19, the split-ring commutator is shown as a metal cylinder that has been split into two halves and, in practice, insulated from each other. The insulation is not shown in the diagram, for the purpose of clarity. One half of the commutator is connected to one side of the armature coil and the other half is connected to the other end of the coil, and the commutator rotates with the armature coil.

Current is supplied to and from the commutator by means of carbon blocks, called **brushes**, shaped to maintain contact with the commutator from the pressure supplied by a pair of springs. Carbon is not

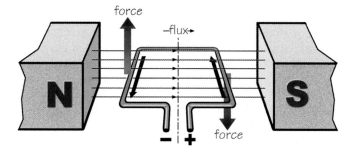

Figure 14.17

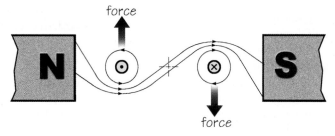

Figure 14.18

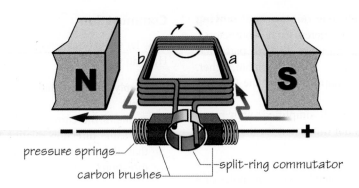

Figure 14.19

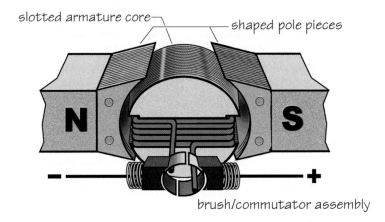

Figure 14.20

only a conductor, but also has the property of being *self-lubricating* – which reduces friction. Carbon also has a negative temperature coefficient of resistance, which means that its resistance falls as its temperature increases which, because of unavoidable arcing, is bound to occur.

In Figure 14.19, the armature current is fed, via the right-hand carbon brush, onto the right-hand half of the commutator, around the coil, and then returned to the external circuit via the left-hand half of the commutator and left-hand carbon brush. In this way, as the armature rotates, current will *always* flow into whichever half of the coil happens to be on the right, and out from whichever half happens to be on the left – thus maintaining a constant current direction relative to the magnetic flux and, therefore, a constant torque direction.

The momentum of the rotating coil moves the rotor assembly past the 'mid-way' point when the armature coil approaches the vertical position, so it does not 'hang' at that point, but keeps rotating instead.

Finally, to maximise the magnetic flux, it's necessary to reduce the size of the air gap between the two magnetic poles. This is achieved by winding the motor's coil in slots cut into a cylindrical armature, typically made of laminated silicon steel, and extending the poles using shaped 'pole pieces', to minimise the air gap between the poles and the armature (see Figure 14.20).

Finally . . .

Now that you have completed this chapter, are you able to achieve the objectives or learning outcomes listed at the beginning of this chapter?

Ask yourself, 'Can I . . .'

1 explain evidence how an electric current is accompanied by a magnetic field.
2 determine the shape and direction of the magnetic field around a current-carrying straight conductor.

3 determine the shape and direction of the magnetic field around a current-carrying single loop conductor.

4 determine the shape and direction of the magnetic field around a current-carrying coil.

5 explain the term 'flux density', specifying its SI unit of measurement.

6 calculate the flux density of simple magnetic fields.

7 sketch the magnetic field resulting from a current-carrying conductor placed in a permanent magnetic field, and use this to explain why the conductor is subject to a force.

8 apply Fleming's Left Hand Rule to determine the direction of force on a current-carrying conductor placed in a magnetic field.

9 calculate the value of the force acting on a current-carrying conductor placed in a magnetic field.

10 describe the principle of operation of a simple d.c. motor.

Online resources

The companion website to this book contains further resources relating to this chapter. The website can be accessed via the following link:

www.routledge.com/cw/waygood

Chapter 15

Magnetic circuits

On completion of this chapter, you should be able to

1 explain the term 'magnetic circuit'.
2 compare a magnetic circuit with an electric circuit.
3 explain the factors that determine:
   • magnetomotive force
   • reluctance
   • magnetic field strength.
4 specify the SI units of measurement for
   • magnetomotive force
   • flux
   • flux density
   • reluctance
   • magnetic field strength.
5 explain the equivalent of 'Ohm's Law' for magnetic circuits.
6 explain the equivalent of 'Kirchhoff's Laws' for magnetic circuits.
7 briefly explain the relationship between absolute permeability (μ), the permeability of free space (μ_o) and relative permeability (μ_r).
8 briefly explain the relationship between absolute permeability (μ) and the flux density and magnetic field strength of a magnetic circuit.
9 explain the reason for the shape of a typical **B-H curve**.
10 describe the main features of a **hysteresis loop**, in particular:
   • residual flux density
   • coercive force
   • saturation points
   • cross-sectional area of the loop.
11 explain '**magnetic leakage**' and '**fringing**'.
12 solve problems on magnetic circuits.

Introduction

In this chapter, we are going to explore the behaviour of **magnetic circuits**. A magnetic circuit is a key part of all electrical machines such as generators, transformers and motors, as well as other electrical devices such as relays and contactors, and a great deal of effort goes into its design.

There are similarities between *magnetic* circuits and *electric* circuits, so it's usual to use an electric circuit as an analogy for explaining how magnetic circuits behave. And that's what we'll be doing throughout this chapter.

Electric circuit analogy

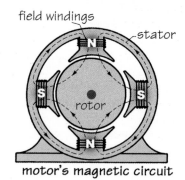

Figure 15.1

A **magnetic circuit** may be defined as one, or more, closed paths within which *lines of magnetic flux are confined.*

An Introduction to Electrical Science, Waygood, ISBN 9780415810029, 2013. © Taylor & Francis

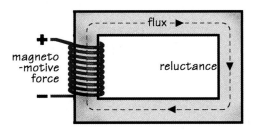 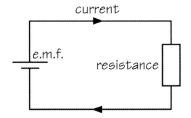

Figure 15.2

Magnetic circuits are manufactured from magnetically 'soft' ferromagnetic metals, such as iron or silicon steel. In the case of a transformer, the magnetic circuit is its 'core'. In the case of a generator or a motor, the magnetic circuit is more complex, and is a combination of the machine's 'stator', 'rotor' and the airgaps between the two, as illustrated in Figure 15.1 (where the broken lines respresent the flux paths set up by the field windings).

As already pointed out, there are certain similarities between magnetic and electric circuits, which will help us understand the behaviour of magnetic circuits.

The lines of **magnetic flux** (symbol: Φ, pronounced 'phi'), confined within a magnetic circuit, are equivalent to the *electric current* in an electric circuit.

In order to establish this magnetic flux, we need a **magnetomotive force** (symbol: F), which is provided by means of a current-carrying winding or coil, and this is equivalent to the *electromotive force* applied across an electric circuit.

The material which forms a magnetic circuit always offers some degree of opposition to the formation of magnetic flux, and we call this opposition **reluctance** (symbol: R_m). This is equivalent to the *resistance* of an electric circuit.

Finally, **magnetic field strength** (symbol: H) of a magnetic circuit, is equivalent to a voltage gradient, or 'voltage per unit length'.

> The original (but a, by far, more descriptive) term for magnetic field strength was '**m.m.f. gradient**', and it's a pity the scientific community hasn't retained that term instead of changing it to '**magnetising force**' and, finally, to '**magnetic field strength**', both of which are rather non-descriptive!

Table 15.1 summarises this

Table 15.1

For a magnetic circuit		For an electric circuit
magnetomotive force (F)	*is equivalent to*	**electromotive force (E)**
magnetic flux (Φ)	*is equivalent to*	**electric current (I)**
reluctance (R_M)	*is equivalent to*	**resistance (R)**
magnetic field strength (H)	*is equivalent to*	**voltage gradient (V/l)**

Series and series-parallel equivalent circuits

In the same way that we can have **series**, **parallel** or **series-parallel** *electric* circuits, we can also have equivalent series and series-parallel (but *not* 'parallel'!) *magnetic* circuits – examples of which are illustrated in Figure 15.3.

Figures 15.3 (a), (b) and (c) are examples of **series magnetic circuits**, together with their electric circuit equivalents. Figures 15.3 (d) and (e) are examples of **series-parallel magnetic circuits**, together with their electric circuit equivalents.

> In **Figure 15.3 (a)**, R_m represents the reluctance of the complete magnetic circuit.
>
> In **Figure 15.3 (b)**, R_{m1} represents the reluctance of the metal part of the magnetic circuit, while R_{m2} represents the reluctance of the air gap.

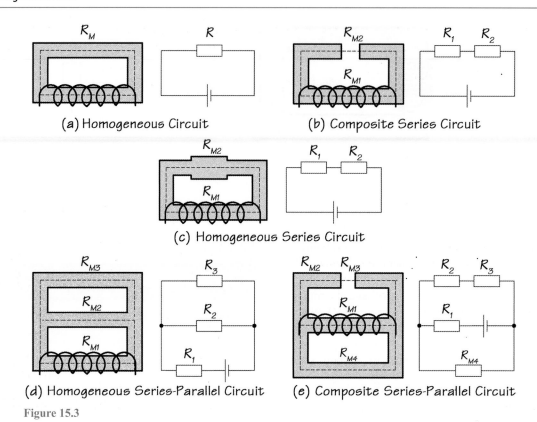

(a) Homogeneous Circuit

(b) Composite Series Circuit

(c) Homogeneous Series Circuit

(d) Homogeneous Series-Parallel Circuit

(e) Composite Series-Parallel Circuit

Figure 15.3

In **Figure 15.3 (c)**, R_{m1} represents the reluctance of the metal part with the smaller cross-sectional area, while R_{m2} represents the reluctance of the metal part with the large cross-sectional area.

In **Figure 15.3 (d)**, R_{m1} represents the reluctance of the lower part of the magnetic circuit, while R_{m2} represents the reluctance of the centre part, and R_{m3} represents the reluctance of the upper part of the magnetic circuit.

In **Figure 15.3 (e)**, R_{m1} represents the reluctance of the centre metal part of the magnetic circuit, while R_{m2} represents the total upper metal part, and R_{m3} represents the reluctance of the airgap in the upper branch, and R_{m4} represents the reluctance of the lower part of the magnetic circuit.

We can further classify magnetic circuits as being either '**homogeneous**' or '**composite**', where:

- **homogeneous** (meaning 'the same throughout') describes a magnetic circuit in which the magnetic flux passes through entirely the same material. Figure 15.3 (a), (c) and (d) are examples of homogeneous magnetic circuits, because the flux is contained entirely within a ferrous-metal core. An example of an homogeneous magnetic circuit is a transformer core.
- **composite** describes a magnetic circuit in which the magnetic flux passes through two or more *different* materials. For example, in Figure 15.3 (b) and (e), the flux not only passes through ferrous metal but also through airgaps. Composite magnetic circuits are used in generators and motors, where there are airgaps between the stationary and rotating parts of the machines.

Now, let's move on and look in more detail at the *quantities* we listed in our comparison table of magnetic and electric circuits.

Magnetomotive force (equivalent to 'electromotive force')

Magnetomotive force (symbol: F), which is often abbreviated to **m.m.f.**, is the term used to describe *the means by which magnetic flux is set up within a magnetic circuit*. In all practical magnetic circuits, this is provided using a current-carrying winding (coil), and is the product of the *current* and the *number of turns in the winding*. Its SI unit of measurement is the **ampere**

(symbol: A) which, in this context, is usually spoken as 'ampere turn' to avoid any confusion with the unit for electric current:

$$F = IN$$

where:

F = m.m.f., in amperes (A)

I = current, in amperes (A)

N = number of turns (no units)

Note! Since the m.m.f is the product of amperes and a simple number (of turns), its SI unit of measurement is the 'ampere' but, to avoid confusion with the unit for current, it is usually spoken or pronounced as '*ampere turn*'. Despite its unit of measurement, it's important to realise that m.m.f. is equivalent to e.m.f. and *not* current!

Magnetic flux (equivalent to 'current')

Magnetic flux (symbol: Φ, pronounced 'phi') describes *the quantity of magnetic flux established within a magnetic circuit*. Its unit of measurement is the **weber** (symbol: Wb) – pronounced '*vay-ber*'. In practice, a weber is a *very* large unit and you are more likely to see flux expressed in milliwebers (mWb) or even microwebers (μWb).

Reluctance (equivalent to 'resistance')

Reluctance (symbol: R_M), is *the opposition a magnetic circuit offers to the formation of magnetic flux*. As we shall explain shortly, its unit of measurement is the **ampere per weber** (symbol: A/Wb) – usually spoken as 'ampere-turn per weber', for the reason already explained.

Just like resistance, reluctance is *directly proportional* to the **length** (symbol: l) of a magnetic circuit, and is *inversely proportional* to its **cross-sectional area** (symbol: A):

$$R_M \propto \frac{l}{A}$$

To change the 'proportional' sign into an 'equals' sign we must, of course, introduce a *constant*. This constant is called '**magnetic reluctivity**', and is directly equivalent to 'resistivity' in an electric circuit. However, it's far more common to use its *reciprocal* instead, which we call '**absolute permeability**' (symbol: μ,

pronounced '*mu*'), which is equivalent to 'conductivity' (the reciprocal of 'resistivity') in an electric circuit.

$$R_M = \left(\frac{1}{\mu}\right) \times \frac{l}{A}$$

$$R_M = \frac{l}{\mu A}$$

We will learn more about absolute permeability a little later in this chapter but, for now, you can think of it as a measure of the *ease* with which a particular material allows the formation of magnetic flux, making it directly equivalent to the '*conductivity*' (the reciprocal of 'resistivity') of an electrical conductor (see Table 15.2).

Table 15.2

For a magnetic circuit		For an electric circuit
$R_M = \dfrac{l}{\mu A}$	*is equivalent to*	$R = \rho \dfrac{l}{A}$
permeability (μ)	*is equivalent to*	conductivity

'Ohm's Law' for magnetic circuits

The relationship between **magnetomotive force**, magnetic **flux** and **reluctance** is *exactly* equivalent to the relationship between potential difference, current and resistance, as established in the earlier chapter on *Ohm's Law*.

Although this relationship is, for obvious reasons, widely known as the '*Ohm's Law for Magnetic Circuits*', strictly speaking, it should be termed '**Hopkinson's Law**', as the relationship was originally determined by the British electrical engineer, John Hopkinson (1849–1898).

We'll discuss whether Ohm's Law is, in fact, a good analogy or not, towards the end of this chapter. However, regardless of what we call it, the relationship is as follows:

$$\Phi = \frac{F}{R_M}$$

where:

Φ = magnetic flux, in webers (Wb)

F = m.m.f., in amperes (A)

R_M = reluctance, in amperes per weber (A/Wb)

Note! By rearranging this equation, and substituting the units of measurement, we can now confirm the unit of measurement of **reluctance**:

$$R_M = \frac{F}{\Phi} = \frac{ampere}{weber} = ampere\ per\ weber$$

Table 15.3

For a magnetic circuit		For an electric circuit
$R_M = \dfrac{F}{\Phi}$	*is equivalent to*	$R = \dfrac{E}{I}$

Worked example 1   A mild steel ring has a coil of 200 turns, carrying a current of 10 A, wound uniformly around it. If a resulting flux of 750 μWb is established within the ring, calculate

a   the m.m.f.
b   the reluctance.

Solution

a $F = IN = 10 \times 200 = 2000$ A (ampere turns)
     Answer (a.)

b   Using 'Hopkinson's Law' for magnetic circuits:

$$R_M = \frac{F}{\Phi} = \frac{2000}{750 \times 10^{-6}}$$

$$= 2.67 \times 10^6\ A/Wb\ or\ 2.67\ MA/Wb\ (Answer\ b.)$$

'Kirchhoff's Laws' for magnetic circuits

You will recall that '**Kirchhoff's Voltage Law**' specifies that, '*in any closed loop, the applied voltage is equal to the sum of the voltage drops around that loop*'. Well, a similar principle applies to magnetic circuits, which is often called '**Kirchhoff's Law for Magnetomotive Force**'; that is, *for any 'closed path' around a magnetic circuit, the applied magnetomotive force is equal to the sum of the m.m.f. drops around that same path.*

Since magnetic field strength is defined as 'm.m.f. per unit length', then an 'm.m.f. drop' must be the product of the magnetic field strength and the length *(Hl)* of the relevant part of the circuit; that is,

$$since\ H = \frac{F}{l} \quad then \quad F = Hl$$

. . . as shown in Figure 15.4 and the equations that follow.

$$F = (H_{iron}\ l_{iron}) + (H_{air}\ l_{air}) \qquad E = U_1 + U_2$$

You will also recall that '**Kirchhoff's Current Law**' specifies that *the sum of the currents approaching a junction must equal the sum of the currents leaving that junction*. Again a similar principle applies to magnetic circuits, which is sometimes called '**Kirchhoff's Law for Flux**'; that is '*the amount of flux approaching a junction must equal the sum of the fluxes leaving*', as shown in Figure 15.5 and the equations that follow.

$$\Phi = \Phi_1 + \Phi_2 \quad I = I_1 + I_2$$

It should be emphasised that these two 'laws' are actually the unnamed 'equivalents' of Kirchhoff's Laws for electric circuits, as Kirchhoff did *not* intend for his laws to apply to magnetic circuits.

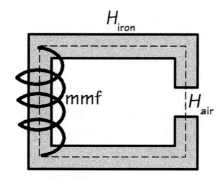
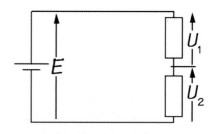

Figure 15.4

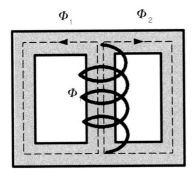
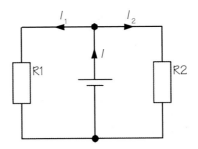

Figure 15.5

Understanding magnetisation (B-H) curves

'**Magnetisation curves**' are graphs that show the relationship between a magnetic material's **flux density** *(B)* and its **magnetic field strength** *(H)*. For that reason, they are also commonly called '*B-H* **curves**'.

Flux density *(B)*

We have already briefly explained magnetic **flux density** in the chapters on *magnetism* and on *electro-magnetism*, but it is sufficiently important for us to remind ourselves what it means here.

Flux density is a measure of the 'intensity' of the magnetic flux at any point within a magnetic circuit, and is defined as '*the flux per unit area*' at that point. Its unit of measurement is, therefore, the weber per square metre which, in SI, is given the special name **tesla** (symbol: **T**) in honour of the remarkable Serbian-American engineer, Nikola Tesla (1856–1943).

So, we can express flux density as. . .

$$B = \frac{\Phi}{A}$$

where:

B = flux density (T)

Φ = flux (Wb)

A = cross-sectional area of magnetic circuit (m²)

> **Worked example 2**   A steel magnetic circuit has a rectangular section measuring 20 mm by 25 mm, and supports a flux of 750 µWb. What is its flux density in the magnetic circuit?

Solution   The cross-sectional area of the magnetic circuit is:

$$A = (25 \times 10^{-3}) \times (20 \times 10^{-3}) = 500 \times 10^{-6}\,\text{m}^2$$

. . . therefore, the flux density:

$$B = \frac{\Phi}{A} = \frac{750 \times 10^{-6}}{500 \times 10^{-6}} = 1.5\ \text{T (Answer)}$$

Magnetic field strength *(H)*

At the beginning of this chapter, we compared magnetic circuits with electric circuits, and described the term **magnetic field strength** as being equivalent to a *voltage gradient* around an electric circuit. While this analogy is accurate, it's a little less obvious, as we don't usually talk in terms of voltage gradients around circuits but it is, nevertheless, a useful analogy!

> We do consider 'voltage gradient' in insulators or dielectrics, where it is known as 'dielectric strength', expressed in volts per metre.

As we have already briefly mentioned, 'magnetic field strength' was originally known as '**m.m.f. gradient**', which is by far the most descriptive and meaningful term for this quantity. It was more recently known as '**magnetising force**', which is just as undescriptive as its present term!

Magnetic field strength (symbol: H) is defined as '*the magnetomotive force per unit length of a magnetic circuit*':

$$H = \frac{F}{l}$$

where:

H = magnetic field strength (A/m)

F = magnetomotive force (A)

l = length of magnetic circuit (m)

Worked example 3   A cast iron ring has a mean diameter of 50 mm, and is uniformly wound with 200 turns of insulated wire. If the current flowing in the wire is 2.5 A, what is the ring's magnetic field strength?

Solution   We start by calculating the magnetomotive force:

$$F = IN = 2.5 \times 200 = 500 \text{ A (ampere-turns)}$$

Next, we determine the mean length of the magnetic circuit:

$$\text{length} = \text{circumference} = \pi D$$
$$= \pi \times (50 \times 10^{-3}) = 157 \times 10^{-3} \text{ m}$$

Now, we can determine the magnetic field strength:

$$H = \frac{F}{l} = \frac{500}{157 \times 10^{-3}} = 3185 \text{ A/m (Answer)}.$$

Now that we've learnt a little more about **flux density (B)** and **magnetic field strength (H)**, let's move on and examine 'magnetisation curves' or 'B-H curves'.

By means of a relatively simple experiment, it is possible to produce a graph which shows the relationship between the two for a sample of ferrous metal. The experiment is illustrated in Figure 15.6.

The sample, in this case a toroid (ring) of iron, is wound with an insulated conductor and supplied, via a variable resistor, from a direct-current supply.

The magnetic field strength is proportional to the current so, by incrementally increasing the current, we are also able to incrementally increase the magnetic field strength. For each incremental increase in magnetic field strength, we can then measure the corresponding value of flux density within the iron ring, using an instrument, called a 'teslameter' (formerly called a 'gaussmeter') to do so.

If we were then to plot the values of magnetic field strength **(H)**, horizontally, and the resulting **flux density (B)**, vertically, on a graph, the result would look something like the curve shown to the right of the schematic diagram in Figure 15.6, or the one illustrated in Figure 15.7.

You may ask why should the resulting curve follow this particular shape instead of being, say, a straight line. Well, to understand this, we must remind ourselves of the '**domain theory of magnetism**' that we learnt about in the chapter on *magnetism*.

A '**domain**', you will recall, is a molecule which behaves just like a tiny permanent magnet. In a sample of *unmagnetised* ferromagnetic material, the domains form closed 'chains', with each north pole 'chasing' the south pole of an adjacent domain. In a sample of *magnetised* ferromagnetic material, the domain 'chains' have been broken, and all or most of the domains realigned to form rows.

So let's examine the magnetisation curve, and see what is happening to the domains within the iron sample as the magnetic field strength is gradually increased. This is illustrated in Figure 15.7.

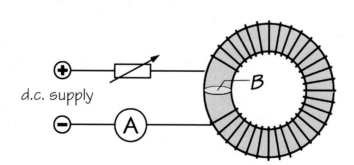

magnetising curve experiment

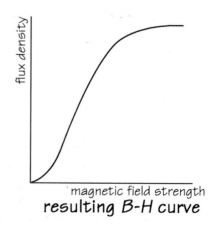

resulting B-H curve

Figure 15.6

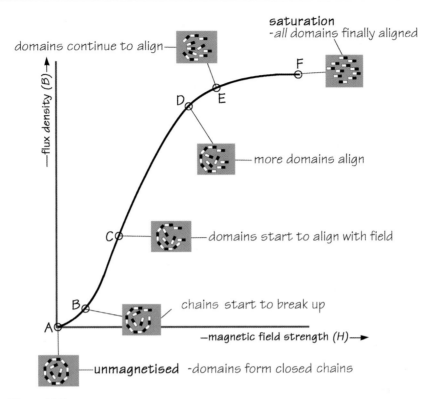

domains continue to align—

saturation
-*all domains finally aligned*

F

D E

—flux density (B)→

more domains align

C

domains start to align with field

B

chains start to break up

A

—magnetic field strength (H)—▶

unmagnetised -*domains form closed chains*

Figure 15.7

- At point **A**, before we actually start to increase the magnetic field strength, the sample is *unmagnetised*, with its domains forming closed 'chains'.
- At point **B**, the increasing magnetic field strength starts to cause the domain chains to break up and realign, creating a weak magnetic field (low flux density).
- At points **C** and **D**, the magnetic field strength continues to increase, causing more and more domains to align with the field and, so, the flux density increases yet further.
- At point **E**, the majority of domains have now realigned, and the sample of iron is starting to approach '**saturation**'. 'Saturation' is the point of maximum flux density.
- At point **F**, *all* the domains have now realigned, and the sample has finally reached '**saturation**' – in other words, *there are no more domains to realign*, and the sample's flux density has reached its maximum possible value.

Once **saturation** has been reached, any further increase in magnetic field strength will have absolutely no effect whatsoever on the sample's flux density.

From the shape of the magnetisation curve, it should now be obvious that the ratio of *B:H* can only be (approximately) constant over the straight part of the

magnetisation curve. Below point **B** on the above curve, and beyond point **D**, the ratio will change considerably. These points are called the 'lower knee' and 'upper knee', respectively (not all *B-H* curves have a noticeable 'lower knee', but they *all* have an 'upper knee').

Absolute permeability: the ratio of flux density to magnetic field strength

The ratio of *flux density to magnetic field strength (B:H)* is important enough to be given its own name, and is called '**absolute permeability**' (symbol: μ, pronounced '*mu*'). Earlier, we described absolute permeability as being equivalent to 'conductance' in an electric circuit.

For air, the value of absolute permeability is a *constant*, and is equal to a figure of $4\pi \times 10^{-7}$ H/m (henry per metre) or, if you prefer, 1.257×10^{-6} H/m. You can think of the absolute permeability for air as being a sort of 'reference' permeability, and it is important enough to be given its own name: the **absolute permeability of free space** (symbol: μ_o).

Strictly speaking 'free space' means a 'vacuum' but the permeability for air and for a vacuum are taken as being the same.

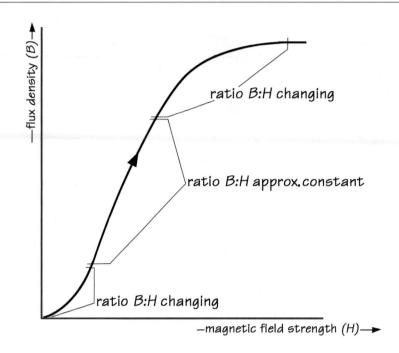

Figure 15.8

But for ferrous metals, the absolute permeability (unlike conductance) is a *variable*, and depends entirely on *where along the metal's magnetisation curve it is measured*. This is because the ratio of *B:H* changes along that curve (Figure 15.8).

Just like the figure for absolute permeability of free space, figures representing absolute permeability are quite awkward and clumsy. So, in an attempt to simplify them, permeabilities of ferrous metals are more usually expressed as a '*relative* permeability' (μ_r), which is a simple ratio. For example, a metal having a relative permeability of, say, 1000, has a permeability that is 1000 times greater than that of air. This is equivalent to an absolute permeability of 1257×10^{-6} H/m, but presented as a much more convenient figure. And this is the way in which you will normally see permeabilities expressed in data tables.

So the relationship between these three 'different' permeabilities is:

$$\mu_r = \frac{\mu}{\mu_o}$$

where:

μ_r = relative permeability (no units)

μ = absolute permeability (H/m)

μ_o = absolute permeability of freespace (H/m)

To summarise this section, we can say that the ratio **B:H** can *either* be expressed in terms of a material's **absolute permeability** (μ):

$$\frac{B}{H} = \mu$$

. . . *or* in terms of the **absolute permeability of free space** (μ_o) *and* **relative permeability** (μ_r):

$$\frac{B}{H} = \mu_o \mu_r$$

Worked example 4   What is the relative permeability, at a magnetic field strength of 1500 A/m, of a sample of ferromagnetic material whose *B-H* curve is reproduced in Figure 15.9?
Solution   For a magnetic field strength of 1500 A/m, the corresponding flux density (obtained from the graph) is 1.45 T, so:

$$\mu_r = \frac{B}{\mu_o H} = \frac{1.45}{4\pi \times 10^{-7} \times 1500} \approx 769 \text{ (Answer)}$$

So, in the above worked example, the permeability of the sample at a magnetic field strength of 1500 A/m is **769** times higher than that for air.

Although the magnetisation curves for *all* ferromagnetic materials follow roughly the same shape, the positions of their lower and upper 'knees' and the gradient of their linear part will vary considerably – *depending on how easily the material is magnetised.*

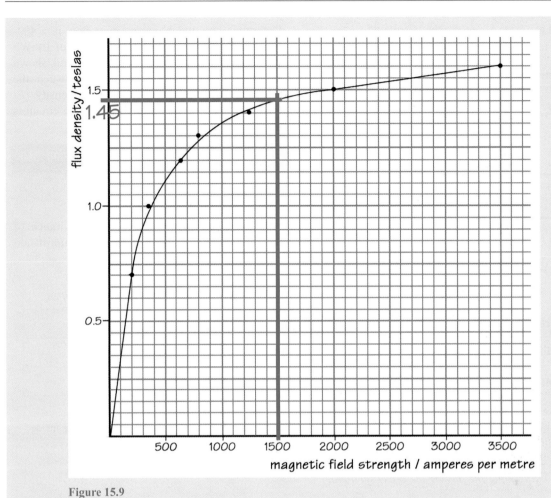

Figure 15.9

Solving problems on magnetic circuits

Most problems on magnetic circuits involve trying to determine the amount of magnetomotive force required to establish a specified amount of flux (or, perhaps, flux density) within part, usually within an airgap, of a magnetic circuit. This has practical relevance to the design of magnetic circuits for rotating machines.

Most problems on magnetic circuits, then, can be worked out methodically, through the use of what is known as the '**results-table method**', in which Table 15.4 is gradually completed, starting with the information supplied in the question.

So, in the following worked example, we will use the '**results-table method**' to solve the problem, showing how the table is completed, step-by-step. Of course, when you actually use this method, you only need to complete the one table; we are repeating the image of the table, here, to demonstrate how each column is completed.

Table 15.4

Part:	Length/m	A/m²	Φ/Wb	B/T	H/A/m	F/A
				$B = \dfrac{\phi}{A}$	$H = \dfrac{B}{\mu}$	$F = Hl$

Worked example 5 (using the 'results-table method')

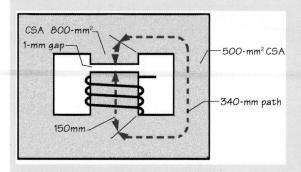

Figure 15.10

Determine how much current must flow in a 400-turn winding, in order to establish a flux of 1 mWb within the airgap of the magnetic circuit shown in Figure 15.10. The relationship between the magnetic field strength *(H)* and flux density *(B)* for the material from which the magnetic circuit is manufactured is as follows:

H/A/m	200	350	650	800	1250	2000	3500
B/T	0.8	1.0	1.3	1.3	1.4	1.5	1.6

Solution The first step is to fill in as much of the results table as you can from the information supplied in the question:

Part:	Length/m	A/m²	Φ/Wb	B/T	H/A/m	F/A
Centre	150×10⁻³	800×10⁻⁶	*1×10⁻³			
Outside	340×10⁻³	500×10⁻⁶	**0.5×10⁻³			
Airgap	1×10⁻³	800×10⁻⁶	1×10⁻³			

*As we are told fringing can be ignored, the flux in the centre limb will be equal to the flux in the airgap.
**As both outside limbs are identical, the amount of flux in each of these outer limbs must be *half* that of the inside limb/airgap.

Next, we can complete the flux density *(B)* column, by *dividing the flux by the cross-sectional area, i.e. $B = \dfrac{\phi}{A}$*.

Part:	Length/m	A/m²	Φ/Wb	B/T	H/A/m	F/A
Centre	150×10⁻³	800×10⁻⁶	1×10⁻³	**1.25**		
Outside	340×10⁻³	500×10⁻⁶	0.5×10⁻³	**1.00**		
Airgap	1×10⁻³	800×10⁻⁶	1×10⁻³	**1.25**		

To complete the magnetic field strength for the **airgap** is straightforward, as it is the flux density divided by the absolute permeability of free space:

$$H = \frac{B}{\mu_o} = \frac{1.25}{4\pi \times 10^{-7}} = 995 \times 10^3 \, \text{A/m}$$

Part:	Length/m	A/m²	Φ/Wb	B/T	H/A/m	F/A
Centre	150×10⁻³	800×10⁻⁶	1×10⁻³	1.25		
Outside	340×10⁻³	500×10⁻⁶	0.5×10⁻³	1.00		
Airgap	1×10⁻³	800×10⁻⁶	1×10⁻³	1.25	**995×10³**	

But to determine the magnetic field strength for the centre and outside limbs, we would normally first have to construct the B/H graph, from the data supplied. However, as you can see, from the supplied data, a flux density of 1 T corresponds to a magnetic field strength of 350 A/m.

For the centre limb, a flux density of 1.25 T corresponds to a magnetic field strength of **720 A/m**; for the outer limbs, a flux density of 1.00 T corresponds to a magnetic field strength of **350 A/m**.

Part:	Length/m	A/m²	Φ/Wb	B/T	H/A/m	F/A
Centre	150×10^{-3}	800×10^{-6}	1×10^{-3}	1.25	**720**	
Outside	340×10^{-3}	500×10^{-6}	0.5×10^{-3}	1.00	**350**	
Airgap	1×10^{-3}	800×10^{-6}	1×10^{-3}	1.25	995×10^{3}	

To complete the final column, the 'magnetomotive force drop' *(F)* is *the product of the magnetomotive force and the length of that part of the magnetic circuit:*

For the centre limb:

$$F = Hl = 720 \times (150 \times 10^{-3}) = 108 \text{ A}$$

For the outer limbs:

$$F = Hl = 350 \times (340 \times 10^{-3}) = 119 \text{ A}$$

For the airgap:

$$F = Hl = 995 \times (1 \times 10^{-3}) = 995 \text{ A}$$

Part:	Length/m	A/m²	Φ/Wb	B/T	H/A/m	F/A
Centre	150×10^{-3}	800×10^{-6}	1×10^{-3}	1.25	720	**108**
Outside	340×10^{-3}	500×10^{-6}	0.5×10^{-3}	1.00	350	**119**
Airgap	1×10^{-3}	800×10^{-6}	1×10^{-3}	1.25	995×10^{3}	**995**

To determine the total magnetomotive force required, we simply *add up the m.m.f. drops in the right-hand column:*

Part:	Length/m	A/m²	Φ/Wb	B/T	H/A/m	F/A
Centre	150×10^{-3}	800×10^{-6}	1×10^{-3}	1.25	720	108
Outside	340×10^{-3}	500×10^{-6}	0.5×10^{-3}	1.00	350	119
Airgap	1×10^{-3}	800×10^{-6}	1×10^{-3}	1.25	995×10^{3}	995
					total mmf:	**1222 A**

Finally, since m.m.f. is the product of the current through the winding, and its number of turns:

$$I = \frac{F}{N} = \frac{1222}{400} = 3.06 \text{ A (Answer)}$$

Magnetic materials

Magnetic materials are classified according to their values of relative permeability, as shown in Figures 15.11 and 15.12.

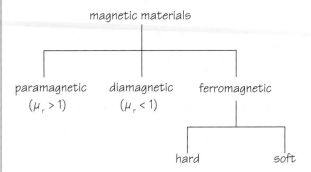

Figure 15.11

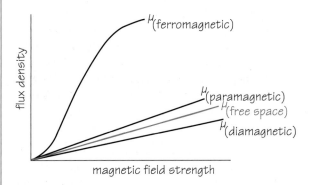

Figure 15.12

Paramagnetic materials have values of relative permeability that are *slightly greater than unity*, which cause a very slight *increase* in the flux density within a sample of that type of material, when placed within a uniform magnetic field. Paramagnetic materials include magnesium and lithium.

Diamagnetic materials have values of relative permeability that are *slightly less than unity*, which actually cause a slight *reduction* in the flux density

within a sample of that material, when placed within a uniform magnetic field. Examples of diamagnetic materials include mercury, silver and copper.

Ferromagnetic materials are in a completely different category to the others, and have values of relative permeability that lie in the *hundreds or thousands*. These materials are further sub-classified as being either 'hard' or 'soft'. 'Hard' ferromagnetic materials (e.g. steel) are relatively difficult to magnetise and demagnetise, whereas 'soft' ferromagnetic materials (e.g. iron) are relatively easy to magnetise and demagnetise. Rather oddly, some ferromagnetic alloys (e.g. heusler alloy) are actually mixtures of entirely paramagnetic or diamagnetic materials!

Paramagnetic and diamagnetic materials are of little practical importance, in comparison with ferromagnetic materials – *all large-scale energy conversion, using generators, transformers and motors, depends entirely on the use of ferromagnetic materials*.

Hysteresis loops

In the magnetisation curve experiment described earlier, we increased the magnetic field strength until the sample became saturated.

But what would happen if we continued the experiment further, by *reducing* the magnetic field strength *back to zero* and, then, *reversing* the direction of the magnetic field strength? Well, let's find out!

First of all, let's repeat the last experiment, so that the sample reaches saturation. So, in Figure 15.14, as we increase the current flowing through the coil, the magnetic field strength increases, and the flux density increases, following the curve until saturation is reached at point **a**.

In Figure 15.15, we gradually reduce the current back towards zero. As we do this, the magnetic field strength reduces to zero as well, reducing the flux density in the sample. However, the reduction in flux density *doesn't follow the original curve* but, instead, it follows the *second* curve *(a–b)*. So when the magnetic

Figure 15.13

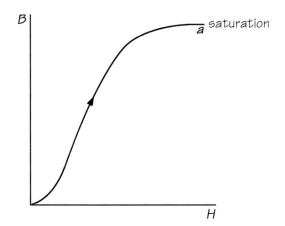

Figure 15.14

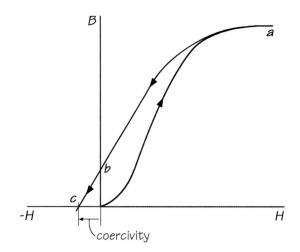

Figure 15.16

field strength eventually reaches zero, the flux density has *lagged behind* and has *not* fallen to zero, but only to point *(b)* on the new curve! In other words, the sample hasn't completely demagnetised, but has *retained* a certain amount of its flux density. We call this remaining flux density '**residual flux density**', or the '**remanence**' of the sample.

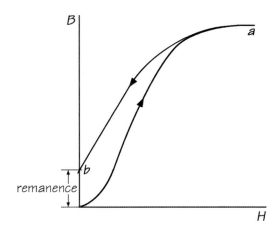

Figure 15.15

To *completely* demagnetise the sample (that is, to entirely remove the residual flux density), it's necessary to *reverse* the direction of the current in the coil, thereby *reversing the direction of the magnetic field strength* and, then, applying just enough to completely demagnetise the sample – i.e. dropping its flux density to zero (point *c*). The amount of 'negative' magnetic field strength required to achieve this is called '**coercivity**' or the '**coercive force**' – as illustrated in Figure 15.16.

On a humorous note, you may think of this property in the following terms: 'it takes more **effort** to get a bus-load of football fans *out* of a pub than the amount of **effort** it took to get them *into* that pub'! In this context, '**effort**' is equivalent to '**magnetic field strength**', and the 'bus-load of football fans' is equivalent to flux density!

In Figure 15.17, we have continued to increase the current in the *reverse* direction, increasing the magnetic field strength in the same direction. This now causes the flux density to increase in the negative direction (i.e. reversing the polarity of the sample) until it, once more, reaches saturation again at point *d*.

In Figure 15.18, we now reduce the current again, reducing the magnetic field strength and causing the flux density to fall, following the curve *d–e*. Again, when the magnetic field strength has fallen to zero, some flux density remains – so, again, a certain amount of residual magnetism remains within the sample. By *reversing* and *increasing* the current (back to its original direction) we can remove this residual magnetism *(e–f)*. As the current continues to increase, the flux density follows curve *(f–a)* back to saturation in the original direction.

From now on, no matter how often we continue to cycle the value of the current through the magnetising coil, the *outside* curve, (a–b–c–d–e–f–a) will *always* result – as illustrated in Figure 15.19.

This curve is called a '**hysteresis loop**', and is particularly important when we apply an alternating current (and, therefore, an alternating magnetic field

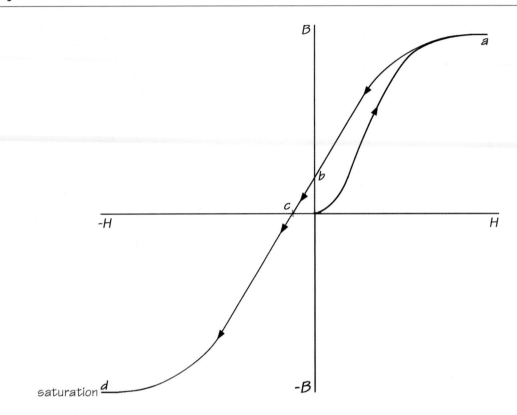

Figure 15.17

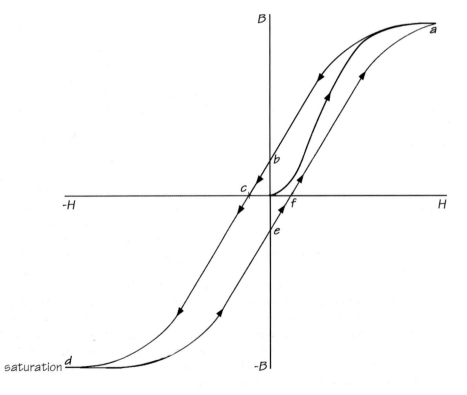

Figure 15.18

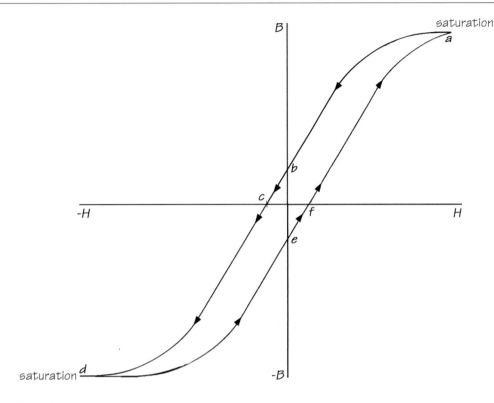

Figure 15.19

strength) to the sample – e.g. in transformers or a.c. motors.

The term '**hysteresis**' was coined by the Scottish engineer and academic Sir James Ewing (1855–1935), who conducted a great deal of research into the magnetic properties of metals in the 1920s, and who is credited with discovering this property. The word itself is derived from a Greek word, meaning *'to lag behind'* and, in this context, describes how a change in **flux density** 'lags behind' any change in **magnetic field strength**.

The cross-sectional area enclosed by the hysteresis loop represents the energy (expressed in 'joules per cubic metre per cycle') required to magnetise and demagnetise the sample – the *greater* this area, the *greater* the energy required to do so. Again, this is particularly important for alternating-current magnetic circuits (i.e. generators, transformers and motors) and accounts for part of the energy losses (called 'hysteresis losses') that occur within the magnetic circuits of a.c. machines.

Different ferromagnetic materials have hysteresis loops of different cross-sectional areas – e.g. 'soft' iron and silicon steel have very *narrow* cross-sections, whereas 'hard' steels have very *wide* cross-sections, as shown in Figure 15.20.

The left-hand figure shows a relatively narrow hysteresis loop, with low remanence (residual flux density). This is typical for a 'soft' ferromagnetic material, which is easily magnetised and demagnetised, and which retains very little residual flux density when the magnetic field strength is removed. This represents a material that is ideal for making *temporary magnets* and magnetic circuits for electrical machines. The low cross-sectional area indicates that relatively little energy is required to magnetise and demagnetise the sample.

The right-hand figure shows a relatively wide hysteresis loop, with high remanence. This is typical for a 'hard' ferromagnetic material, which is difficult to magnetise, and just as difficult to demagnetise. This represents a material that is ideal for making *permanent magnets*, but would be no use whatsoever for making the magnetic circuits for electrical machines. The large cross-sectional area indicates a relatively large amount of energy is required to magnetise and demagnetise the sample.

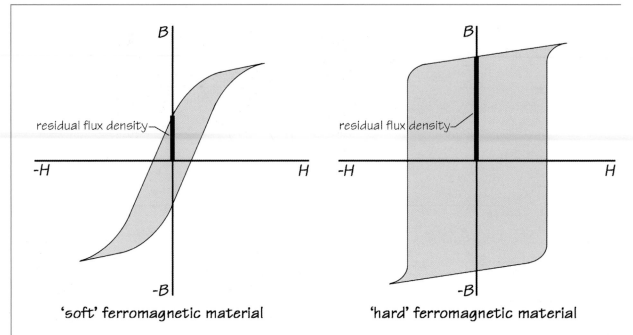

'soft' ferromagnetic material   'hard' ferromagnetic material

Figure 15.20

Practical applications of magnetic circuits

As briefly mentioned at the beginning of this chapter, the main application for magnetic circuits is to provide a low-reluctance 'path' for the magnetic flux utilised in electrical machines.

Transformers

In the case of a simple, two-winding **transformer**, its core is an example of a *homogeneous* magnetic circuit, the purpose of which is to efficiently link the

magnetic flux created by its primary winding with its secondary winding. There are two different types of transformer core design: 'core type' (Figure 15.21, left) and 'shell type' (Figure 15.21, right). By using a low-reluctance ('soft') ferromagnetic material (typically, silicon steel), maximum flux density is assured, with minimum remenance, and with a minimum energy requirement – thereby maximising the efficiency of the transformer.

Rotating machines

Electrical rotating machines, such as generators and motors, use a *composite* magnetic circuit in the

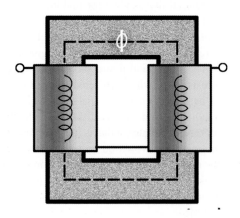
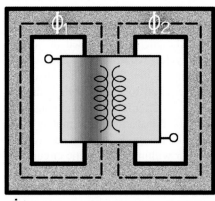

Figure 15.21

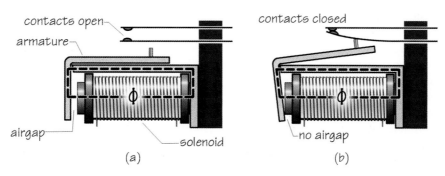

contacts open
armature
airgap
solenoid
(a)

contacts closed
no airgap
(b)

Figure 15.22

construction of their stators (the stationary part of the machine) – again, usually laminated silicon steel. These magnetic circuits (see Figure 15.1) are 'composite' because the magnetic path includes airgaps. The magnetic circuits of rotating machines comprise silicon steel and air (the airgap) and have *two* purposes. The first is to *link the field windings and armature winding*. The second is to *concentrate and maximise the flux density within the airgap* between the stator and rotor, through which the armature windings move, to ensure a maximum induced e.m.f. in those windings.

Relays, contactors, etc.

A **relay** or **contactor** (a heavy-duty relay) consists of a *solenoid* and a spring-loaded, hinged, *armature* – as illustrated in Figure 15.22. The function of the armature is to open or close contacts which are used to control an external circuit. The solenoid and armature form part of a composite magnetic circuit (shown by the broken line, below), which includes an airgap between the face of the solenoid and the armature – as illustrated in Figure 15.22a.

When the solenoid is initially energised, the magnetic circuit includes the airgap and, so, the reluctance of the magnetic circuit is relatively high, although the flux is sufficient for the solenoid to attract the armature towards its face.

As soon as the armature makes contact with the face of the solenoid (Figure 15.22b), the airgap disappears and the reluctance of the magnetic circuit falls appreciably – so far less flux is required to 'hold' the relay closed, than is needed to cause it to close in the first place.

Magnetic leakage and fringing

Only in an 'ideal' magnetic circuit will *all* the magnetic flux be contained entirely within the medium of the circuit itself. In all *practical* magnetic circuits (transformers, motors, generators, etc.), part of the flux produced by the magnetomotive force 'bypasses' the magnetic circuit. This is called '**magnetic leakage**'.

Similarly, repulsion between adjacent lines of magnetic force within any airgaps (in the case of motors, generators and measuring instruments) cause the outermost lines of flux to bulge outwards, essentially reducing the flux density (flux per unit area) within the airgap. This is called '**magnetic fringing**'.

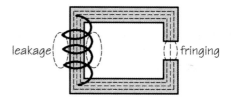

leakage | fringing

Figure 15.23

Magnetic leakage and fringing are relatively minor within a well-designed magnetic circuit, but nevertheless act to reduce the theoretical flux density achievable within the magnetic circuit.

Limitations of the 'electric circuit' analogy

At the beginning of this chapter, we learnt that there are useful similarities between **magnetic circuits** and **electric circuits**, which we then used to help us understand the behaviour of magnetic circuits.

This analogy is a widely used approach for teaching students about magnetic circuits, but it is, nevertheless, an *analogy* (a likeness) and, like all analogies, *it has its limitations*.

So we will finish this chapter by briefly looking at some of the significant *differences* between magnetic circuits and electric circuits.

First of all, an electric current is a *movement*, or *drift*, of electric charges (free electrons, in the case of metal conductors) *around* a circuit. Magnetic flux is *not a* movement, or drift, of anything at all! It is simply the 'shape' assumed by lines of magnetic flux contained within the magnetic circuit.

As we learnt in the chapter on *resistance*, *'the consequence of resistance is heat'*. Whenever current passes through resistance, energy is dissipated through heat transfer away from the circuit. A magnetic circuit's equivalent of resistance is **reluctance**, but *absolutely no energy is dissipated by reluctance due to the formation of magnetic flux.*

Although we described absolute permeability as being 'equivalent' to conductance, unlike conductance its value is not constant as it depends on the ratio of flux density to magnetic field strength, which varies according to where the ratio is measured along its magnetisation curve.

With the exception of insignificantly small 'leakage currents' through a conductor's insulation, electric current is confined within its circuit. Magnetic flux, on the other hand, is subject to both '**leakage**' and '**fringing**'. 'Leakage' describes magnetic flux which is created by a magnetomotive force but which, then, partially or entirely 'bypasses' the magnetic circuit, and 'fringing' describes the reduction in flux density whenever flux passes through an airgap in the magnetic circuit. Because of the difficulty is coping with leakage and fringing, most magnetic-circuit calculations you will be expected to perform include the words, *'ignore any leakage or fringing'.*

Most, although not all, metal conductors are 'ohmic' or 'linear' and, therefore, obey Ohm's Law; in other words the ratio of voltage to current (resistance) is usually *constant* for wide variations in voltage. However, this is most definitely *not* the case for magnetic circuits, which are exclusively '**non-linear**' – that is their ratios of magnetomotive force to flux (reluctance) are *not* constant. As well as the length and cross-sectional area of a magnetic circuit, reluctance also depends upon the absolute permeability of that circuit's material. Relative permeability, in turn, depends upon the ratio of flux density to magnetic field strength *(B/H)*, which can vary considerably for different values of magnetising current.

Finally, when the potential difference is removed from an electric circuit, no current can continue to flow in that circuit. However, in a magnetic circuit, when the magnetomotive force is removed, some (residual) magnetic flux usually remains, due to the circuit's **remanence**.

To summarise, then. Continue to use the analogy between magnetic circuits and electric circuits to understand the behaviour of magnetic circuits. But be aware that it is *only* an analogy and, like most analogies, it tends to ignore any significant differences!

Summary

A **magnetic circuit** is the path that encloses magnetic flux in air, or in a ferromagnetic material.

There are similarities between **magnetic circuits** and *electric circuits*, where:

- **magnetomotive force** *(F)* is equivalent to *electromotive force.*
- **magnetic flux** (Φ) is equivalent to *electric current.*
- **reluctance** *(R_m)* is equivalent to *resistance.*
- **magnetic field strength** *(H)* is equivalent to *voltage gradient.*

These quantities are related by the magnetic equivalent of Ohm's Law:

$$\Phi = \frac{F}{R_m}$$

A **magnetomotive force** is created when an electric current flows through a coil, and is the product of the current and number of turns:

$$F = IN \text{ amperes}$$

Reluctance is directly proportional to the *length* of the magnetic circuit, and inversely proportional to its *cross-sectional area:*

$$R_m = \frac{l}{\mu_o \mu_r A}$$

Absolute permeability *(μ)* is equivalent to the *conductivity* of an electric circuit, and is the product of the **absolute permeability of free space** *(μ_o)* and **relative permeability** *(μ_r)*, where:

- **absolute permeability of free space**, $\mu_o = 4\pi \times 10^{-7}$ H/m.
- **relative permeability**, μ_r, varies from material to materal, and can be as high as 100 000 (no units).

Flux density *(B)* is defined as *the flux per unit area*, measured in teslas (T):

$$B = \frac{\Phi}{A}$$

Magnetic field strength *(H)* is defined as *the magnetomotive force per unit length of magnetic circuit*, measured in amperes per metre (A/m):

$$H = \frac{F}{l}$$

Absolute permeability is equal to the ratio of flux density to magnetic field strength:

$$\mu_o \mu_r = \frac{B}{H}$$

The **relative permeability** for any particular material is not constant because the ratio of *B:H* is not constant, because it varies according to the shape of the *B-H* curve (Figure 15.24).

A **hysteresis loop** shows the variation in *flux density* as the magnetic field strength is varied in a positive and negative sense. The resulting change in flux density 'lags' behind changes in magnetic field strength – which is what 'hysteresis' means (Figure 15.25).

'**Soft**' ferromagnetic materials (iron, silicon steel) have narrow hysteresis loops which exhibit little residual flux density when the magnetic field strength is removed, and require little energy to magnetise and demagnetise the materials.

'**Hard**' ferromagnetic materials have wide hysteresis loops which exhibit large residual flux densities when the magnetic field strength is removed, and require larger amounts of energy to magnetise and demagnetise the materials.

The magnetic circuits of electrical machines need to be manufactured from '**soft**' ferromagnetic materials.

'**Leakage**' describes flux that is not contained within the magnetic circuit.

'**Fringing**' describes flux within a motor's or generator's magnetic circuit airgap which does not link with the rotor.

Leakage and **fringing** both act to reduce the theoretical effective flux within a magnetic circuit.

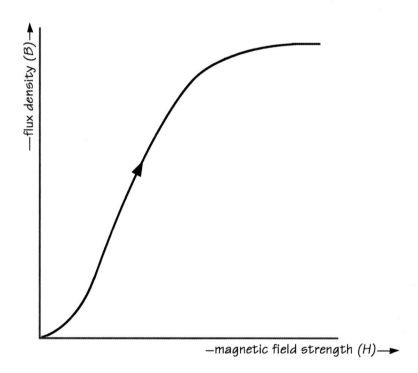

Figure 15.24

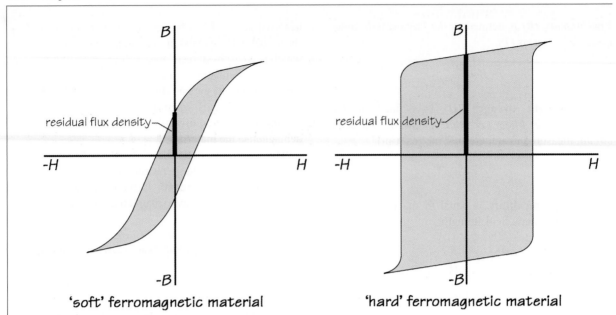

'soft' ferromagnetic material   'hard' ferromagnetic material

Figure 15.25

Finally . . .

Now that you have completed this chapter, are you able to achieve the objectives or learning outcomes listed at the beginning of this chapter?

Ask yourself, 'Can I . . .'

1 explain the term 'magnetic circuit'.
2 compare a magnetic circuit with an electric circuit.
3 explain the factors that determine:
   • magnetomotive force
   • reluctance
   • magnetic field strength.
4 specify the SI units of measurement for
   • magnetomotive force
   • flux
   • flux density
   • reluctance
   • magnetic field strength.

5 explain the equivalent of 'Ohm's Law' for magnetic circuits.
6 explain the equivalent of 'Kirchhoff's Laws' for magnetic circuits.
7 briefly explain the relationship between absolute permeability (μ), the permeability of free space (μ_o) and relative permeability (μ_r).
8 briefly explain the relationship between absolute permeability (μ) and the flux density and magnetic field strength of a magnetic circuit.
9 explain the reason for the shape of a typical **B-H curve**.
10 describe the main features of a **hysteresis loop**, in particular:
   • residual flux density
   • coercive force
   • saturation points
   • cross-sectional area of the loop.
11 explain '**magnetic leakage**' and '**fringing**'.
12 solve problems on magnetic circuits.

Online resources
The companion website to this book contains further resources relating to this chapter. The website can be accessed via the following link:
www.routledge.com/cw/waygood

Chapter 16

Electromagnetic induction

On completion of this chapter, you should be able to

1 explain the result of moving a permanent magnet towards, or away from, a coil.
2 describe the effect of moving a straight conductor perpendicularly through a magnetic field.
3 determine the magnitude and direction of the potential difference induced into a straight conductor moved through a magnetic field.
4 demonstrate the application of Fleming's Right-Hand Rule for generator action.
5 briefly explain the effects of self-induction.
6 explain Faraday's and Lenz's Laws for electromagnetic induction.
7 list the factors that affect the self-induction of a coil.
8 describe the effect of self-induction on the growth and decay of direct currents in inductive circuits.
9 solve simple problems on the growth and decay of direct currents in inductive circuits.
10 explain the terms 'mutual inductance' and 'coupled circuits'.
11 describe the behaviour of the mutual induction between two inductive circuits.
12 solve simple problems on mutual induction.
13 describe the construction and operation of an 'ideal' transformer.
14 solve simple problems on 'ideal' transformers.
15 describe how energy is stored in magnetic fields.
16 solve simple problems on the energy stored in magnetic fields.
17 recognise and give simple examples of the functions of inductors.
18 solve simple problems on inductors in series and in parallel.

Important! Conventional current drift (positive to negative) is assumed throughout this chapter.

Introduction

One of the most important advances in the science of electrical engineering was made in 1831, when **Sir Michael Faraday** (1791–1867) discovered that whenever there is relative motion between lines of magnetic flux and a conductor, a potential difference is 'induced' into the conductor.

So, as predicted by scientists such as Øersted, it was indeed possible to use *movement* to produce electricity!

We call this behaviour **electromagnetic induction** and, within a year of Faraday's discovery, it was discovered quite independently by the American physicist **Joseph Henry** (1797–1878). Rightly, both men have been credited with this discovery.

This discovery directly led to the invention and development, by Faraday, of the **generator** and the **transformer** without which large-scale generation, transmission and distribution of electrical energy would be quite impossible.

Faraday's experiments

Like so many other great scientific breakthroughs, this behaviour was discovered quite by accident while Faraday was conducting a completely separate experiment, and it led him to develop more experiments with which to further investigate the behaviour. The basis of the experiments conducted by Faraday is explained as follows.

With the ends of an insulated coil connected to a sensitive measuring instrument, called a 'galvanometer', we have a closed circuit (Figure 16.1). Of course, as things stand, no current will drift, because there is no *electromotive force*.

However, if a permanent magnet is now suddenly plunged *towards* the coil, the galvanometer's needle

An Introduction to Electrical Science, Waygood, ISBN 9780415810029, 2013. © Taylor & Francis

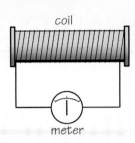

Figure 16.1

will be seen to respond by giving a sharp 'kick', or sharp deflection, in one direction or the other (Figure 16.2).

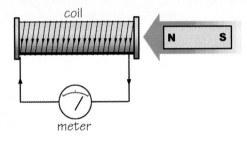

Figure 16.2

If the magnet is then suddenly *withdrawn* from the coil, the galvanometer will again register a sharp 'kick' – but, this time, in the *opposite* direction (Figure 16.3).

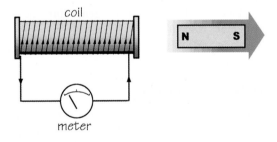

Figure 16.3

The reaction of the galvanometer confirmed the presence of a potential difference across the ends of that coil, causing a current to drift through the coil and be registered by the galvanometer.

Faraday described these potential differences as being '**induced**' into the coil.

It's important to understand that it's a **potential difference** that's being induced into the coil,

not a current. A current only flows if the coil is connected to a load, such as the galvanometer. There is no such thing as an 'induced current'!

The experiment also demonstrated that the *direction* of the induced potential difference *reverses whenever the direction of movement of the magnet is reversed*.

Importantly, the experiment also confirmed that this potential difference is induced *only when there is movement by the magnet*. If the magnet is held stationary, *no* electromotive force is induced into the coil (Figure 16.4).

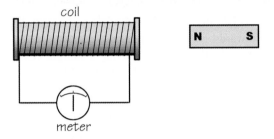

Figure 16.4

Actually, it doesn't really matter whether *the magnet is moved relative to the coil*, or *the coil is moved relative to the magnet*. Providing there is *relative movement between the two*, then a potential difference will be induced into the coil.

From this experiment Faraday was able to conclude that . . .

Whenever there is relative motion between a conductor and a magnetic field, a potential difference will be induced into that conductor. The direction of the induced potential difference depends upon the direction of motion of the conductor relative to the lines of magnetic flux.

Faraday repeated his experiments, while changing the various conditions involved, and discovered that the magnitude of the induced voltage varied according to

- the **speed** at which the magnet was moved
- the **flux density** of the bar magnet
- the **number of turns** wound on the coil.

Increasing *any* of these factors caused the magnitude of the induced voltage to increase. We shall return to the effect of these variations, a little later in this chapter.

Faraday also found that replacing the magnet with an electromagnet would achieve the same effect, not by moving the electromagnet, but by *varying the current through it*. Increasing the current through the electromagnet produced exactly the same result as moving a magnet *towards* the coil, while decreasing the current produced exactly the same effect as moving a magnet *away* from the coil.

What Faraday had discovered was that it was possible to induce a voltage into a coil, through no physical movement whatsover! He had, in fact, invented a primitive **transformer**.

Figure 16.5 shows Faraday's experimental set-up with the electromagnet: two coils, wound around a common 'core' (apparently, he used a wooden core!). One coil (the electromagnet), termed the 'primary winding', is connected to a battery, via a switch, while the other, termed the 'secondary winding', is connected to a galvanometer.

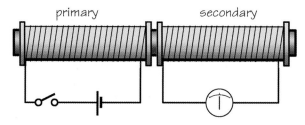

Figure 16.5

Whenever Faraday closed the switch in the primary circuit, he noticed a sharp deflection by the secondary's galvanometer. With a steady current drifting through the primary circuit, the galvanometer indicated zero. When he then re-opened the switch, he noticed another sharp deflection by the galvanometer but, this time, in the opposite direction.

Faraday went on to vary, rather than to switch, the current in the primary circuit, and noticed that as the primary current varied, the galvanometer deflected in one direction when the primary current was increased, but in the opposite direction when the primary current was reduced.

Generator action

Let's now move on to examine the **magnitude** and **direction** of an induced potential difference – initially

using the simple case of a straight conductor that is forced to move perpendicularly (at right angles) through a permanent magnetic field set up between the poles of a magnet (Figure 16.6).

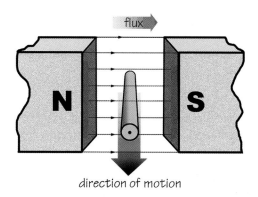

direction of motion

Figure 16.6

As the conductor is forced vertically downwards through the magnetic field, we say that it is *'cutting'* the lines of magnetic flux. The term 'cutting', in this context, isn't meant literally; it simply means that the conductor is *'moving across'* or *'passing through'* the magnetic field.

Because there is now relative motion between the conductor and the magnetic field, a potential difference is induced into that conductor.

For a conductor cutting the magnetic flux perpendicularly, the **magnitude** of this induced potential difference is given by the following equation:

$$E = Blv$$

where:

E = potential difference (V)

B = flux density (T)

l = length of conductor (m)

v = velocity of conductor (m/s)

> **Important!** By 'length of conductor', what we *really* mean is that length of the conductor that lies *within the magnetic field – not* the entire length of the conductor!

Actually, this equation is *only* true when the conductor *cuts the flux at right angles*. For any other angle, the above equation needs to be modified as follows.

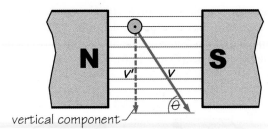

conductor cutting flux at an angle θ

Figure 16.7

If the conductor cuts the flux at an angle, *θ* (pronounced *'theta'*) at a velocity represented by vector *v*, then we must find the perpendicular component of that velocity vector, *v'* – as represented by the broken line in Figure 16.7.

This can be determined from the sine ratio, as follows:

$$\sin\theta = \frac{\text{opposite}}{\text{hypotenuse}} = \frac{v'}{v}$$

$$v' = v\sin\theta$$

So we can now modify the original equation so that it can be used for conductors moving at *any* angle, *θ*, through the flux:

$$E = Blv\sin\theta$$

where:

E = potential difference (V)

B = flux density (T)

l = length of conductor (m)

v = velocity of conductor (m/s)

θ = angle cutting flux (°)

Worked example 1   A conductor moves through a permanent magnetic field of flux density 250 mT at a velocity of 20 m/s. If the length of conductor within the field is 175 mm, calculate the voltage induced into the conductor, if it cuts the flux (a) perpendicularly, and (b) at 60°.

Solution

$$E = Blv\sin\theta$$
$$= (250\times10^{-3})\times(175\times10^{-3})\times20\times\sin90°$$
$$= 0.875 \text{ V (Answer a.)}$$

$$E = Blv\sin\theta$$
$$= (250\times10^{-3})\times(175\times10^{-3})\times20\times\sin60°$$
$$= 0.875\times0.866$$
$$= 0.758 \text{ V (Answer b.)}$$

Of course, no potential difference will be induced in the conductor should that conductor be moved *parallel* to the lines of magnetic flux, because it will not be 'cutting' them.

Fleming's Right-Hand Rule

The **direction** of this induced potential difference may be determined by using (for conventional flow) **Fleming's Right-Hand Rule** for 'generator action', which works as follows.

The *thumb, first finger (index finger)* and *second finger* of the right hand are held at right angles to each other, as shown in Figure 16.8.

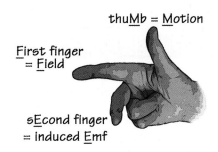

Figure 16.8

Then:

- the **thuMb** indicates the direction of **M**otion of the *conductor relative to the field*
- the **First finger** (index finger) indicates the direction of the magnetic **F**ield (i.e. north to south)
- the **sEcond finger** indicates the direction of the induced **E**.m.f. (Potential difference).

Note that, for Fleming's Right-Hand Rule to apply, the direction of motion *always* refers to the movement of the *conductor* relative to the magnetic field, *never* the other way around!

So if we apply **Fleming's Right-Hand Rule** to the downward-moving conductor, shown previously, you will find that the resulting induced potential difference will act *towards* you (i.e. *out* from the page). In other words, the nearer end of the conductor

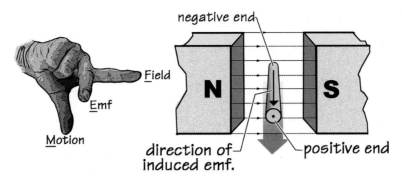

Figure 16.9

will be the positive end, and the far end will be the negative end.

Now, if this conductor is connected to a closed *external circuit*, the induced potential difference will cause a current to drift in the same direction as that induced potential difference – i.e. in this case, *towards* you as the reader, as indicated by the 'dot' convention shown in Figure 16.9.

> **Reminder**: For conventional current, a **dot** represents current drifting *towards* you, and a **cross** represents current drifting *away* from you.

As already pointed out in an earlier chapter, when we talk about the 'direction of current', we are *always* referring to the *direction of current through the load*; **never** *through the voltage source itself* – so, in this particular case, the current will be leaving the moving conductor from its positive (nearest) end, drifting through the load, and re-entering the conductor at its negative (far) end.

The *plan* (downward) view of this arrangement (shown in Figure 16.10) should help clarify this.

As you can see, with the downward-moving conductor acting as a 'voltage source' for the external load, the resulting load current drifts from positive to negative (while the current drift *within* that moving conductor is from negative to positive).

Action and reaction: Lenz's Law

In the previous section, we learnt that if the conductor moving through a magnetic field is connected to a load, then the induced potential difference will cause a current to drift through any external circuit.

> This current is sometimes referred to as an *'induced current'*. This, however, is incorrect. It's the *potential difference* that is induced, *not* the resulting current! Current will *only* flow as a result of this induced potential difference, provided the moving conductor is connected to an external load.

This current, of course, is capable of expending energy in the external load – e.g. if the load were, say, a lamp,

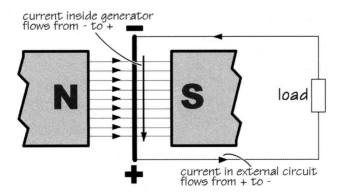

Figure 16.10

then the current would heat its filament. If the **Law of the Conservation of Energy** *('energy can neither be created nor destroyed, but only changed from one form to another')* is to be maintained, then this current can only expend energy *if work has been done to produce it in the first place!*

The *work expended* in moving the conductor through a magnetic field is the product of the **force** applied to that conductor and the **distance** through which it is moved. And, according to **Newton's Third Law**, *this force must be opposed by an equal and opposite force* called a '**reaction**' *('for every force, there is an equal and opposite reaction').*

And this reaction *can only come from a magnetic field which the current itself produces!* That is, the magnetic field set up around the conductor by the load current flowing through it.

This reasoning led the Russian physicist, **Emil Lenz** (1804–1865), to conclude that . . .

> *The current resulting from the induced potential difference, due to the motion of a conductor through a magnetic field, must act in such a direction that its own magnetic field will then oppose the motion that is causing that induced potential difference'.*

This statement is one interpretation of what is known as **Lenz's Law**, and we will meet other interpretations later.

You may find it necessary to re-read this statement through a couple of times, as it can be rather confusing the first time around!

In the meantime, let's see if we can make Lenz's Law more understandable, by considering it step by step.

Firstly, let's see if we can explain it in terms of **Fleming's Right-** and **Left-Hand Rules**.

Figure 16.11 shows the lower side of a continuous rectangular loop of wire being pushed downwards through a magnetic field.

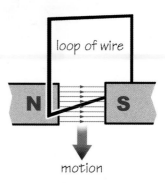

Figure 16.11

If you apply **Fleming's Right-Hand Rule** (for 'generator' action) to this conductor, you will see that the resulting induced potential difference acts towards you, causing a current to drift in the *same* direction – as illustrated in Figure 16.12.

If we now apply **Fleming's Left-Hand Rule** (for 'motor' action) to this current we see that the resulting force (i.e. the 'reaction') due to that current will act *upwards*. That is, it will react *against* the direction of downward motion (Figure 16.13).

In case you also found *this* explanation rather confusing, let's try yet another approach . . . This time, we'll ignore Fleming's Rules, and concentrate on the magnetic flux set up around the conductor due to the current drifting through it

In Figure 16.14, the conductor is forced downwards through the magnetic field, and a potential difference is induced into the conductor.

If the conductor is connected to an external load, then a current will flow in the direction shown (towards you).

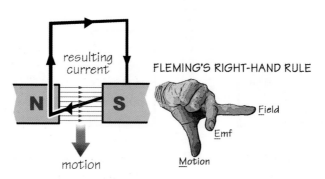

Figure 16.12

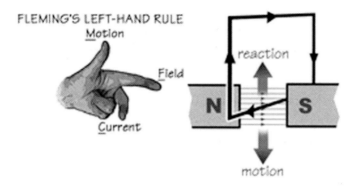

Figure 16.13

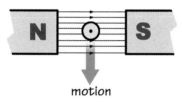

motion

Figure 16.14

In Figure 16.15, we see the magnetic field surrounding the conductor, caused by the current drifting through it. For clarity, we have *not* shown the permanent field.

field due to 'induced' current

Figure 16.15

In Figure 16.16, we show how the conductor's magnetic field *and* the permanent field combine, and react with each other to create an upward-acting *reaction* against the force responsible for the conductor's downward motion.

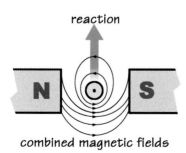

combined magnetic fields

Figure 16.16

If we extend the same logic (for every action there must be a reaction) to Faraday's magnet and coil experiment described earlier, then we must conclude that if we move the magnet *towards* the coil, then the direction of the induced potential difference must be such that the resulting current causes the polarity of the coil-end nearest the magnet to be the same as the nearest pole of the magnet – thus *opposing* (reacting against) the force moving the magnet *towards* the coil (like poles repel). And when we move the magnet *away* from the coil, the induced potential difference and resulting current reverses, causing the coil's polarity to *reverse* and oppose the force moving the magnet *away from* the coil (unlike poles attract)!

> This **action-reaction** effect is a ***very*** important concept, and you will meet it again when you study both generators and motors. *So it is highly recommended that you re-read this section again – perhaps several times – so that you fully understand the important principle of Lenz's Law.*

Self-inductance

In this section we are going to learn how, whenever the current drifting through a coil *changes in either magnitude or direction*, the resulting change in its magnetic field will induce a potential difference into that *same* coil, and the direction of this induced potential difference will *always* act *to oppose that change in current.*

As we now know, there are two *laws* which define the effect of changing the current drifting through a coil: **Faraday's Law** and **Lenz's Law**.

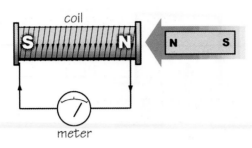
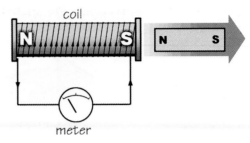

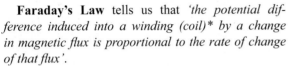

polarity of coil opposes
movement of magnet *towards* coil

polarity of coil opposes
movement of magnet *away* from coil

Figure 16.17

Faraday's Law tells us that *'the potential difference induced into a winding (coil)* by a change in magnetic flux is proportional to the rate of change of that flux'.*

We can write this as:

$$u \propto \frac{\text{change in flux}}{\text{change in time}}, \quad \text{written as:} \quad u \propto \frac{\Delta \Phi}{\Delta t}$$

[*Actually, this also applies to a single, straight, conductor. But its effect is practically insignificant compared with that of a coil.]

In the above expression, we use a lower-case *'u'* to represent a *changing*, as opposed to a continuous, value of induced voltage; and the Greek letter Δ ('delta') which mathematicians use to mean 'change in'. So the above expression simple means that *the value of the induced potential difference at any instant is proportional to the rate of change of flux at that same instant.*

Lenz's Law, tells us that *'the direction of an induced potential difference is such that it always opposes the change producing it'*. This is the 'action-reaction' process already described.

By *combining* these two laws, we can rewrite the above expression as:

$$u \propto -\frac{\Delta \Phi}{\Delta t}$$

The negative sign simply indicates the *sense*, or the *direction*, in which the induced potential difference is acting; it has nothing to do with the polarity of electric charge.

As always, to change a *proportional* sign into an *equals* sign, we must introduce a *constant*. For a coil, this constant is the **number of turns** *(N)* linked by

the magnetic flux. So, the above expression may be rewritten as:

$$u = -N\frac{\Delta \Phi}{\Delta t} \qquad \text{—equation (1)}$$

where:

u = potential difference (V)

N = number of turns

$\Delta \phi$ = change in flux (Wb)

Δt = change in time (s)

Now, if you refer back to the chapter on *magnetic circuits*, you will recall that

$$\text{flux density} = \frac{\text{flux}}{\text{area}} \quad i.e: \quad B = \frac{\Phi}{A} \quad \text{so} \quad \Phi = BA$$

For any given coil, its cross-sectional area, *A*, is fixed (a 'constant'), so if we substitute for Φ in equation (1), we have:

$$u = -NA\frac{\Delta B}{\Delta t} \qquad \text{—equation (2)}$$

If you again refer back to the chapter on *magnetic circuits*, you will recall that:

$$H = \frac{IN}{l} \text{ and } \mu_o \mu_r = \frac{B}{H} \quad \text{so} \quad B = \mu_o \mu_r H = \mu_o \mu_r \frac{IN}{l}$$

So, if we substitute for *B* in equation (2), we have:

$$u = -\left(\frac{NA\mu_o\mu_r N}{l}\right) \times \frac{\Delta I}{\Delta t} = -\left(\frac{N^2 A\mu_o\mu_r}{l}\right) \times \frac{\Delta I}{\Delta t}$$

—equation (3)

The entire expression appearing *inside the brackets* is known, simply, as '**self-inductance**' (symbol: L), and its SI unit of measurement is the **henry** (symbol: H). So, the final equation becomes:

$$u = -L\frac{\Delta I}{\Delta t} \qquad \text{—equation (4)}$$

where:

u = induced voltage (V)

L = inductance (H)

ΔI = change in current (A)

Δt = change in time (s)

From this equation, we should now be able to understand the significance of the definition of the **henry**:

*A circuit has an inductance of one **henry** when a potential difference of **one volt** is induced into it when its current changes at a uniform rate of **one ampere per second**.*

Although you are not expected to remember this definition, you should be able to see how this relates to equation (4).

Worked example 2   The current drifting through a coil of self-inductance 0.2 H is increased from 0 A to 5 A in 0.01 s. What is the value of the potential difference induced into the coil?

Solution   In this example, the change in current can be determined from:

$$\Delta I = \left(I_{final} - I_{initial}\right) = (5-0) = 5\text{ A}$$

Substituting for ΔI in equation (4):

$$u = -L\frac{\Delta I}{\Delta t} = -0.2\times\frac{(5)}{0.01} = -100\text{ V (Answer)}$$

Note that, in this worked example, the negative sign indicates that the induced potential difference is acting to **oppose** *the increase in current.*

Worked example 3   The current drifting through a coil of self-inductance 0.25 H is reduced from 5 A to 0 A in 0.01 s. What is the value of the potential difference induced into the coil?

Solution   In this example, the change in current can be determined from:

$$\Delta I = \left(I_{final} - I_{initial}\right) = (0-5) = -5A$$

Notice, here, that the current is getting smaller, hence the negative sign.

Substituting for δI in equation (4):

$$u = -L\frac{\Delta I}{\Delta t} = -0.25\times\frac{(-5)}{0.01} = -0.25\times(-500)$$
$$= +125\text{ V (Answer)}$$

This time, the positive sign indicates that the induced potential difference is acting in the *same* direction as the current – i.e. it is trying to *oppose its* **collapse** – or, to put in another way, it is acting to *maintain the current.*

Factors affecting the self-inductance of a coil

The list of factors that affect the **self-inductance** *(L)* of a coil were developed in equation (3), above:

$$L = \frac{N^2 A \mu_o \mu_r}{l} \qquad \text{—equation (5)}$$

where:

L = self-inductance (H)

N = number of turns

A = area of coil (m^2)

μ_o = permeability of free space

μ_r = relative permeability

l = length of coil (m)

Figure 16.18 summarises these relationships:

• *doubling* the **number of turns** will *quadruple* its inductance
• *doubling* its **cross-sectional area*** will *double* its inductance
• *doubling* its **length** will *halve* its inductance
• inserting a 'soft' **ferromagnetic core** will increase its inductance by *hundreds* or even *thousands of times*

[*this is the csa of the **coil**, *not* the conductor]

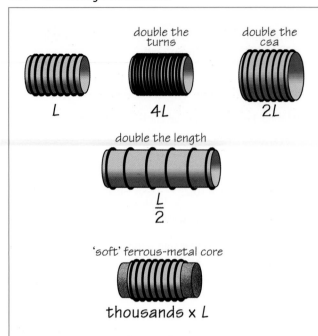

double the
turns

double the
csa

L

$4L$

$2L$

double the length

$\dfrac{L}{2}$

'soft' ferrous-metal core

thousands × L

Figure 16.18

You will recall, from the chapter on *magnetic circuits* that 'soft' ferrous metals (iron, silicon steel, etc.) will have a **relative permeability** (μ_r) that is *thousands* of times greater than the **absolute permeability of free space** (i.e. that of a hollow coil), so inserting a ferromagnetic core inside the coil will have, *by far*, the greatest effect on increasing a coil's self-inductance.

In an earlier chapter, we learnt that the absolute permeability (μ) of a magnetic material only remains (approximately) constant over the linear portion of its **B-H curve**. Accordingly, the inductance value of an iron-cored inductor applies only when that inductor is operating within the linear portion of its *B-H* curve, and the core of an inductor is always designed in such a way that its flux density never, for example, approaches its saturation level. Therefore, manufacturers of inductors normally specify the range of currents for which the specified inductance value applies.

Another way of expressing **self-inductance, L,** is as follows.

If you refer back to the chapter on *magnetic circuits*, you will recall that a magnetic circuit's reluctance is given by the equation:

$$R_m = \frac{l}{\mu_o \mu_r A}$$

(the equivalent of $R = \rho \dfrac{l}{A}$ for an electric circuit).

Most of the variables in the equation for reluctance also appear in the equation for inductance. In fact, if you separate N^2 from the rest of the equation, you will notice that the remaining part is equivalent to the *reciprocal* of reluctance:

$$L = \frac{N^2 A \mu_o \mu_r}{l} \text{ which we can rewrite as}$$

$$L = N^2 \left(\frac{A \mu_o \mu_r}{l} \right) = N^2 \times \frac{1}{R_m}$$

$$L = \frac{N^2}{R_m}$$

where:

$$N = \text{number of turns}$$

$$R_m = \text{reluctance}$$

Worked example 4   A ferromagnetic ring has a mean circumference of 3 m, a circular cross-sectional area of 2000 mm², and a relative permeability of 2000. If it is wound with a coil of 1000 turns, calculate (a) its reluctance, and (b) its inductance.

Solution

$$R_m = \frac{l}{\mu_o \mu_r A} = \frac{3}{(4\pi \times 10^{-7}) \times 2000 \times (2000 \times 10^{-6})}$$

$$= \frac{3}{5.03 \times 10^{-6}} = 597 \text{ kA/Wb (Answer a.)}$$

$$L = \frac{N^2}{R_m} = \frac{1000^2}{597 \times 10^3} = 1.675H \text{ (Answer b.)}$$

Behaviour of d.c. inductive circuits

Figure 16.19 represents a circuit of inductance L and resistance R, supplied from a d.c. source via a two-way switch.

The circuit could represent an **inductor** of inductance, L, and negligible *resistance*, in series with a **resistor** of resistance, R. On the other hand, it could also be an equivalent circuit, representing an **inductor** having an *inductance, L,* and a *resistance R*.

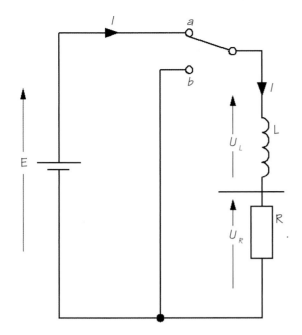

Figure 16.19

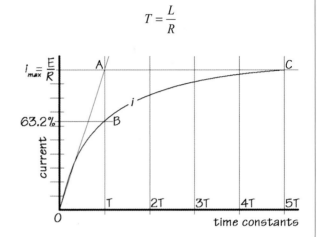

The increasing load current initially follows the line **O-A**, as shown in Figure 16.20. If it *were* to continue to follow that line, then it would reach its maximum value (determined by the potential difference divided by resistance) in a period of time, called a **time constant** (symbol: *T*), measured in seconds, where:

$$T = \frac{L}{R}$$

Figure 16.20

At the instant the switch is moved to position *(a)*, the external d.c. supply voltage, *E,* is suddenly applied to the inductive circuit, and a resulting load current *(i)* starts to increase in value. The maximum value *(I_{max})* this current can reach is limited by the resistance *(R)* of the circuit, where:

$$I_{max} = \frac{E}{R}$$

Now, if this were a purely resistive circuit, this current would reach its maximum value *instantaneously*. But this *cannot* happen in an inductive-resistive circuit, as explained by **Faraday's** and **Lenz's Laws**, which tell us that, due to its self-inductance, a changing current induces a potential difference *(u_L)* into that circuit, the direction of which *will always oppose that change in current*.

So it's important to understand that while this induced voltage *opposes* the *growth* in current, *it does **not** prevent it from eventually reaching its maximum value*. In other words, it acts to *reduce* its rate of growth, but does not act to *prevent* its growth.

A circuit's **resistance** will determine the maximum current in a circuit; whereas its **self-inductance** will *reduce its rate of growth*.

But, instead of continuing to follow line **O-A**, and reaching its maximum value in that time period it, instead, follows curve **O-C** and, so, actually only reaches approximately 63.2 per cent of its final value (position **B** on the graph).

The time it *actually* needs to reach its maximum value (point **C** on the curve) is approximately *five* time constants:

$$\text{time needed to reach maximum value} \approx 5\frac{L}{R}$$

To summarise, then. For an inductive-resistive circuit suddenly connected to a d.c. voltage, the current *cannot* rise to its maximum value instantaneously but, instead, *will always follow the curve shown above*. Once the current has eventually reached its maximum value, it becomes constant, so is no longer opposed by the inductive-effect of the circuit, and will remain at this value.

The *exact* value of current at *any* point along this graph line can be determined using calculus – which is beyond the scope of this text. So, you should bear in mind that the figures of '**63.2%**' and '**5 × time constants**' used in this chapter are very *good approximations* of the actual values derived by using calculus.

Now, *with a constant current flowing through the circuit,*
Figure 16.21 shows what happens if we suddenly move
the switch from position *(a)* to position *(b)*.

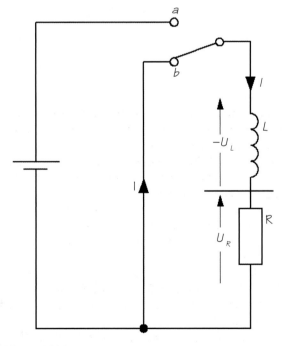

Figure 16.21

With the switch in position *(b)*, the external voltage
supply is removed and the inductive circuit is short-
circuited, so the current will start to collapse although
its *direction* will remain the same.

If this were a purely resistive circuit, then the
current would simply collapse to zero instantaneously.
However, the collapsing current induces a voltage
($-u_L$) which acts to *oppose* that collapse of current
or, to put it another way, *tries to maintain that
current.*

Once again, it's important to understand that the
induced voltage does not *prevent* the current from
collapsing to zero but, rather, it acts to prevent it
reaching zero *instantaneously,* by acting to *try to
maintain the current!* So the collapse of current follows
the curve shown in Figure 16.22. As you will notice,
this curve has *exactly* the same shape as before, except
that it is now inverted.

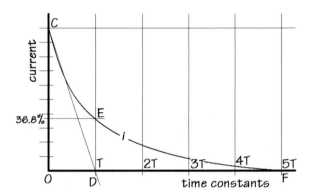

Figure 16.22

Initially, the current follows the line **C-D** and, if it
continued to follow that line, it would reach zero in a
period (**O-D**) called the time constant, *T*, measured in
seconds, where:

$$T = \frac{L}{R}$$

But, instead of reaching zero in that time period,
it *actually* follows the curve **C-F** and only falls
by approximately 63.2 per cent of its initial value
(position **E** on the graph), reaching **36.8 per cent**
of i_{max}.

And the time it actually takes to reach zero (point
F on the graph) is, again, approximately *five* time
constants:

$$\text{time needed to reach zero} \simeq 5\frac{L}{R}$$

Inductance and arcing

For circuits, such as the one in Figure 16.23, when the switch is closed onto a highly inductive load, the resulting current will, of course, increase to its maximum value, following the curve described above.

However, when the switch is opened, and the current starts to collapse, an **arc** can occur across the gap between the switch's contacts as the load's self-induced voltage tries to maintain the current across that gap.

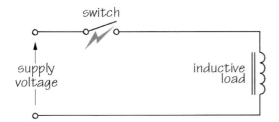

Figure 16.23

The temperature of an electric arc is very high and may actually cause the switch contacts to partially melt. For this reason, *it can be hazardous to break the supply to any load that has a substantial amount of self-inductance – such as the field winding of a d.c. machine.*

There are various ways of preventing, or at least minimising, this arcing problem. One such method is to use a special type of 'make-before-break' switch which, when operated, temporarily connects a **discharge resistor** in parallel with the inductive load before it actually breaks the supply to that load. This discharge resistor then provides a route for the collapsing current to pass through, rather than by arcing across the switch contacts.

Energy stored in a magnetic field

When a current drifts through a purely inductive circuit (i.e. a circuit without any resistance) increasing in magnitude, it does **work** by *expanding*, and storing energy in, the magnetic field. You will remember from the chapter on *magnetism* that lines of magnetic flux are considered to have an 'elastic' property, and work must be done to *stretch* them.

The energy, thus stored in the magnetic field, is later returned to the circuit when the magnetic field collapses as it attempts to maintain current drift.

An analogy can be made, here, to winding up the rubber-band 'motor' of a model aeroplane. The energy thus stored in the twisted rubber band, when released, will then drive the propellor until all that stored energy is released.

The expression for the work done in storing energy in an inductive circuit's magnetic field is:

$$W = \frac{1}{2}LI_{max}^2$$

where:

$$W = \text{work (J)}$$
$$L = \text{inductance (H)}$$
$$I_{max} = \text{steady-state current (A)}$$

Worked example 5   How much energy will be stored in the magnetic field of a coil of self-inductance 0.05 H and resistance 3 Ω, when connected to a 12-V d.c. supply?

Solution   We first need to find out the steady-state current (I_{max}) that will drift in the coil:

$$I_{max} = \frac{E}{R} = \frac{12}{3} = 4 \text{ A}$$

Now, we can determine the energy stored:

$$W = \frac{1}{2}LI_{max}^2 = \frac{1}{2} \times 0.05 \times 5^2$$
$$= 0.625 \text{ J or } 625 \text{ mJ (Answer)}$$

Mutual inductance

Mutual inductance (symbol: *M*) exists between *two* separate self-inductive circuits when a change in the flux set up by one circuit induces a potential difference into the second circuit – just like Faraday's experiment with the coil and electromagnet, explained towards the beginning of this chapter.

Although it is common to use the term 'inductance', when we strictly mean 'self-inductance', we should *never* use 'inductance' whenever we mean 'mutual inductance'.

In Figure 16.24, the self-inductive circuit connected, via a switch, to the battery is called the **primary** circuit, and the self-inductive circuit connected to the resistive load is called the **secondary** circuit.

In Figure 16.24(a), when the switch is closed, a current drifts through the primary winding in the direction

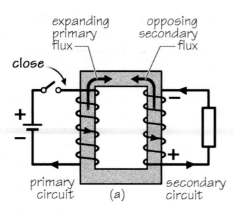
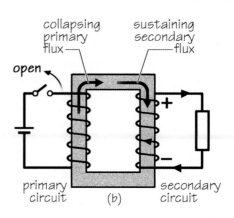

Figure 16.24

shown. As we have already learnt, this current *cannot* rise *instantaneously* to its maximum value, due to the **self-inductance** of the primary winding. However, as the current increases towards its maximum value (determined by the primary circuit's resistance) it creates, in this example, a correspondingly expanding magnetic flux in a *clockwise* direction around the core. You can confirm the direction of this flux by applying the '**right-hand grip rule**' to the primary winding (coil).

Most of this changing flux will be contained within the core, and link with the secondary winding, inducing a secondary potential difference into that winding by **mutual induction**.

Since the secondary circuit is connected to a load, the secondary induced potential difference will cause a current to drift through the secondary circuit. In accordance with Lenz's Law, the direction of this secondary current must create a secondary flux which will act *counterclockwise* around the core – in order to *oppose the increase in the primary flux*.

In Figure 16.24(b), the switch is opened and the primary flux starts to collapse (remember, the primary current is collapsing, *not reversing direction*, so its flux still acts *clockwise* around the core). This collapse in primary flux again induces a potential difference into the secondary winding, by **mutual induction** – this time, in the *opposite* direction to the originally induced secondary potential difference. The resulting reverse in the secondary current now causes the secondary flux to reverse, which will now act clockwise around the core, trying *to sustain the primary flux*.

It's very important to understand that the behaviour we are describing *only* occurs for *changing* currents. When a current in the primary circuit

becomes steady or constant, the associated magnetic flux also becomes steady, and cannot induce a potential difference into the secondary circuit.

We say that the primary and secondary circuits, in the above example, are 'coupled by **mutual induction**' (symbol: **M**) which, just like self-inductance, is measured in **henrys**.

The value of the potential difference induced into the secondary circuit is given by the expression:

$$e_s = -M \frac{\Delta i_p}{\Delta t}$$

. . . where $\dfrac{\Delta i_p}{\Delta t}$ means the *rate of change* in the primary current, and the *negative* sign indicating that, by **Lenz's Law**, the potential difference induced into the secondary circuit acts in a direction that will *oppose* the change in primary current.

From the above expression, we can say that '*two circuits will have a mutual inductance of one henry if a uniform change in current of one ampere per second in one circuit causes a potential difference of one volt to be induced into the other*'.

Worked example 6   Two self-inductive circuits, **A** and **B**, have a mutual inductance of 0.05 H. If the current in circuit **A** increases from zero to 10 A in 0.2 s, what will be the value and direction of the potential difference (u_B) induced into circuit **B**?

Solution   In this example the change in current in circuit **A** may be found from:

$$\Delta I = i_{initial} - i_{final} = 10 - 0 = 10 \text{ A}$$

$$\text{so, } u_B = -M\frac{\Delta I}{\Delta t}$$

$$= -0.05 \times \frac{10}{0.2} = -0.05 \times 50 = -2.5 \text{ V (Answer)}$$

Once again, the negative sign indicates that the direction of the induced voltage is acting to oppose the change in current in the circuit *A*.

Expression for mutual inductance

For two circuits, having self-inductances of L_1 and L_2 respectively, and which are linked by *all* of the flux created by the current in the primary circuit, it can be shown that the **mutual inductance**, *M*, is given by:

$$M = \sqrt{L_1 L_2}$$

Coupled circuits

In practice, not *all* the flux created by the primary current will actually link with the secondary circuit. That flux which does *not* link with the secondary circuit is called **leakage flux**, or is simply referred to as *'leakage'*.

If two self-inductive circuits (such as a pair of coils), are placed so that there is mutual induction between them, then we say that the two circuits are magnetically **coupled**.

The magnitude of the circuits' mutual inductance depends upon their 'magnetic closeness'. By this, we *don't* mean how physically close the two coils are but, rather, the degree by which the magnetic flux is able to link the two circuits, or *how closely the circuits are magnetically 'coupled'*:

- If the circuits are linked by *only part* of the flux, then they are said to be **loosely coupled**.
- If the circuits are linked by *most* of the flux, then they are said to be **closely coupled**.

In Figure 16.25, the pair of coils to the left are 'loosely coupled' whereas the pair of coils to the right, while being the same *physical* distance apart, are 'closely coupled' – thanks to the iron core that links them.

The degree of **looseness** or **closeness** of coupling is represented by a number called the '**coupling coefficient**' or '**coefficient of coupling**' (symbol: *k*), the value of which depends upon the *degree of coupling*.

For two circuits that are *perfectly coupled* (no leakage), the coupling coefficient is **unity** (1); for two circuits that are so separated that there *is no mutual inductance between them whatsoever*, the coupling coefficient is **zero**.

So the **mutual induction** between the two self-inductive circuits, expressed above, must be modified, as follows:

$$M = k\sqrt{L_1 L_2}$$

where:

$$M = \text{mutual inductance (H)}$$
$$k = \text{coupling coefficient}$$
$$L_1 = \text{primary circuit self-inductance (H)}$$
$$L_2 = \text{secondary circuit self-inductance (H)}$$

Worked example 7   Two circuits, of 8 mH and 16 mH respectively, are closely coupled with a coupling coefficient, $k = 0.75$. What is the mutual inductance between the two circuits?

Solution

$$M = k\sqrt{L_1 L_2}$$
$$= 0.75\sqrt{8 \times 16}$$
$$= 0.75\sqrt{128} = 0.75 \times 11.31 = 8.48 \text{ mH (Answer)}$$

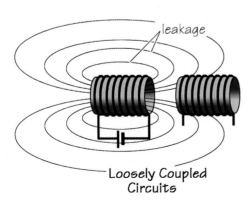

Loosely Coupled Circuits

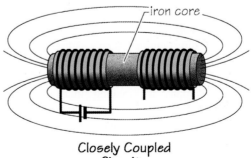

Closely Coupled Circuits

Figure 16.25

Worked example 8 What is the coupling coefficient between two circuits, having self-inductances of 240 μH and 360 μH, if their mutual inductance is 250 μH?

Solution

$$M = k\sqrt{L_1 L_2}$$

$$\text{so, } k = \frac{M}{\sqrt{L_1 L_2}}$$

$$= \frac{250}{\sqrt{240 \times 360}} = \frac{250}{\sqrt{86\,400}} = \frac{250}{294} = 0.85$$

Two practical examples of mutually inductive circuits are a motor car's **ignition coil**, and the **transformer**.

The **ignition coil** is used to step up the 12-V d.c. battery supply to the thousands of volts required by the car's spark plugs. In order to provide the changing-current in the primary coil, necessary to induce the high-voltage into the secondary coil (which has many more turns than the primary coil), the primary circuit is continuously switched on and off by a rotary switch driven by the car's engine.

Transformers

A **transformer** is a device that transfers electrical energy from one circuit (the 'primary' circuit) to another (the 'secondary' circuit) through mutually inductive coupled conductors: the transformer's coils or 'windings', with practically 100 per cent efficiency. Its purpose is to either 'step up' (increase), or to 'step down' (decrease), the voltage applied to its primary circuit.

For a transformer to function, there *must* be a continuously changing flux linking the primary and secondary circuits. For this reason, *transformers are alternating-current machines*.

An '**induction coil**' is a transformer which will work from a direct-current supply, such as a battery. This is achieved by continuously inter-rupting the d.c. current in the primary circuit using, for example, a vibrating mechanical contact. Induction coils are used in vehicles, for providing the high-voltage output necessary to operate spark plugs. In older vehicles, the nec-essary interruption to the current in the primary circuit was provided by a set of contacts operated by a cam in the distribution head (these days, it is done electronically).

Because it has no moving parts, a practical transformer is very close to being 100 per cent efficient. For the purpose of this introduction, we will assume that we are dealing with an 'ideal' transformer that *is* 100 per cent efficient.

A transformer, then, consists of two windings (coils), termed the **primary winding** and the **secondary winding**, which are wound concentrically around each other as well as around a silicon-steel magnetic circuit, termed a **core**. By definition, the primary winding is the winding connected to the supply, while the secondary winding is the winding connected to the load.

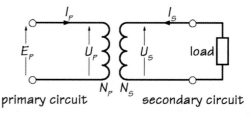

Figure 16.26

Although, in practice, the two windings are wound around each other as well as around the transformer's core, they are shown separately in schematic diagrams, as illustrated in Figure 16.26.

Earlier in this chapter, we learnt that the voltage *(u)* induced into a coil, by self-induction, is given by the equation:

$$u = -N\frac{\Delta\Phi}{\Delta t}$$

So, for a given amount of magnetic flux (Φ), the voltage induced into the primary winding, by self-induction, is proportional to the number of turns (N_P) in the primary winding.

Since, for an 'ideal' transformer, every turn in the secondary winding must link with exactly the same flux, the voltage induced into the secondary winding must be proportional to the number of turns (N_S) in the secondary winding.

It follows, therefore, that *the ratio of primary voltage (U_P) to secondary voltage (U_S)* must be exactly the same as *the ratio of the number of primary turns (N_P) to the number of secondary turns (N_S)*, that is:

$$\frac{U_P}{U_S} = \frac{N_P}{N_S}$$

We call the ratio of U_P to U_S, the transformer's '**voltage ratio**', and the ratio of N_P to N_S, the transformer's '**turns ratio**'. So, for an 'ideal transformer', its voltage ratio is exactly the same thing as its turns ratio.

Worked example 9   An ideal transformer has a primary winding of 500 turns and a secondary winding of 2000 turns. If the primary voltage is 230 V, calculate the transformer's secondary voltage.

Solution

$$\frac{U_P}{U_S} = \frac{N_P}{N_S}$$

$$U_S = U_P \frac{N_S}{N_P} = 230 \times \frac{2000}{500} = 920 \text{ V (Answer)}$$

In the above worked example, the transformer is a '**step-up**' transformer.

We can consider an 'ideal' transformer to be 100 per cent efficient, then we can say that the primary power must equal the secondary power. In other words, when the transformer is supplying a load, the product of its primary voltage and primary current must be equal to the product of its secondary current and its secondary (load) current:

$$P_P = P_S$$
$$U_P I_P = U_S I_S$$

Rearranging this equation:

$$\frac{U_P}{U_S} = \frac{I_S}{I_P}$$

So *the ratio of primary voltage (U_P) to secondary voltage (U_S)* is exactly the same as *the ratio of the number of secondary current (I_S) to the primary current (I_P)*.

Worked example 10   A load draws a secondary current of 0.5 A from an ideal transformer. If the transformer's primary voltage is 230 V and its secondary voltage is 12 V, what will be its primary current?

Solution

$$\frac{U_P}{U_S} = \frac{I_S}{I_P}$$

$$I_P = I_S \frac{U_S}{U_P} = 0.5 \times \frac{12}{230} \approx 26 \text{ mA (Answer)}$$

We can combine the equations listed above, as follows:

$$\frac{U_P}{U_S} = \frac{N_P}{N_S} = \frac{I_S}{I_P}$$

Inductors

All circuits have some degree of *natural* self-inductance – even a single conductor has some self-inductance, albeit extremely low compared to that of a coil.

However, we can *modify* a circuit's natural self-inductance using circuit components called **inductors** – in just the same way in which we can modify a circuit's resistance by using resistors.

Inductors have a great many applications. At one end of the scale, they are used in low-power electronic circuits while, at the other end of the scale, they are used with the 400-kV supergrid transmission system! So they come in an enormous range of shapes, and in sizes measuring from a few millimetres to several metres.

The terms '**reactor**', '**choke**' and '**ballast**' are often used instead of 'inductor' but, strictly speaking, these terms describe the *application* of inductors, rather than the devices themselves.

- '**Reactors**' are inductors used, for example, in electrical transmission systems to limit the magnitude of switching and fault currents.
- The term '**ballast**' is applicable to resistors as well as to inductors, and describes their use in limiting the rise of alternating currents – e.g. to limit the current flow through a fluorescent lamp once its gas has been ionised.
- '**Chokes**' are inductors used to limit the flow of alternating currents while allowing the passage of direct currents, or to block ('filter out') high frequency currents (e.g. telecommunication signals) from low frequency currents (e.g. mains current).

Despite their differences in size, inductors are essentially *all* constructed in very much the same way, consisting of a **coil** of insulated wire wound around a 'soft' ferromagnetic laminated **core**, such as silicon steel.

Like resistors, inductors can be connected in series, parallel, series-parallel and complex. In this section, we will only examine inductors that are connected in **series** and in **parallel**. You will notice that inductors are handled in exactly the same way as resistors, as we shall describe below.

Inductors in series

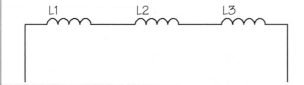

Figure 16.27

Ignoring any *mutual* inductance that may exist, the **equivalent self-inductance** for series self-inductance is given by:

$$L = L_1 + L_2 + L_3 + etc.$$

> **Worked example 11** What is the total self-inductance if four inductors, of self-inductance 2 mH, 3 mH, 4 mH and 5 mH, are connected in series.
>
> **Solution**
>
>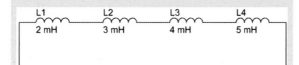
>
> Figure 16.28
>
> $$L = L_1 + L_2 + L_3 + L_4 = 2+3+4+5 = 14 \text{ mH (Answer)}$$

Inductors in parallel

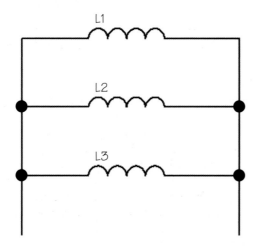

Figure 16.29

Ignoring any *mutual* inductance that may exist, **the equivalent self-inductance** for parallel self-inductance is given by:

$$\frac{1}{L} = \frac{1}{L_1} + \frac{1}{L_2} + \frac{1}{L_3} + etc.$$

> **Worked example 12** What is the total self-inductance if three inductors of 6 mH, 18 mH and 36 mH, are connected in parallel (Figure 16.30)?
>
>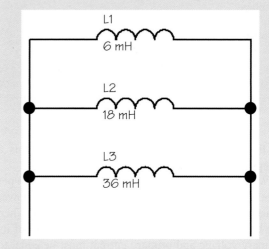
>
> Figure 16.30
>
> **Solution**
>
> $$\frac{1}{L} = \frac{1}{L_1} + \frac{1}{L_2} + \frac{1}{L_3}$$
>
> $$\frac{1}{L} = \frac{1}{6} + \frac{1}{18} + \frac{1}{36} = \frac{6+2+1}{36} = \frac{9}{36}$$
>
> $$L = \frac{36}{9} = 4 \text{ mH (Answer)}$$

Unintentional coupling between adjacent inductors

In each of the examples above, we specified 'ignoring any mutual induction that might exist'. But what happens if we can't ignore it?

If adjacent inductors are located too close together in the same circuit, then the magnetic field of one inductor can affect the other and *vice versa*. In other words, mutual inductance can occur *unintentionally* between pairs of inductors if they are located too close together.

We call this effect '**cumulative**' or '**differential**' coupling.

For a pair of inductors connected in series, the effect of this mutual inductance can be to either increase or to decrease the total inductance of the circuit – depending on how they are wound relative to each other.

For example, suppose two inductors, L_1 and L_2, are connected in series close enough, and in such a way that their fields *reinforce* each other ('cumulatively coupled'), then their total inductance will be given by the equation:

$$L = L_1 + L_2 + 2M$$

If, on the other hand, the two inductors are connected in series close enough, and in such a way that their fields *oppose* each other ('differentially coupled'), then their total inductance will be given by the equation:

$$L = L_1 + L_2 - 2M$$

For each of these equations, the mutual inductance, M, is given by:

$$M = k\sqrt{L_1 L_2}$$

. . . where k is the coefficient of coupling.

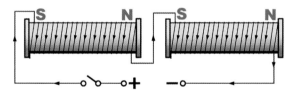

Figure 16.31

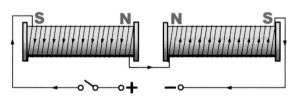

Figure 16.32

Worked example 13   Two inductors, of inductance 20 mH and 30 mH are connected in series and close enough to each other such that they are cumulatively coupled. If the coefficient of coupling is 0.25, what is the total inductance of the circuit?

Solution

$$L = L_1 + L_2 + 2M = L_1 + L_2 + 2k\sqrt{L_1 L_2}$$
$$= 20 + 30 + 2 \times 0.25\sqrt{20 \times 30}$$
$$= 50 + 0.5\sqrt{600}$$
$$\simeq 50 + 12.25$$
$$\simeq 62.25 \text{ mH (Answer)}$$

Worked example 14   The same two inductors are connected in series with each other such that they are differentially coupled, with the same coefficient of coupling. What will be the total inductance of the circuit?

Solution

$$L = L_1 + L_2 - 2M = L_1 + L_2 - 2k\sqrt{L_1 L_2}$$
$$= 20 + 30 - 2 \times 0.25\sqrt{20 \times 30}$$
$$= 50 - 0.5\sqrt{600}$$
$$\simeq 50 - 12.25$$
$$\simeq 37.75 \text{ mH (Answer)}$$

The worst case scenario for this condition is for the mutual induction between a pair of inductors to either *double* the series inductance of the inductors or to completely *cancel* the series inductance of the two inductors!

Finally . . .

Now that you have completed this chapter, are you able to achieve the objectives or learning outcomes listed at the beginning of this chapter?

Ask yourself, 'Can I . . .'

1 explain the result of moving a permanent magnet towards, or away from, a coil.
2 describe the effect of moving a straight conductor perpendicularly through a magnetic field.
3 determine the magnitude and direction of the potential difference induced into a straight conductor moved through a magnetic field.
4 demonstrate the application of Fleming's Right-Hand Rule for generator action.
5 briefly explain the effects of self-induction.
6 explain Faraday's and Lenz's Laws for electromagnetic induction.

7 list the factors that affect the self-induction of a coil.

8 describe the effect of self-induction on the growth and decay of direct currents in inductive circuits.

9 solve simple problems on the growth and decay of direct currents in inductive circuits.

10 explain the terms 'mutual inductance' and 'coupled circuits'.

11 describe the behaviour of the mutual induction between two inductive circuits.

12 solve simple problems on mutual induction.

13 describe the construction and operation of an 'ideal' transformer.

14 solve simple problems on 'ideal' transformers.

15 describe how energy is stored in magnetic fields.

16 solve simple problems on the energy stored in magnetic fields.

17 recognise and give simple examples of the functions of inductors.

18 solve simple problems on inductors in series and in parallel.

Online resources

The companion website to this book contains further resources relating to this chapter. The website can be accessed via the following link:

www.routledge.com/cw/waygood

Chapter 17

Capacitors and capacitance

On completion of this chapter, you should be able to

1 explain the primary function of a capacitor.
2 describe the essential components of any capacitor.
3 describe the charging/discharging process of a capacitor.
4 describe what is meant by 'the charge on a capacitor'.
5 describe how electric charge behaves with capacitors connected in series.
6 recognise the circuit symbols for different types of capacitor.
7 state the unit of measurement for capacitance.
8 list the factors that affect capacitance, and their relationship.
9 describe how a capacitor's dielectric affects capacitance of that capacitor.
10 describe the relationship between absolute permittivity, the permittivity of free space, and relative permittivity.
11 determine the energy stored by a capacitor.
12 describe the basic construction of practical fixed-value and variable-value capacitors.
13 solve simple problems on the time constant of a resistive-capacitive circuit.
14 solve simple series, parallel and series-parallel capacitive circuits.

Introduction

In 1745, barely three years before his death, a German cleric and physicist Ewald Georg von Kleist (1700–1748) invented a device for 'storing static electricity' for the purpose of his experiments. It consisted of a glass jar, coated both inside and out with metal foil (often, gold leaf), and with its inner coating connected, via a metal chain, to a brass rod that passed through an insulated wooden stopper.

A year later, in 1746, before details of von Kleist's invention had even been published, a Dutch physicist working at the University of Leiden, Pieter van Musschenbroek (1692–1761), independently invented an almost identical device which became known as a 'Leiden Jar' or '**Leyden Jar**', the forerunner of what we know, today, as a **capacitor**.

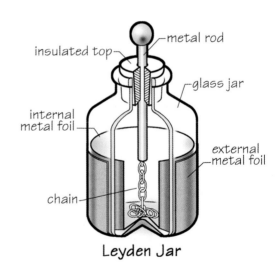

Figure 17.1

Although invented well over 250 years ago, the Leyden and Kleistien Jars each have *exactly* the same components as a modern capacitor: two conducting '**plates**' (its separate inner and outer metal foils), separated by an insulator or **dielectric** (glass).

Prior to the invention of these devices, physicists studying electricity were only able to store separated electrical charges on physically large, electrically isolated conductors; the Leyden Jar offered a more compact, convenient and portable alternative.

Until around the 1950s, in the English-speaking world, capacitors were called '**condensers**' from the Italian word, *'condensatore'*, coined by Alessandro Volta, meaning to 'squeeze' or to 'compress', because of their ability to store a *greater density* of separated electrical charge than the larger normally isolated conductors used at that time. Incidentally, this is the word by which capacitors are *still* known by engineers and electricians who live, for example, in France *('condensateur')* and in Germany *('kondensator')*.

In fact, in the English-speaking automotive industry, the capacitors used in ignition systems are still widely known as 'condensers'!

Capacitors

In this chapter we will learn about the behaviour of **capacitors**, and about a naturally occuring electrical quantity, called **capacitance**, which exists in most circuits.

Throughout this chapter, we will be using **electron flow**, as the use of conventional flow will, otherwise, make this particular topic unreasonably confusing.

Let's begin by looking at what a capacitor *does*.

> A **capacitor** is a circuit component that will *temporarily store electrical energy.*

We must *not* interpret this statement as meaning that a capacitor offers a practical means of storing useful amounts of energy for later use, rather in the same way a cell or battery does! Rather, this ability gives a capacitor some very useful properties which, as we will learn later, are used by both d.c. and a.c. applications.

A common misconception that we need to dispel right from the very beginning is that 'capacitors store *charge*'! Unfortunately, a great many textbooks state this to be the case but, as we shall learn, this is both misleading and prevents us from understanding their behaviour.

So it's worth repeating that capacitors store **energy**, *not* charge.

In their simplest form, *all* capacitors consist of two thin, parallel metal sheets (or foils), called '**plates**', placed very close together, but separated by an insulating material which we call a '**dielectric**'. In Figure 17.2, the dielectric is simply air, but another

insulating material could be used instead, including mica, plastics, oil-impregnated paper, etc. In fact, capacitors are named according to their dielectric, hence; 'mica capacitor', 'paper capacitor', etc.

As already explained, these components are exactly equivalent to the interior and exterior metal foil coatings and the glass dielectric of the original Leyden Jar.

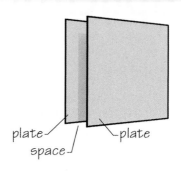

plate — plate
space

Figure 17.2

If the capacitor's plates are connected to an external source of d.c. potential difference, such as that provided by a battery, electrons will be transferred from one plate to the other. The resulting deficiency of electrons on the first plate will cause that plate to acquire a positive potential, while the excess of electrons on the other plate will cause that plate to acquire a negative potential – creating a potential difference across the plates, and an electric field between them. And *it is within this electric field that a capacitor stores energy.*

We call this process of charge transfer between its plates 'charging the capacitor', and it is explained as follows.

Before charging takes place

With the switch open, each plate is electrically *neutral* (Figure 17.3) – i.e. they each contain equal quantities of protons and electrons (represented in the illustrations by the small positive and negative signs).

Remember, only the negatively charged *free electrons* are able to move, as their corresponding positively charged protons are bound within the nucleii of their immovable atoms.

Charging action

When the switch is closed, the positive terminal of the battery will immediately attract electrons away from the left-hand plate, while its negative terminal will drive electrons onto the right-hand plate (Figure 17.4).

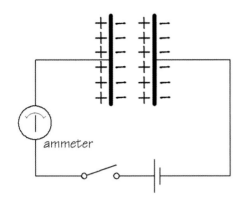

Figure 17.3

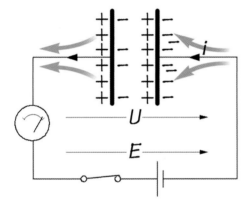

Figure 17.4

We call this movement of free electrons through the capacitor's external circuit, a '**charging current**'.

As more and more free electrons arrive at the right-hand plate, the amount of negative charge on that plate continues to increase while the loss of these electrons from the left-hand plate means that the amount of positive charge on that plate increases by the same amount, causing the potential difference *(U)* between the plates to rapidly increase.

The direction of this increasing potential difference, of course, is *opposite* that of the battery *(E)*, and eventually:

$$U = E$$

During the process, the charging current follows the curve shown in Figure 17.5. When the battery is first connected across the capacitor, the resulting current is high. But as the opposing potential difference *(U)*

builds up across the plates of the capacitor, the resulting charging current falls towards zero.

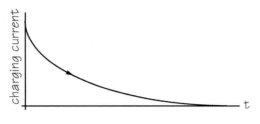

Figure 17.5

As we shall learn later, if the resistance of the circuit is negligible, then the charging current will fall to zero in just milliseconds or microseconds.

> The potential difference across the plates of a capacitor *will always rise to equal the potential difference of the external source*, but will act in the opposite sense – that is, the polarities of a capacitor's plates will always match those of the external potentials to which they are connected.

It's important to understand that, during the 'charging' process, all that is happening is that *electric charge is being transferred from one plate to the other*. There is no change to the *net amount* of charge on those plates – i.e. the battery is *not* forcing any additional charges onto the plates of the capacitor.

So, when we refer to the amount of 'charge' on a capacitor, we are actually referring *either* to the amount of positive charge on the positive plate, *or* to the identical amount of negative charge on the negative plate – *not to the sum of these two charges!* However, in keeping with the electron theory, it is convenient to express the 'charge on a capacitor' in terms of *the amount of negative charge, expressed in coulombs, accumulated on the negative plate of the capacitor.*

> **Important!**
> By convention, the 'charge on a capacitor' is the amount of negative charge, expressed in coulombs, that has accumulated on the negative plate of that capacitor. It is *not* the sum of the positive and negative charges accumulated on *both* plates.

With the switch open

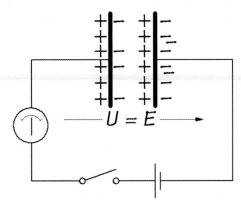

Figure 17.6

If the switch is now opened (Figure 17.6), the capacitor will *retain the potential difference that has built up across its plates*. This is because the open switch has broken the only direct path by which those electrons which have accumulated on the right-hand plate can return to the left-hand plate – other than through the dielectric.

In practice, there is no such thing as a *perfect* dielectric. So, *some* of the excess electrons on the right-hand plate do, in fact, manage to find their way back to the left-hand plate through the dielectric. This is termed **leakage current**, and results is *a very gradual reduction in the potential difference across the capacitor's plates*.

Discharge action

If the battery is now removed from the circuit (Figure 17.7), and replaced by a load (in this example, a resistor), the excess electrons crowded onto the

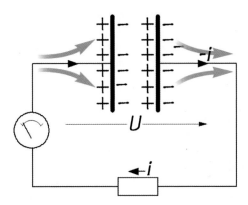

Figure 17.7

right-hand plate now have a direct route back around the circuit to the left-hand plate, and the fully charged capacitor will begin to **discharge**. This discharge current will follow exactly the same-shaped curve as the charging current but, of course, *in the opposite direction* (see Figure 17.8).

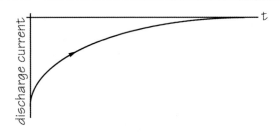

Figure 17.8

Again, as we shall learn later, if the resistance of the load is negligible, then the capacitor will completely discharge in milliseconds or microseconds.

Charge distribution on capacitors in series

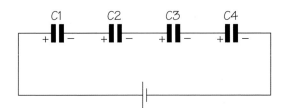

Figure 17.9

We earlier described the charge on a capacitor as the 'amount of negative charge accumulated on its negative plate'. So, what happens if a number of capacitors are connected in *series?*

If several capacitors are connected in series across a d.c. supply, as illustrated in Figure 17.9, free electrons will be transferred from the left-hand plate of capacitor C_1, counterclockwise, through the circuit to the right-hand plate of capacitor C_4.

The negative charge acquired by the right-hand plate of capacitor C_4 will then repel electrons from its left-hand plate, driving them onto the right-hand plate of capacitor C_3. This action is repeated for capacitor C_2, and the resulting charge distribution will be as indicated by the positive and negative signs shown in Figure 17.9.

However, the **total charge** *on all four capacitors will simply be the original amount of negative electric* charge transferred from the left-hand plate of capacitor C_1 onto the right-hand plate of capacitor C_4.

In other words, the total charge will remain the same, *regardless of how many capacitors are connected in series between the outer plates of the outermost two capacitors!* We will return to this topic, for further explanation, a little later.

Water-flow analogy of a capacitor

Imagine a hollow, water filled, chamber divided in half by a flexible rubber diaphragm, and connected to a water pump via water-filled pipework (Figure 17.10).

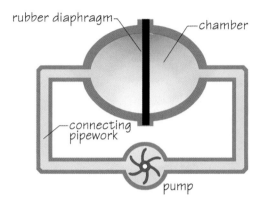

Figure 17.10

The volume of water on either side of the diaphragm is equivalent to the balanced charge on the plates of an uncharged capacitor, and the diaphragm itself is equivalent to the electric field that will appear between the two plates.

When the pump starts, water will flow (in this case) counterclockwise around the system and start to push against the diaphragm (see Figure 17.11).

Initially, there will be a sudden rush of water around the system. But, as the diaphragm stretches, it will increasingly oppose the flow of water until, eventually, it becomes fully stretched, its opposition will be equal and opposite to the pump's pressure and the flow will stop altogether (Figure 17.12). At the same time, the work done in stretching the diaphragm means that the diaphragm has gained energy. The **diaphragm** *is the equivalent of a capacitor's electric field*, and the action described is equivalent of *a capacitor being charged.*

The increased volume of water to the right of the diaphragm represents the negative charge that has built up on the negative plate of the capacitor. But, although the volume of water to the right of the diaphragm has increased, the volume to the left has decreased by exactly the same amount, so there has been no change in the overall volume of water within the chamber, yet energy has been stored in the stretched diaphragm.

Imagine, now, that the pump is allowed to freewheel (Figure 17.13). The stretched diaphragm will now relax and return to its normal shape, forcing water to flow in the *reverse* direction (clockwise) around the system. At the same time, the energy stored in the fully stretched diaphragm is allowed to dissipate. Initially, there will be a sudden rush of water, but this flow will reduce as the diaphragm returns to its normal shape. *This action is equivalent to a capacitor being discharged.*

If we think of the volume of water to the right of the diaphragm, within the chamber, as being equivalent to the 'charge transferred' to the negative plate of a capacitor, it should become apparent that, as no overall change in the volume of water has taken place, then it should be equally clear that no 'additional electric charge' has been 'stored' on the plates of a capacitor during the 'charging' process. What *has* been 'stored' in the capacitor is **energy**, and this energy has been stored in the its **electric field**.

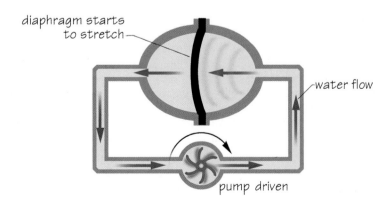

Figure 17.11

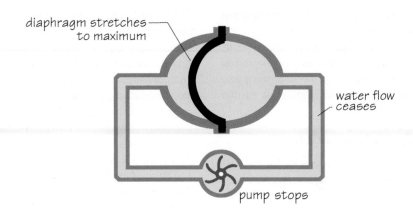

diaphragm stretches to maximum

water flow ceases

pump stops

Figure 17.12

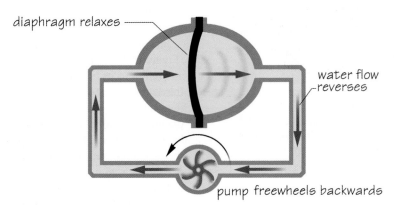

diaphragm relaxes

water flow reverses

pump freewheels backwards

Figure 17.13

So we can say that **a capacitor is a circuit component that stores** *energy*.

Perhaps we should replace the terms, 'charging' and 'discharging' with 'energising' and 'de-energising'? Well, unfortunately, the terms 'charging' and 'discharging' are too well established for us to change them at this stage!

Despite this, many textbooks continue to insist that a capacitor 'stores charge', but this is misleading because, as we have learnt, there is no increase in the overall charge on the plates whenever a capacitor is charged. All that has happened is that charge has been *separated*, and has been moved from one plate across to the other; *no additional charge has been introduced or stored*.

Actually, these textbooks aren't really 'wrong' because what they *mean* (as opposed to what they *say*!) is that a capacitor is a circuit component which

'stores *separated* charges' which, of course, is quite correct!

Circuit symbols for capacitors and capacitance

Capacitors may be **fixed value** or **variable value**. A special type of variable resistor is the **trimmer** type; this is a variable resistor that has been pre-set to a particular value of capacitance and not intended to be adjusted further.

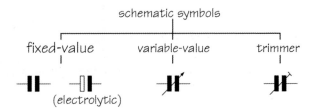

schematic symbols

fixed-value   variable-value   trimmer

(electrolytic)

Figure 17.14

'Electrolytic' capacitors are fixed-value capacitors which are 'polarised', which means that it is important which way around a potential difference is applied to their plates according to their markings (for the circuit symbol, shown in Figure 17.14, the solid black plate represents its negative plate).

Figure 17.15

You should be aware that the US circuit symbol of a capacitor is quite different from the European version, and can be confusing for the following reason.

What appears to Europeans as being the symbol for a capacitor is, in fact, the US circuit symbol for a set of normally open electrical **contacts** (normally *closed* contacts use the same symbol, but with a diagonal line through it). The American symbol for a capacitor uses a *curved* line to represent one of its plates (the negative plate, in the case of a polarised capacitor).

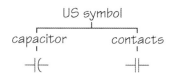

Figure 17.16

So care must be taken when reading US-sourced schematic diagrams, to ensure that you do not confuse the two symbols.

Capacitance

During the charging process, as free electrons accumulate on the negative plate of a capacitor, this causes the potential of that plate, measured with respect to the positive plate, to increase.

The resulting potential difference *(U)* between the two plates is directly proportional to the amount of charge *(Q)* accumulating on the negative plate, and can be expressed as follows:

$$U \propto Q$$

To change the proportional sign to an equals sign, we need to introduce a constant:

$$U = \text{constant} \times Q$$

Expressed another way, we can say:

$$\text{constant} = \frac{Q}{U}$$

We call this constant the **capacitance *(C)*** of the capacitor, which is defined as *'the ratio between the amount of electric charge on the negative plate, and the potential difference between the two plates'*.

$$C = \frac{Q}{U} \qquad\qquad \text{—(equation 1)}$$

where:

C = capacitance, in farads

Q = charge, in coulombs

U = potential difference, in volts

The SI unit of measurement for capacitance is equivalent to a 'coulomb per volt' which, in SI, is given the special name: the **farad** (symbol: F).

> The **farad** (symbol; **F**) is defined as *'the capacitance of a capacitor, between the plates of which there appears a difference in potential of 1 volt, when it is charged to 1 coulomb'*.

The farad, in practical terms, is absolutely enormous so, in practice, capacitance is usually expressed in **microfarads** (μF), in **picofarads** (pF) or in **nanofarads** (nF).

> Interestingly, one of the earliest units for capacitance was the **'jar'** (named after the 'Leyden Jar'). This unit was used well into the 1930s, where a capacitance of one microfarad is equivalent to 900 jar.

It's important to understand that the ratio of charge to potential difference merely tells us *what the capacitance happens to be for that particular ratio*, in much the same way as the ratio of voltage to current merely tells us what the resistance happens to be for that particular ratio.

The capacitance is not *determined* by that ratio, any more than resistance is determined by the ratio of voltage to current.

So what factors actually determine the capacitance of a capacitor?

Factors affecting capacitance

The capacitance of a capacitor is, to some extent, dependent upon the physical size of that capacitor, being *directly proportional* to the **area of overlap** (symbol; *A*) of its plates, and *inversely proportional* to the **distance** (symbol; *d*) between the plates.

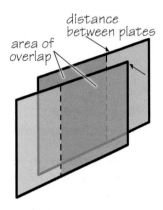

Figure 17.17

$$C \propto \frac{A}{d}$$

In other words, the capacitance *increases* with a *larger* area of overlap, and *decreases* if the plates are moved further apart. Again, we can change the proportional symbol, in the above expression, to an equals symbol by introducing a constant; in this case, we call the constant the **absolute permittivity** (symbol: ε, pronounced '*epsilon*'), which is a physical property of the dielectric.

The capacitance of a capacitor, then, is expressed as:

$$C = \varepsilon \frac{A}{d} \qquad \text{—(equation 2)}$$

> **Important**! It is *not* the total area of either plate that affects the capacitance but, rather, the area by which the two plates *overlap* each other.

Changing the dielectric can have a *significant* affect on altering a capacitor's capacitance.

This is summarised in Table 17.1.

How a dielectric affects capacitance

The capacitor's dielectric has *three* important functions. It must

- keep the two plates apart
- insulate the plates from each other
- improve the capacitance of the capacitor.

Table 17.1

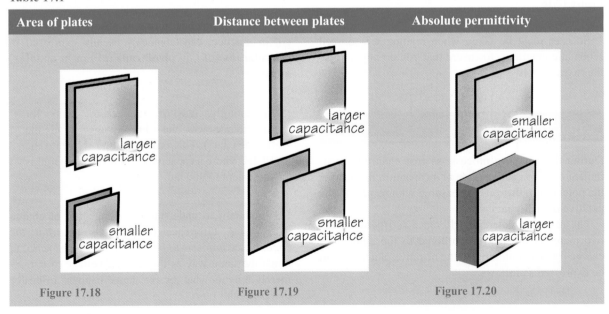

Area of plates	Distance between plates	Absolute permittivity

Figure 17.18   Figure 17.19   Figure 17.20

The first point is obvious; the plates mustn't be allowed to come into contact with each other or they will short-circuit.

The second point is that the insulating property of the dielectric must not be allowed to fail when the rated voltage of the capacitor is applied across its plates. We express this property in terms of its **dielectric strength**, which is measured in volts per metre (in practice, kilovolts per millimetre). An ideal dielectric should have a very high dielectric strength.

The third point is less obvious but, as we shall see, it can have a dramatic effect on increasing the capacitance of the capactor. So, *why does the dielectric affect a capacitor's capacitance?* Well, we know that there are relatively few free electrons in an insulating material, with the overwhelming majority of electrons being strongly tied within valence shells surrounding their atoms' nucleii. When they are not exposed to an electric field, their shells behave quite normally – as illustrated by the three atoms represented in the left-hand diagram in Figure 17.21.

In the right-hand diagram in Figure 17.21, however, these same atoms are subject to the electric field due to the potential difference between the plates of a charged capacitor. This field acts to 'stretch', or to distort, the electron shells so that they become 'elongated' along the direction of the lines of electric flux, with the 'positive centre' of each atom biased towards the negative plate, and the 'negative centre' biased towards the positive plate. We say that the dielectric's atoms have become 'polarised'.

The amount of polarisation – i.e. the amount by which the electron orbit 'stretches' – depends entirely on the amount of potential difference applied across the dielectric. Whenever this potential difference changes, the amount of polarisation changes – constituting what amounts to a momentary current, which we call a '**displacement current**'.

In Figure 17.21, only three dielectric atoms are shown for the sake of clarity. In reality, of course, this behaviour applies to *each* of the billions of atoms within the dielectric.

When the capacitor is fully charged, all the polarised atoms within the dielectric will have orientated themselves so that their positive centres are attracted towards the negatively charged plate, while their negative centres are attracted towards the positively charged plate. So the direction of electric field *within each polarised atom itself* effectively acts in the opposite sense (direction) to the *main* electric field between the two plates.

The effect of this is to weaken the main field and, therefore, to *reduce the voltage* across the plates *without affecting the charge* on the plates – *the resulting increase in the ratio of charge to voltage, therefore, acts to cause a corresponding increase in the capacitance of the capacitor.*

The amount of increase in capacitance depends on the material from which the dielectric is manufactured, with some having a far greater effect than others.

The 'reference dielectric material' is 'free space' (a term used by scientists to describe a vacuum or, in practice, air), whose permittivity is termed the '**permittivity of free space**' (symbol: ε_o). This is numerically equal to (8.85×10^{-12}) F/m.

All other dielectric materials are compared with the dielectric of free space – for example, mica has five

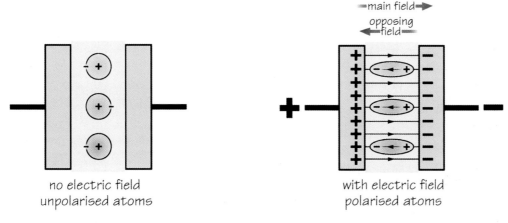

no electric field
unpolarised atoms

with electric field
polarised atoms

electric field within polarised dielectric atoms opposes main electric field
–reducing effective potential difference between plates

Figure 17.21

times the permittivity of free space, so we say that mica has a '**relative permittivity**' (symbol: ε_r) of 5.

> An older, alternative, name for **relative permittivity** is '**dielectric constant**'.

So, **absolute permittivity** (ε) is the product of the **permittivity of free space** (ε_o) and **relative permittivity** (ε_r). So we can now rewrite equation (2) in its final form, as follows:

$$C = \varepsilon_o \varepsilon_r \frac{A}{d} \qquad \text{—(equation 3)}$$

where:

C = capacitance, in farads

ε_o = permittivity of free space (farads per metre)

ε_r = relative permittivity (no units)

A = overlapping area of plates (square metres)

d = distance between plates (metres)

> **Worked example 1**   The plates of a capacitor measure 25 mm square, and are placed 1 mm apart. Calculate (a) its capacitance if the dielectric is air, and (b) its capacitance if the dielectric is mica. Assume that the relative permittivity of air is 1, and the relative permittivity of mica is 5.
>
> **Solution**   We first need to determine the cross-sectional area of the capacitor's plates:
>
> $$A = (25 \times 10^{-3}) \times (25 \times 10^{-3})$$
> $$= 25 \times 25 \times 10^{-6} = 625 \times 10^{-6}\,\text{m}^2$$
>
> a   For an **air** dielectric:
>
> $$C = \varepsilon_o \varepsilon_r \frac{A}{d} = \left(1 \times 8.85 \times 10^{-12}\right) \times \frac{625 \times 10^{-6}}{1 \times 10^{-3}}$$
> $$= 5.53 \times 10^{-12} = 5.53\,\text{pF Answer (a)}$$
>
> b   For a **mica** dielectric:
>
> $$C = \varepsilon_o \varepsilon_r \frac{A}{d} = \left(5 \times 8.85 \times 10^{-12}\right) \times \frac{625 \times 10^{-6}}{1 \times 10^{-3}}$$
> $$= 27.66 \times 10^{-12} = 27.66\,\text{pF Answer (b)}$$

Relative permittivity versus dielectric strength

Table 17.2 compares the average relative permittivities with the average dielectric strengths of a range of dielectrics commonly used in the manufacture of capacitors.

You will notice from the table that there is absolutely *no* correlationship whatsoever between a dielectric's **relative permittivity**, and its **dielectric strength**. For example, barium-strontium titanate (termed a 'ferroelectric ceramic') has a *huge* relative permittivity of 7500, but a *very low* dielectric strength of just 3 kV/mm which is no better than that of air.

This can be a problem for manufacturers who need to design capacitors, particularly those for use in high voltage systems, who find themselves unable to use a dielectric with a high relative permittivity if its dielectric strength is too low. So the selection of an appropriate dielectric is an exercise in compromise – i.e. deciding on which is the most important for a particular application; its relative permittivity or its dielectric strength!

Table 17.2 Examples of average relative permittivity and dielectric strength

Dielectric material	Relative permittivity	Dielectric strength (kV/mm)
Air	1.0006	3
Teflon	2	60
Paper	2.5	20
Rubber	3	28
Transformer oil	4	16
Mica	5	200
Glass	6	120
Porcelain	6	8
Bakelite	7	16
Barium-strontium titanate	7500	3

Absolute permittivity vs absolute permeablity

In the chapter on *magnetic circuits*, we met the term '**absolute permeability**' and learnt that it was a measure of the ease with which a medium (air or a ferromagnetic material) allows the formation of magnetic flux.

We also learnt that absolute permeability was the ratio of **magnetic flux density** *(B)* to **magnetic field strength** *(H), that is:*

$$\mu = \frac{B}{H}$$

An electric circuit's 'equivalent' of magnetic field strength, you will recall, is a 'voltage gradient' which, when applied to a dielectric, is called its **dielectric strength** *(D)*.

Absolute permittivity (symbol: ε) compares with absolute permeability, as it is the ratio of **electric flux density** (symbol: *D*) to **dielectric strength** (symbol: *E*), that is:

$$\varepsilon = \frac{D}{E}$$

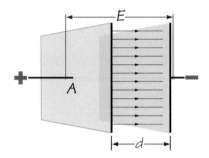

Figure 17.22

If you imagine every line of electric flux represented to eminate from a charge of one coulomb, then the unit of measurement for electric flux density will be a **coulomb per square metre** (C/m²). And, of course, dielectric strength will be measured in **volts per metre** (V/m).

For *air* (or, more strictly, for a vacuum), absolute permittivity corresponds to the **permittivity of free space** (ε_o), which is a constant and equal to:

8.85 × 10⁻¹² F/m (farads per metre)

To find the absolute permittivity for any other dielectric, we must multiply the absolute permittivity of free space by the relative permittivity of that particular dielectric:

$$\varepsilon = \varepsilon_r \varepsilon_o$$

This equates with the following for magnetic circuits:

$$\mu = \mu_r \mu_o$$

Voltage rating of capacitors

Unless specifically stated otherwise, a capacitor's **voltage rating** is *always* a **d.c. rating**. So if, for example, the voltage rating of a capacitor is 12 V, then any applied voltage in excess of 12 V might cause its dielectric to break down and fail.

Care, therefore, needs to be taken if the same capacitor is to be used in an *a.c.* circuit, to ensure the circuit's a.c. *peak value* of voltage does not exceed the capacitor's d.c. voltage rating.

Unfortunately, a.c. voltages are *always* expressed as *root-mean-square* (*r.m.s.*) values, never as peak values – where, $E_{peak} = 1.414\ E_{rms}$.

> The '230 V' which supplies your residence is an r.m.s. value, *not* a peak value; its peak value is 325 V!

This means that an a.c. voltage, having an r.m.s. value of 12 V, will actually *peak* to very nearly 17 V, which is significantly more than the dielectric of a capacitor rated at 12 V is designed to withstand.

We will learn about r.m.s. values in a later chapter on *Alternating Current*, but for now it's good enough to understand that an a.c. peak value is 1.414 *Erms*.

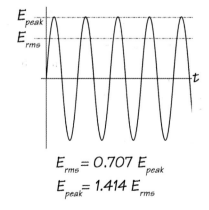

Figure 17.23

So, installing a 12-V capacitor into a 12-V a.c. circuit will subject that capacitor to nearly 17 V which will likely damage that capacitor!

So, the safe a.c. (r.m.s.) voltage which can be applied to a capacitor is no more than 0.707 × the capacitor's rated d.c. voltage.

> **Important!** Unless otherwise stated, a capacitor's rated voltage is a d.c. voltage. Care must, therefore, be taken when using a capacitor in an a.c. circuit, because a.c. voltages are always root-mean-square voltages, which peaks to 1.41× the r.m.s. value.

> **Worked example 2**   Can a capacitor rated at 250-V(d.c.) be used in parallel with a 230-V(a.c.) load for the purpose of power factor correction?
>
> **Solution**   Remember, the peak value of the a.c. voltage must not exceed the rated d.c. value of the capacitor, or there is a risk that the capacitor's dielectric may be damaged. So, in this case, the supply's peak value must not exceed 250 V.
>
> Since 230 V is the supply's r.m.s voltage,
>
> $$E_{rms} = 0.707\, E_{max}$$
>
> $$\text{so, } E_{max} = \frac{E_{rms}}{0.707} = \frac{230}{0.707} = 325 \text{ V}$$
>
> So, this capacitor *cannot* be connected across the 230-V load without the risk of its dielectric failing.

Energy stored by a capacitor

We learnt, in the chapter on *potential and potential difference*, that whenever we isolate an electric charge, there must be an equal charge, of opposite polarity, left behind, and that an electric field is set up between them.

The **energy** stored within a capacitor's electric field originates from the work done establishing that electric field.

As the potential difference *(U)* across a capacitor's plates is directly proportional to the charge *(Q)*, if we were to draw a graph of potential difference against charge throughout the charging process, the result would be a straight-line graph, as illustrated in Figure 17.24.

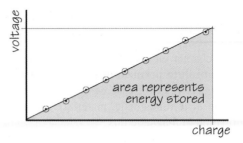

Figure 17.24

The triangular area (shaded grey in Figure 17.24) under the graph line represents the quantity of energy stored in that field. The area of a triangle, you will recall, is half its base times its height; so the equation for the energy stored in the field may be expressed as:

$$W = \frac{1}{2} U Q \qquad \text{—(equation 4)}$$

If we apply simple dimensional analysis to this, it should become apparent *why* this area represents the stored energy:

$$U \times Q = \text{volt} \times \text{coulomb} = \left(\frac{\text{joule}}{\text{coulomb}}\right) \times \text{coulomb} = \text{joule}$$

So, as you can see, the product of voltage and charge is **energy**, expressed in joules.

If we now substitute for Q in equation (4), we can rewrite the equation as follows:

$$W = \frac{1}{2} U (CU)$$

$$W = \frac{1}{2} CU^2 \qquad \text{—(equation 5)}$$

where:

$$W = \text{energy, in joules}$$
$$C = \text{capacitance, in farads}$$
$$U = \text{voltage, in volts}$$

> **Worked example 3**   A fully charged 250 μF capacitor has a potential difference of 100 V across its plates. What is the amount of charge transferred?
>
> **Solution**
>
> $$Q = CU = \left(250 \times 10^{-6}\right) \times 100 = 2500 \times 10^{-6}$$
> $$= 25 \text{ mC (Answer)}$$

Worked example 4   How much energy is stored within the electric field of the above capacitor when it is fully charged?

Solution

$$W = \frac{1}{2}CU^2 = \frac{1}{2} \times (250 \times 10^{-6}) \times 100^2$$
$$= 1.25 \text{ J (Answer)}$$

Construction of practical capacitors

Up until now, we've rather assumed that a capacitor consists of two *flat* metal plates, separated by a dielectric – just like its circuit symbol. However, this does *not* reflect the actual construction of practical capacitors.

First of all, as we have already learnt, there are basically *two* types of capacitor; **fixed value capacitors** and **variable value capacitors**.

As the names indicate, fixed-value capacitors are manufactured with a fixed value of capacitance, while variable-value capacitors allow their capacitance to be varied by the user.

As we have also learnt, the plates of a capacitor assume the same polarities as the external potentials to which they are attached, so it doesn't really matter how they are connected to their voltage source. The exception to this rule is a type of fixed-value capacitor, called an 'electrolytic capacitor' – this capacitor *must* be connected according to the polarity marks printed on its container. Failure to do so will result in the destruction of the electrolytic capacitor.

Fixed-value capacitors

Capacitors are manufactured in a huge range of physical sizes, ranging from the size of a match head to the size of a distribution transformer.

Capacitors are generally named after the material from which their dielectrics are made. These include *paper capacitors*, *mica capacitors*, *ceramic capacitors*, *oil capacitors* and *electrolytic capacitors*.

Paper capacitors typically consist of two long strips of metal foil, interlaced with two strips of oil-impregnated paper dielectric, rolled together rather like a 'Swiss roll', and sealed inside an aluminium or plastic protective tubular container. These days, plastics have tended to replace paper, but there are still lots of paper capacitors about.

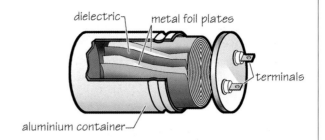

Figure 17.25

Electricians are most likely to come across **oil-impregnated paper capacitors**. This type of capacitor exists in a great many shapes and sizes, from relatively small types used to start some types of single-phase a.c. motor, to much larger types used for **power factor improvement** (see the chapter on *power factor improvement*) in industrial installations.

Capacitors used for these industrial purposes are very much larger (as illustrated in Figure 17.26), partly because they often have to operate at relatively high voltages, so their plates need to be relatively far apart and consist of interleaved rectangular plates, connected in parallel, and immersed in a mineral oil.

Figure 17.26

Variable-value capacitors

A **variable-value capacitor** is one designed to allow its capacitance to be varied over a given range. One of the most common applications for a variable capacitor, of the type shown in Figure 17.27, is the tuning control of a radio. It consists of a number of fixed vane-shaped plates, interleaved with moving plates controlled by a rotating knob. The capacitance is varied according to the area of overlap of the two sets of plates.

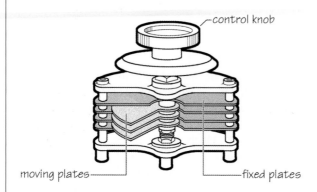

Figure 17.27

Behaviour of capacitors in d.c. circuits

In the circuit shown in Figures 17.28 and 17.29, a capacitor is connected in series with a resistor across a battery of voltage, E volts.

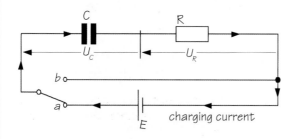

Figure 17.28

With the switch in position **a**, the battery will act to *charge* the capacitor. As the capacitor charges, its voltage, U_C, 'gradually' increases until it equals (but opposes) the supply voltage – following the curve shown to the right. Had the capacitor continued to charge at its initial rate, then its voltage would have followed the dotted line – reaching its fully charged voltage in CR seconds. This is known as the circuit's **time constant**, expressed in seconds (s):

$$\text{time constant} = CR$$

where:

$$C = \text{capacitance (farads)}$$

$$R = \text{resistance (ohms)}$$

However, by following the curve, it actually reaches its fully charged voltage in **5 time constants**:

$$\text{time to fully discharge} = 5\,CR$$

The chain line represents the **charging current,** which is maximum *before* the potential difference across the capacitor (U_C) starts to build up. In fact, *the charging current is **directly proportional to the rate of change of** U_C*, which means that it is greatest when the voltage's curve is at its steepest.

We will return to the fact that a capacitor's charging current is proportional to the rate of change of voltage when we study a.c. circuits, later in this book.

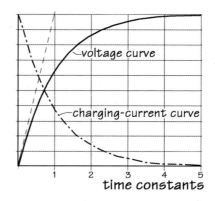

Figure 17.29

With the switch moved to position **b**, the capacitor will discharge. As it discharges, its voltage, U_C, 'gradually' decreases until it reaches zero volts – following the curve shown in Figure 17.31.

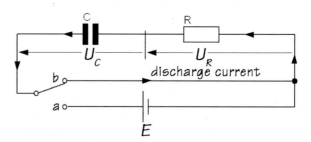

Figure 17.30

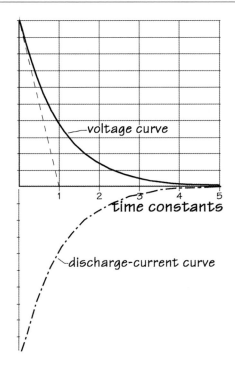

Figure 17.31

Had the capacitor continued to discharge at its *intial* rate then, again, it would fully discharge in CR seconds – again, known as the circuit's **time constant**. However, by following the actual curve, the capacitor will fully discharge in **5 time constants**.

$$\text{time to fully discharge} = 5\,CR$$

*The **time constant** is unaffected by the circuit's voltage, and depends solely upon the capacitance and resistance of the circuit.*

The chain line, again, represents the **discharge current** which, of course, acts *in the opposite direction to the charging current*.

So from this we can see that, in the same way that an inductor will oppose any change in *current*, a **capacitor always acts to oppose any *change* in voltage.**

Worked example 5   A 50 μF capacitor is connected in series with a 100-Ω resistor. If this circuit is suddenly connected across a 100-V (d.c.) supply, how long will the capacitor take to fully charge?

Solution   The 100-V (d.c.) supply is irrelevant to this question, as the time taken for the capacitor to reach its fully charged state will be five time constants. So,

$$\begin{aligned}\text{time to fully charge} &= 5\,CR\\ &= 5\times(50\times10^{-6})\times100\\ &= 25000\times10^{-6}\\ &= 25\text{ ms (Answer)}\end{aligned}$$

Worked example 6   A circuit, comprising a fully charged 50 μF capacitor in series with a 100-MΩ resistor, is suddenly short-circuited. How long will the capacitor take to fully discharge?

Solution

$$\begin{aligned}\text{time to fully discharge} &= 5CR\\ &= 5\times(50\times10^{-6})\times(100\times10^{6})\\ &= 25000\text{s (Answer)}\\ \text{or}\quad \frac{25000}{60} &= 416\min 40\text{ s (Answer)}\end{aligned}$$

As the above examples show, a capacitor's charging and discharging times can vary considerably, depending on the resistance of the circuit.

Natural capacitors

It's not just **capacitors** that exhibit **capacitance**. Both **overhead lines** and **underground cables** behave as 'natural capacitors'!

In the case of overhead lines, adjacent conductors (and each conductor and the earth) behave like the plates of a capacitor, and the air acts as the dielectric.

In the case of underground cables (in fact, *any* long length of cable), adjacent conductors (or individual conductors and earth) act like plates, while the conductors' insulation acts as the dielectric.

The 'capacitance' of cables is significantly higher than for overhead lines, due to the extreme closeness of their conductors ('plates'), and **care must be taken to fully discharge long lengths of cable that may have acquire appreciable charge during an insulation test using a Megger** (a high voltage test instrument), in order to avoid a shock hazard!

Electrical cables used for residential wiring have a capacitance of around 100 pF per metre length and the resulting capacitive currents can be responsible for some odd behaviours that occasionally occur in electrical installations, such as CFLs (compact fluorescent lamps) which continue to flicker *after* they have been switched off.

Capacitors in series and parallel

Capacitors in series

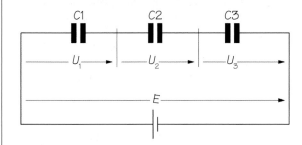

Figure 17.32

In accordance with Kirchhoff's Voltage Law, the sum of the voltage drops across each capacitor will equal the supply voltage:

$$E = U_1 + U_2 + U_3 \qquad \text{—(equation 6)}$$

But we know that voltage is charge divided by capacitance, so we can substitute for voltage in equation (6):

$$\frac{Q}{C} = \frac{Q}{C_1} + \frac{Q}{C_2} + \frac{Q}{C_3} \qquad \text{—(equation 7)}$$

As we learnt earlier, the charge on capacitors in series is simply the amount of charge accumulated on the negative plate of the outermost capacitor (C_3, in the above example), so it is the same *regardless of the number of capacitors*.

So, we can divide equation (7) throughout by Q, giving:

$$\frac{\cancel{Q}}{\cancel{Q}C} = \frac{\cancel{Q}}{\cancel{Q}C_1} + \frac{\cancel{Q}}{\cancel{Q}C_2} + \frac{\cancel{Q}}{\cancel{Q}C_3}$$

$$\frac{1}{C} = \frac{1}{C_1} + \frac{1}{C_2} + \frac{1}{C_3}$$

To summarise for a **series** circuit:

$$U = U_1 + U_2 + U_3$$

$$Q = Q_1 = Q_2 = Q_3$$

$$\frac{1}{C} = \frac{1}{C_1} + \frac{1}{C_2} + \frac{1}{C_3}$$

Worked example 7   Three capacitors, of 25 µF, 50 µF and 75 µF, are connected in series across a 100-V D.C. supply. Calculate each of the following:

a   the circuit's total capacitance
b   the total charge
c   the voltage drop across each capacitor
d   the total energy stored.

Solution   As always, you should start by sketching the circuit (Figure 17.33), and inserting all the values given in the question.

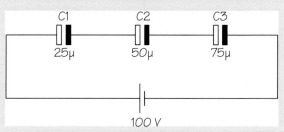

Figure 17.33

a $$\frac{1}{C} = \frac{1}{C_1} + \frac{1}{C_2} + \frac{1}{C_3}$$

$$\frac{1}{C} = \frac{1}{25} + \frac{1}{50} + \frac{1}{75}$$

$$\frac{1}{C} = \frac{6+3+2}{150} = \frac{11}{150}$$

$$C = \frac{150}{11} = 13.64 \text{ µF (Answer a.)}$$

b $$Q = CE = 13.64 \times 10^{-6} \times 100 = 1.364 \text{ mC}$$
(Answer b.)

c $$U_1 = \frac{Q}{C_1} = \frac{1.364 \times 10^{-3}}{25 \times 10^{-6}} = 54.56 \text{ V (Answer c.1)}$$

$$U_2 = \frac{Q}{C_2} = \frac{1.364 \times 10^{-3}}{50 \times 10^{-6}} = 27.28 \text{ V (Answer c.2)}$$

$$U_3 = \frac{Q}{C_3} = \frac{1.364 \times 10^{-3}}{75 \times 10^{-6}} = 18.19 \text{ V (Answer c.3)}$$

check your answer:

$$\left(U_1 + U_2 + U_3 = 54.56 + 27.28 + 18.19 \approx 100 \text{ V}\right)$$

d $W = \dfrac{1}{2}CE^2$

$$= \frac{1}{2} \times 13.64 \times 10^{-6} \times 100^2 = 68200 \times 10^{-6}$$

$$= 68.2 \text{ mJ (Answer)}$$

Capacitors in parallel

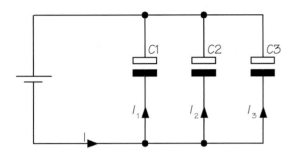

Figure 17.34

In accordance with Kirchhoff's Current Law, the sum of the individual branch currents will equal the supply current:

$$I = I_1 + I_2 + I_3 \qquad \text{—(equation 8)}$$

This time, the total charge must be the sum of the negative charge accumulated on the right-hand plate of each of the individual capacitors:

$$Q = Q_1 + Q_2 + Q_3 \qquad \text{—(equation 9)}$$

Since charge is the produce of capacitance and voltage, we can rewrite equation (9) as follows:

$$CE = C_1U + C_2U + C_3U$$

Since the supply voltage is common to each branch of a parallel circuit, we can now divide throughout by voltage:

$$\frac{C\cancel{E}}{\cancel{E}} = \frac{C_1\cancel{U}}{\cancel{U}} + \frac{C_2\cancel{U}}{\cancel{U}} + \frac{C_3\cancel{U}}{\cancel{U}}$$

$$C = C_1 + C_2 + C_3$$

To summarise for a **parallel** circuit:

$$I = I_1 + I_2 + I_3$$
$$Q = Q_1 + Q_2 + Q_3$$
$$C = C_1 + C_2 + C_3$$

Worked example 8 Three capacitors, of 25 µF, 50 µF and 75 µF, are connected in parallel across a 100-V D.C. supply. Calculate each of the following:

a the circuit's total capacitance
b the charge on each capacitor
c the voltage drop across each capacitor
d the energy stored on capacitor C_1.

Solution As always, you should start by sketching the circuit (see Figure 17.35), and inserting all the values given in the question.

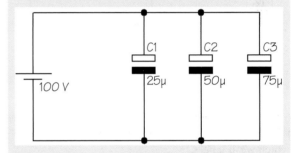

Figure 17.35

a $C = C_1 + C_2 + C_3 = 25 + 50 + 75 = 150 \text{ µF}$
(Answer a.)

b $Q = CU = 150 \times 10^{-6} \times 100 = 15000 \times 10^{-6}$
$= 15 \text{ mC (Answer b.)}$

c $Q_1 = C_1U = 25 \times 10^{-6} \times 100 = 2.5 \text{ mC}$
(Answer c.1)
$Q_2 = C_2U = 50 \times 10^{-6} \times 100 = 5.0 \text{ mC}$
(Answer c.2)
$Q_3 = C_3U = 75 \times 10^{-6} \times 100 = 7.5 \text{ mC}$
(Answer c.3)

check your answer:

$$\left(Q_1 + Q_2 + Q_3 = 2.5 + 5.0 + 7.5 = 15.0 \text{ mC}\right)$$

d $W = \dfrac{1}{2}C_1U^2$

$$= \frac{1}{2} \times 25 \times 10^{-6} \times 100^2 = 125 \text{ mJ (Answer)}$$

Capacitors in series-parallel

We resolve series-parallel capacitive circuits in much the same way we learnt to logically solve resistors in series-parallel. By way of example, let's solve the series-parallel circuit shown in Figure 17.36.

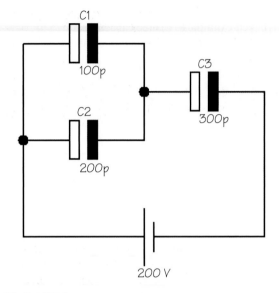

Figure 17.36

We start by finding the equivalent capacitance of the two capacitors, C_1 and C_2 that are connected in parallel, which we'll label C_A:

$$C_A = C_1 + C_2 = 100 + 200 = 300 \text{ pF}$$

This leaves C_A and C_3 in series, from which we can find the equivalent capacitance of the complete circuit, C:

$$\frac{1}{C} = \frac{1}{C_A} + \frac{1}{C_3} = \frac{1}{300} + \frac{1}{300} = \frac{2}{300}$$

$$\therefore C = \frac{300}{2} = 150 \text{ pF} \quad \text{(Answer)}$$

Now, let's find the total charge of the circuit:

$$Q = CU = 150 \times 10^{-12} \times 200 = 30 \times 10^{-9} \text{ C} \quad \text{(Answer)}$$

This charge will appear on capacitor C_3 and on the parallel part of the circuit. So, we can now work out the

voltage drops. First of all, the voltage drop, U_2, across C_3 can be found from:

$$U_2 = \frac{Q}{C_3} = \frac{30 \times 10^{-9}}{300 \times 10^{-12}} = 100 \text{ V} \quad \text{(Answer)}$$

So, $U_1 = E - U_2 = 200 - 100 = 100 \text{ V}$ (Answer)

Finally, we can determine the charge on each of the capacitors C_1 and C_2:

$$Q_1 = U_1 C_1 = 100 \times 100 \times 10^{-12} = 10 \times 10^{-9} \text{ C (Answer)}$$
$$Q_2 = U_1 C_2 = 200 \times 100 \times 10^{-12} = 20 \times 10^{-9} \text{ C (Answer)}$$

(We can confirm our answer from; $Q_1 + Q_2 = Q$)

Finally . . .

Now that you have completed this chapter, are you able to achieve the objectives or learning outcomes listed at the beginning of this chapter?

Ask yourself, 'Can I . . .'

1 explain the primary function of a capacitor.
2 describe the essential components of any capacitor.
3 describe the charging/discharging process of a capacitor.
4 describe what is meant by 'the charge on a capacitor'.
5 describe how electric charge behaves with capacitors connected in series.
6 recognise the circuit symbols for different types of capacitor.
7 state the unit of measurement for capacitance.
8 list the factors that affect capacitance, and their relationship.
9 describe how a capacitor's dielectric affects capacitance of that capacitor.
10 describe the relationship between absolute permittivity, the permittivity of free space, and relative permittivity.
11 determine the energy stored by a capacitor.
12 describe the basic construction of practical fixed-value and variable-value capacitors.
13 solve simple problems on the time constant of a resistive-capacitive circuit.
14 solve simple series, parallel and series-parallel capacitive circuits.

Online resources

The companion website to this book contains further resources relating to this chapter. The website can be accessed via the following link:
www.routledge.com/cw/waygood

Chapter 18

Introduction to alternating current

On completion of this chapter, you should be able to:

1 explain how the voltage induced into a conductor varies according to the angle at which the conductor moves through a magnetic field.
2 apply Fleming's Right-Hand Rule to determine the direction of the voltage induced into a conductor moving through a magnetic field.
3 explain why a conductor or loop, rotating within a magnetic field, generates a sinusoidal voltage.
4 explain each of the following terms specifying, where applicable, their SI units of measurement:
   a amplitude
   b instantaneous value
   c period
   d wavelength
   e cycle
   f frequency.
5 given the peak value of an a.c. voltage or current, calculate its r.m.s. (or 'effective') value, and *vice versa*.

Introduction

Electricity generation, transmission and distribution systems are almost* exclusively **alternating current** (a.c.) systems. The primary reason for this is because high voltages are essential for the transmission and distribution of electrical energy, and a.c. voltages can be easily and efficiently changed using transformers.

*High-voltage, **direct current** transmission systems do exist and are used for very long transmission lines (to reduce losses), to interconnect independent grid systems (to avoid frequency synchronisation difficulties), and for undersea cables (to avoid large capacitive currents).

For any given load, the *higher* the supply voltage, the *lower* the resulting load current, that is:

$$I_{load} = \frac{P_{load}}{E_{supply}}$$

So, **high voltages** are *essential* for the transmission and distribution of electrical energy if we are to avoid (a) enormous voltage drops along the lines, (b) conductors with unrealistically high cross-sectional areas and weights, and (c) unacceptably high line losses.

For these reasons, in the UK, electricity is generated at 11 kV or 25 kV (practical alternators are limited to around 30-kV output voltage), but is transmitted at 400 kV and gradually reduced, at load centres (typically located at the outskirts of towns and cities), through 275 kV, 132 kV, 33 kV, 11 kV, until it finally reaches consumers at 400/230 V. The UK's generation/transmission/distribution system is illustrated in Figure 18.1.

As we shall learn in a later chapter, **three-phase** systems are used throughout the transmission and distribution system, both high voltage and low voltage, because for a given load, less volume of copper is required

An Introduction to Electrical Science, Waygood, ISBN 9780415810029, 2013. © Taylor & Francis

Transmission & Distribution Networks

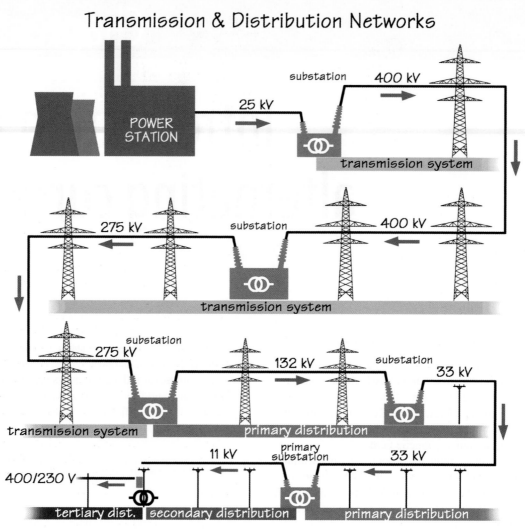

Figure 18.1

for a three-phase system than an equivalent single-phase system, making three-phase more economical compared with single phase.

Generation of a sine wave

From earlier chapters, you will recall that whenever there is relative movement between a conductor and a magnetic field, a potential difference (voltage) is *induced* into that conductor.

Faraday's Law tells us that the *magnitude* of that induced voltage is *directly proportional to the rate at which the lines of magnetic flux are cut by the conductor*. For any given velocity, the maximum rate will occur when the conductor cuts the flux perpendicularly (at right angles); on the other hand, if the conductor runs

parallel with the flux, then no flux is cut, and no voltage will be induced into the conductor.

If the conductor cuts the flux at an angle, θ (pronounced *'theta'*) at a velocity represented by vector v, then we must find the *perpendicular* component of that velocity vector, v' – as represented by the broken line in Figure 18.2.

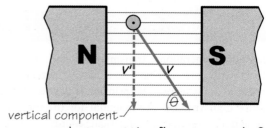

conductor cutting flux at an angle θ

Figure 18.2

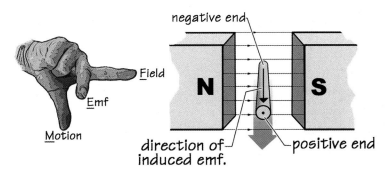

negative end

Field

Emf

Motion

direction of induced emf.

positive end

Figure 18.3

As we learnt in the chapter on *electromagnetic induction*, the general equation for the voltage induced into a conductor that is cutting lines of magnetic flux at an angle, θ, is give by:

$$E = Blv \sin \theta$$

where:

E = potential difference (V)

B = flux density (T)

l = length of conductor (m)

v = velocity of conductor (m/s)

θ = angle cutting flux (°)

The direction of the induced voltage can be deduced by applying **Fleming's Right-Hand Rule**. For example, in Figure 18.3, when the conductor moves vertically downwards through the magnetic field, Fleming's Right-Hand Rule tells us that the positive end of the conductor is at the nearest end, and (conventional) current will enter the external circuit from that end.

Alternators: basic constructional features

A **simple a.c. generator**, or **alternator**, consists of a single loop of wire (called an **armature**), pivoted so that it can rotate within a magnetic field – as illustrated in Figure 18.4.

When it is rotated by an external force, provided by a **prime mover** (steam turbine, a diesel engine, a water turbine, a wind turbine, etc.), a potential difference is induced into each side of the loop. If we apply Fleming's Right-Hand Rule to each side of the loop, we will see that the potential difference induced into the left-hand side will act *towards* us, while the potential difference induced into the right-hand side will act *away* from us – as they are in series with each other, the two act to reinforce each other.

So if the voltage induced into one side is *e* volts, then the voltage induced into the complete loop must be **2e** volts.

In a more practical generator, of course, the armature will be a coil, *not* a single loop. So, for an armature winding with N turns, this voltage will be increased N times. However, for the sake of simplicity, we'll continue for now to illustrate it as a single loop.

When this single loop generator is connected to an external circuit, the resulting (conventional) load current will flow through the armature in the counterclockwise direction shown in Figure 18.4.

Connecting the rotating loop of our simple generator to its external load is achieved using a pair of **slip rings** and spring-loaded **carbon brushes**, as illustrated in Figure 18.5.

One end of the armature winding is connected to one slip ring, while the other end is connected to the second slip ring. As the generator's loop rotates past the mid-way point (i.e. when running parallel with the flux), the voltages induced into each side of the loop reverse direction. So, the output voltage from a generator equipped with slip rings also changes direction, and produces (in theory at least) a *sinusoidal alternating voltage*.

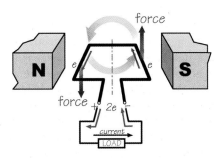

force

force

current

LOAD

Figure 18.4

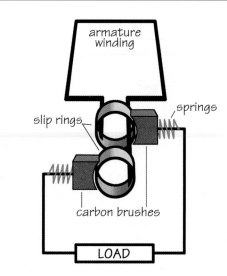

armature
winding

slip rings

springs

carbon brushes

LOAD

Figure 18.5

Action and reaction

In the earlier chapter on *electromagnetism*, we learnt that a current-carrying conductor placed between the poles of a magnet will *experience a force acting at right angles to the field's magnetic flux*. In the same chapter, we also learnt that this force is *proportional to the current* – so, the greater the current, the greater the force.

If we were to apply Fleming's Left-Hand Rule to our simple **generator** when it is supplying a load current, we would find that the resulting force due to this 'motor action' always acts *against the direction in which the conductor is moving!*

In other words, whenever the generator is supplying current to the load, the armature loop behaves *both as a generator and as a motor simultaneously!* The greater the load current, the greater will be the 'motor effect' which acts to oppose the movement of the armature loop. This 'action' and 'reaction' effect plays a very important part in the operation of a generator, requiring it to *react to changes in load* – and explains why, as the load increases (i.e. draws more current), the prime mover 'tends' to slow down and when the load decreases, the speed of the prime mover 'tends' to increase.

In the last sentence, we've deliberately enclosed the word 'tends' in quotation marks for a good reason. This is because the speed of a practical generator's prime mover is actually controlled by a **governor**. So, if the generator does 'tend' to slow down or 'tend' to speed up, for the reasons described in the previous paragraph, then the governor will raise or lower the speed of the prime mover to compensate. For example, if the prime mover was a steam turbine, then the governor would control the steam inlet valve, increasing or decreasing the steam flow through the turbine.

This action/reaction behaviour explains why, when the electrical load demands more energy *from* the generator, more energy must be supplied *to* the generator by the prime mover to *drive* the generator. This is why, for example, leaving unnecessary electrical accessory-loads operating in a car (lights on in daylight, the radio when it is not being listened to, etc.) will actually increase its fuel consumption.

For the types of a.c. generator that supply very large currents and generated voltages (typical power station alternators supply currents of the order 1000 A at 11–25 kV), the use of slip rings is quite impractical (they would be unable to handle such large currents, and the heavy arcing involved would require regular shutdowns for maintenance).

To avoid this problem, the *armature windings are installed on the stator* (the stationary part of the machine) while the *field windings are installed on the rotor* (the rotating part of the machine). In other words, the construction of real alternators is *exactly* **opposite** *to that of our simple generator!*

Because only the field windings rotate, the necessary slip rings need only supply the relatively low current (typical power station alternators utilise field currents of the order 350 A at 400 V d.c.) to the windings that produce the magnetic field, while the fixed armature windings can supply the heavy current directly to the load. As a result, the arcing at the slip rings is insignificant, and the machine's maintenance cycle can be significantly increased and its downtime reduced.

Furthermore, practical alternators have **multiple pairs** of armature windings, not just *two* like our simple generator and, so, are called 'multiple-pole' machines. These machines have significantly more efficient magnetic circuits, but the number of poles affects the frequency of the output voltage – in the case of a **two-pole machine**, for example, *one revolution of the rotor will generate one complete cycle*; whereas in a **four-pole machine** *one revolution will generate two complete cycles*, and so on. Machines with a large number of poles, such as those driven by water turbines, can therefore run more slowly to provide the required output frequency.

The relationship between a machine's **speed**, **number of poles** and its output **frequency** is given by:

$$f = pn$$

where:

f = frequency (hertz)

p = number of *pairs* of poles

n = speed (revolutions per second)

Worked example 1   At what speed must a two-pole alternator run in order to generate a voltage at 50 Hz?

Solution   A two-pole machine has just one *pair* of poles, so:

$$n = \frac{f}{p} = \frac{50}{1} = 50 \text{ rev/s}$$

As it is more usual to express a machine's speed in *revolutions per minute*, so if the machine runs at 50 rev/s, then in one minute (60 s), it will need to run at:

$$n = 50 \times 60 = 3000 \text{ rev/min} \quad \text{(Answer)}$$

Interestingly, as two poles are obviously the *minimum* number of poles it is possible to have, *the highest speed at which a 50-Hz alternator can be allowed to operate is at 3000 rev/min.*

Worked example 2   The standard nominal mains frequency in Canada is 60 Hz. At what speed must a two-pole alternator run in order to generate a voltage at 60 Hz?

Solution   A two-pole machine has just one *pair* of poles, so:

$$n = \frac{f}{p} = \frac{60}{1} = 60 \text{ rev/s}$$

As it is more usual to express a machine's speed in *revolutions per minute*, so if the machine runs at 60 rev/s, then in one minute (60 s), it will need to run at:

$$n = 60 \times 60 = 3600 \text{ rev/min} \quad \text{(Answer)}$$

Worked example 3   At what speed must a six-pole alternator run in order to generate a voltage at 50 Hz?

Solution   A six-pole machine has three *pairs* of poles, so:

$$n = \frac{f}{p} = \frac{50}{3} = 16.67 \text{ rev/s}$$

As it is more usual to express a machine's speed in *revolutions per minute*,

$$n = 16.67 \times 60 = 1000 \text{ rev/min} \quad \text{(Answer)}$$

In the case of power station alternators, in order to maintain the strict frequency requirements laid down, in the United Kingdom, by the **Electricity, Safety, Quality and Continuity Regulations (2002)**, which specify that consumers are to be supplied at a frequency at **50 Hz ± 1%** (i.e. between 49.5 Hz and 50.5 Hz) over a 24-h period, the rotational speed of these alternators is strictly controlled. The machines' terminal voltage, therefore, *cannot be adjusted by changing the speed of the machine* but, instead, by adjusting the flux density of the field, by controlling the field current.

Generation of a sinusoidal voltage

Let's now return to our simple model of an a.c. generator, to remind ourselves of how an a.c. voltage can be generated in a rotating armature loop. In Table 18.1, a conductor follows the counterclockwise circular path, shown, cutting the magnetic flux set up from the north towards the south magnetic poles. We'll assume that, when the conductor cuts the flux at right angles, a voltage of 1 V is induced into the conductor.

So, as the conductor (or, in practice, a **loop**) moves through its circular path between the magnetic poles, the value of voltage induced into the conductor continually varies from zero, to some maximum value in one direction, then back to zero and, finally, to the same maximum value but in the opposite direction.

In practice, the armature conductors of an alternator are part of its stator and remain stationary, while the magnetic field rotates, being produced by a field winding wound around the rotor – i.e. exactly *opposite* to that illustrated in Table 18.1. However, this is very difficult to illustrate clearly, and the actual waveform is exactly the same as shown in the sequence of illustrations!

Let's assume, for a moment, that the *maximum* voltage induced into the conductor is just 1 V. Figure 18.19 shows what we will find if we draw the waveform to scale, and check the values of instantaneous voltages at, say, 30° intervals from 0° to 360°.

Table 18.1

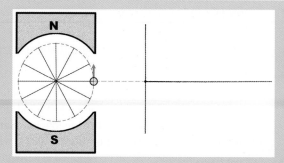

Figure 18.6

In this position, the conductor is moving parallel to the magnetic flux, so no voltage is induced into the conductor.

$$e = 1\sin\theta = 1\sin 0°$$
$$= 0\ \text{V}$$

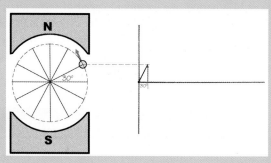

Figure 18.7

The conductor has moved along its circular path by 30°, and is cutting the flux at 30°, so a voltage is starting to be induced into the conductor.

$$e = 1\sin\theta = 1\sin 30°$$
$$= 0.5\ \text{V}$$

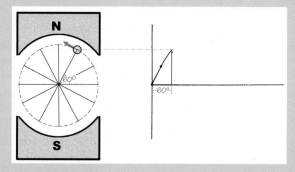

Figure 18.8

The conductor has now moved along its circular path at 60°, and is now cutting the flux by 60°, and an even greater voltage is induced into the conductor.

$$e = 1\sin\theta = 1\sin 60°$$
$$= 0.866\ \text{V}$$

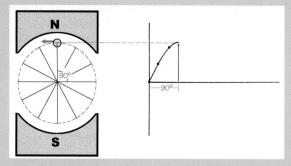

Figure 18.9

The conductor has now moved along its circular path by 90°, and is now cutting the flux at right angles, so the maximum voltage is induced into the conductor.

$$e = 1\sin\theta = 1\sin 90°$$
$$= 1.0\ \text{V}$$

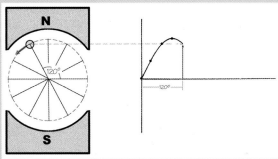

Figure 18.10

The conductor has now moved along its circular path by 120°, and is now cutting the flux at 60°, so the induced voltage is starting to fall.

$$e = 1\sin\theta = 1\sin 120°$$
$$= 0.866 \text{ V}$$

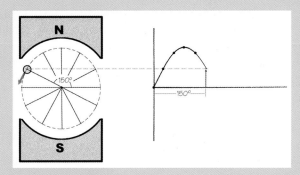

Figure 18.11

The conductor has now moved along its circular path by 150°, and is now cutting the flux at 30°, so the induced voltage has fallen further.

$$e = 1\sin\theta = 1\sin 150°$$
$$= 0.5 \text{ V}$$

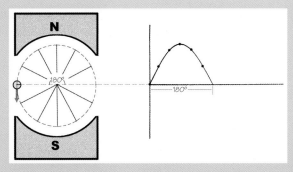

Figure 18.12

The conductor has now moved along its circular path by 180° and, once again, the conductor is moving parallel to the flux, so no voltage is induced into it.

$$e = 1\sin\theta = 1\sin 180°$$
$$= 0 \text{ V}$$

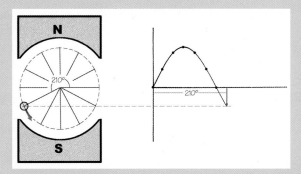

Figure 18.13

The conductor has now moved along its circular path by 210°, and the conductor is cutting the flux at 30° – but this time, in the opposite direction. So the induced voltage is now acting in the opposite sense.

$$e = 1\sin\theta = 1\sin 210°$$
$$= -0.5 \text{ V}$$

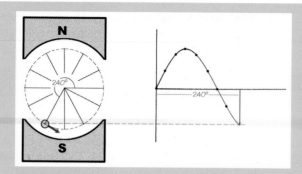

Figure 18.14

The conductor has now moved along its circular path by 240°, and the conductor is cutting the flux at 60°, so the induced voltage is now increasing in the opposite sense.

$$e = 1\sin\theta = 1\sin 240°$$
$$= -0.866 \text{ V}$$

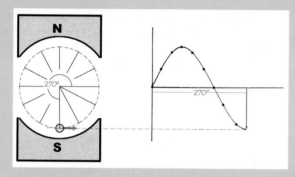

Figure 18.15

The conductor has now moved along its circular path by 270°, and is now cutting the flux at right-angles (but in the opposite direction compared with the 90° position), so the maximum voltage is induced into the conductor.

$$e = 1\sin\theta = 270\sin 30°$$
$$= -1.00 \text{ V}$$

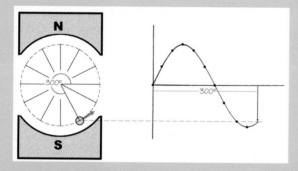

Figure 18.16

The conductor has now moved along its circular path by 300°, and is now cutting the flux at 60°, so the induced voltage is starting to fall again.

$$e = 1\sin\theta = 1\sin 300°$$
$$= -0.866 \text{ V}$$

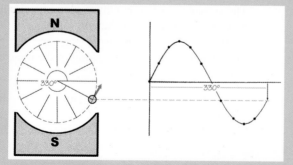

Figure 18.17

The conductor has now moved along its circular path by 330°, and is now cutting the flux at 30°, so the induced voltage is starting to fall further.

$$e = 1\sin\theta = 1\sin 330°$$
$$= -0.5 \text{ V}$$

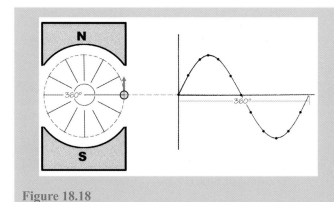

The conductor has now moved through 360° and is, once again, moving parallel to the flux and no voltage is induced into the conductor.

$$e = 1\sin\theta = 1\sin 360°$$
$$= 0.00 \text{ V}$$

Figure 18.18

The angle, expressed in electrical degrees, of any of these instantaneous values of voltage, is termed the **displacement angle** *(θ)*. These values of each of these instantaneous voltage correspond to 1 V times the sine of the corresponding displacement angle, i.e:

*instantaneous voltage = maximum voltage × **sine** of the displacement angle*

instantaneous voltage = maximum voltage × sine of the displacement angle

$$e = E_{max} \sin\theta$$

For this reason, the generated waveform is called a **sine wave**.

So, the **instantaneous voltage** (symbol: *e*) at any point along the sine wave, then, is given by the equation:

$$e = E_{max} \sin\theta$$

where:

$$e = \text{instantaneous voltage}$$
$$E_{max} = \text{peak voltage}$$
$$\theta = \text{displacement angle}$$

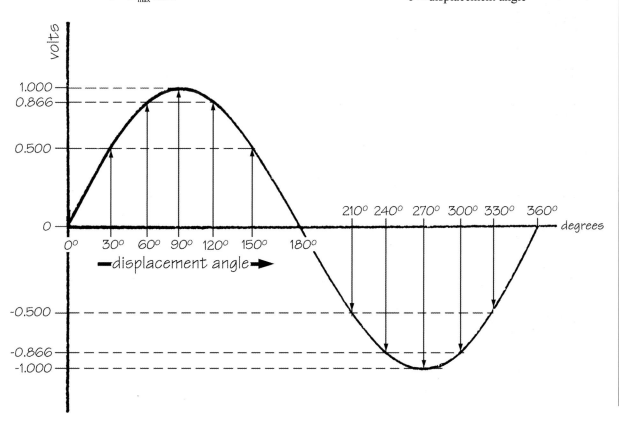

Figure 18.19

Since current is directly proportional to voltage, so we can also state:

$$i = I_{max} \sin \theta$$

where:

i = instananeous current

I_{max} = peak current

θ = displacement angle

Worked example 4   Calculate the instantaneous voltages at displacement angles of 15°, 40°, 70°, 100° and 240°, if the peak value of voltage is 10 V.

Solution

at 15°: $e = E_{max} \sin \theta = 10 \times \sin 15° = 10 \times 0.259$
$$= 2.59 \text{ V } (Answer)$$

at 40°: $e = E_{max} \sin \theta = 10 \times \sin 40 = 10 \times 0.643$
$$= 6.43 \text{ V } (Answer)$$

at 70°: $e = E_{max} \sin \theta = 10 \times \sin 70 = 10 \times 0.94$
$$= 9.40 \text{ V } (Answer)$$

at 100°: $e = E_{max} \sin \theta = 10 \times \sin 100 = 10 \times 0.985$
$$= 9.85 \text{ V } (Answer)$$

at 240°: $e = E_{max} \sin \theta = 10 \times \sin 230 = 10 \times (-0.766)$
$$= -7.66 \text{ V } (Answer)$$

Terminology

The **terminology** shown in Table 18.2 is used to describe any sine wave (voltage or current).

Table 18.2

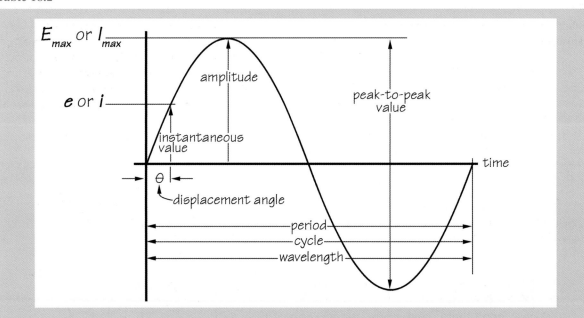

Figure 18.20

Amplitude:	The peak value of an a.c. waveform, in either the positive *or* in the negative sense. Symbols: E_{max} or I_{max}.
Peak-to-peak value:	Twice the amplitude of an a.c. waveform. Symbols: E_{p-p} or I_{p-p}.
Instantaneous value:	The value of voltage or current, at any specified displacement angle, during a complete cycle. Symbols: e or i.

Displacement angle:	The displacement of any instantaneous value of voltage or current, expressed in electrical degrees and measured from the origin of the sine wave. Symbol: θ (the Greek letter, '*theta*').
Period, or **Periodic time:**	The time taken to complete one cycle, measured in seconds. Symbol: T.
Wavelength:	The distance between two displacements of the same phase, measured in metres. Symbol: λ (the Greek letter, '*lambda*').
Cycle:	One complete set of changes in the values of a recurring variable quantity.
Frequency:	The number of complete cycles per unit time, measured in hertz. Symbol: f, where: $$f = \frac{1}{T}$$

Measuring sinusoidal values

Because the value of a sinusoidal voltage or current is continually changing in both magnitude and direction, how do we assign any meaningful value to them? Well, we *could* simply use the *peak value* – but as the waveform only reaches this value twice during any complete cycle, it is hardly representative of the entire waveform's variation.

What about the *average* value? Well, if we work out the waveform's average value over a complete cycle, it will work out at *zero* – because the average value over the *positive* half-cycle will be cancelled out by its average value over the *negative* half-cycle!

> In fact, average values are sometimes assigned to sine waves but only over a half cycle, so are usually applied to rectified a.c. values.

If neither the **peak value**, nor the **average value**, represents a meaningful way of measuring a sinusoidal voltage or current over a complete cycle, *how do we proceed?*

The problem is a little like trying to compare the speed of a reciprocating saw to that of a circular saw! The reciprocating saw blade is continuously oscillating up and down, whereas the circular saw is continuously rotating – so it's rather difficult to compare the two in terms of their 'speeds', as they are measured in completely different ways. Instead, we could compare them in terms of *the rate at which they each cut timber*. In other words, we can say that if the reciprocating saw

cuts timber at exactly the same rate as the circular saw then its reciprocating speed must be '*equivalent*' to (rather than the '*same*' as!) the rotational speed of the circular saw.

We do a similar thing when we measure a.c; that is, *we compare the* **work** *an a.c. current does with the* **work** *that a d.c. current does*. We know that if a voltage is applied to a resistive circuit, the resulting current will cause the temperature of the resistor to rise, and this will happen *regardless* of whether the current is direct current or alternating current, so we make use of this property.

Let's examine the simple experiment shown in Figure 18.21.

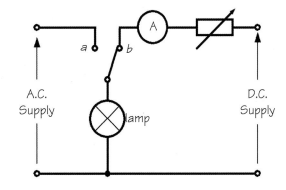

Figure 18.21

With the switch in position *a*, the incandescent lamp is connected to an a.c. supply, and the brightness (or, more accurately, 'luminance') of the lamp is measured using an appropriate instrument (a 'luminance meter'). Next, the switch is moved to position *b*, and the variable resistor is adjusted until *exactly the same brightness is achieved*. The ammeter now indicates the value of d.c. current that has produced *exactly the same brightness* (in other words, produces *exactly the same heating effect*) as that produced by the a.c. current. So if the ammeter indicates a direct current of, say, 0.5 A when identical brightness is achieved, then the effective value of a.c. is also considered to be 0.5 A.

> So, we can say that *'the **effective value** of an alternating current is measured in terms of the direct current that produces exactly the same heating effect in the same resistance'*.

If we were now to compare the a.c. current's effective value with its peak value (using an oscilloscope), we would find (for a sine wave) that the effective value is equal to 0.707× its peak value.

So, the relationship between the effective value of this a.c. current and its peak value would be:

$$I_{effective} = 0.707 \times I_{max}$$

Since voltage and current are directly proportional to each other:

$$E_{effective} = 0.707 \times E_{max}$$

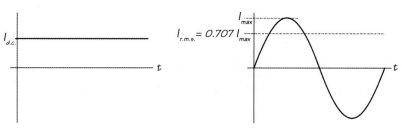

$I_{r.m.s.}$ **does exactly the same work as** $I_{d.c.}$
so, $I_{r.m.s.}$ **is exactly equivalent to** $I_{d.c.}$

Figure 18.22

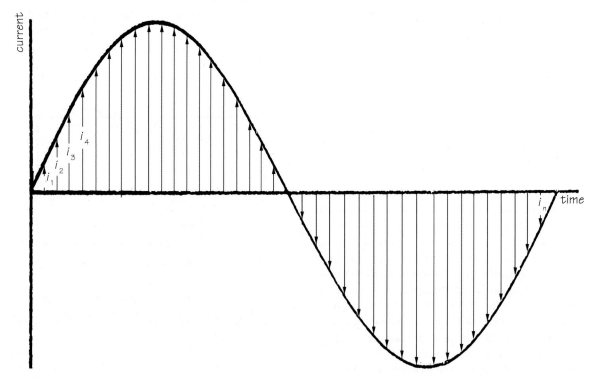

Figure 18.23

The **effective value** of an a.c. current or voltage is more commonly known as its '**r.m.s. value**'. The abbreviation '**r.m.s.**' stands for '**root-mean-square**', a term derived from the *mathematical proof* for obtaining this value of 0.707. We don't need to go into this proof in any great detail in this chapter, but you should be aware of *why* the term **root-mean-square** is used, and *where* it comes from.

In the above explanation of effective value, we said it is based on the *work done* by a current. We know that the equation for work is: $W = I^2Rt$. So, if we start by dividing the a.c. current waveform up into lots and lots of *instantaneous values of current*: i_1, i_2, i_3, i_4, etc. (the more the better!) and apply this equation to each value of instantaneous current (Figure 18.23), we'll end up with the following equation:

$$I_{effective}^2 Rt = i_1^2 Rt + i_2^2 Rt + i_3^2 Rt + i_4^2 Rt + etc.$$

(for the whole waveform)

By dividing throughout by Rt, we can eliminate both R and t (both being constants) from the equation:

$$I_{effective}^2 \frac{\cancel{Rt}}{\cancel{Rt}} = i_1^2 \frac{\cancel{Rt}}{\cancel{Rt}} + i_2^2 \frac{\cancel{Rt}}{\cancel{Rt}} + i_3^2 \frac{\cancel{Rt}}{\cancel{Rt}} + i_4^2 \frac{\cancel{Rt}}{\cancel{Rt}} + etc.$$

(for the whole waveform)

$$I_{effective}^2 = i_1^2 + i_2^2 + i_3^2 + i_4^2 + etc. \text{ (for the whole waveform)}$$

Next, we find the **mean** (average) for all the individual **squared** instantaneous currents:

$$I_{effective}^2 = \frac{i_1^2 + i_2^2 + i_3^2 + i_4^2 + etc. \text{ (for the whole waveform)}}{n}$$

(where n represents the number of individual instantaneous currents)

Finally, to eliminate the squares, we can find the **square root** of both sides of the equation:

$$I_{effective} = \sqrt{\frac{i_1^2 + i_2^2 + i_3^2 + i_4^2 + etc. \text{ (for the whole waveform)}}{n}}$$

So, as you can now see, the **effective value of current** is equal to the **square root** of the **mean** (i.e. 'average') of the **squares** of each of the individual instantaneous currents. Hence, the term: '**root-mean-square**'. If you were to insert actual values into the above equation then, for a *sine wave* it would *always* work out to be **0.707 I_{max}** (it would be different for other shaped waveforms).

In practice, this calculation is performed using calculus, rather than by the technique described above – but this is beyond the scope of this text.

To summarise:

> The **effective**, or **r.m.s. value** of a.c. current is given by:
>
> $$I_{rms} = 0.707\ I_{max}$$

And, since voltage and current are proportional to each other:

> The **effective**, or **r.m.s. value** of a.c. voltage is given by:
>
> $$E_{rms} = 0.707\ E_{max}$$

It's important to understand that **a.c. currents and voltages are *always* expressed in effective, or r.m.s., values *unless otherwise specified***. For example, voltmeters and ammeters are calibrated to indicate r.m.s. values. So, we do *not* normally need to add the subscript 'rms' to the symbols for voltage or current – in other words, we usually simply write '*E*' or '*I*', rather than 'E_{rms}' or 'I_{rms}'.

> **Important!**
> Unless otherwise specified, a.c. values are ***always*** quoted in r.m.s. values, and ***all*** a.c. ammeters and voltmeters are calibrated to output their readings in r.m.s. values. Because of this, you will rarely see the subscript 'r.m.s.' used, unless clarity is necessary.

Worked example 5 What is the peak value of a 230-V a.c. residential supply?

Solution   (For clarity, we'll retain the subscript 'r.m.s.' in this example.)

Since *all* a.c. values are normally expressed as r.m.s. values, then 230 V is an r.m.s. value, so:

since $E_{rms} = 0.707\ E_{max}$

then $E_{max} = \dfrac{E_{rms}}{0.707} = \dfrac{230}{0.707} \approx 325$ V (Answer)

Worked example 6 Using an oscilloscope, we measure the peak value of a sine-wave voltage across a resistor as 250 mV. What is its r.m.s. value?

Solution (Again, for clarity, we'll retain the 'r.m.s.' subscript in this example.)

$$E_{rms} = 0.707\ E_{max}$$
$$= 0.707 \times (250 \times 10^{-3}) \simeq 177 \times 10^{-3}\ V$$
$$\text{or} \quad 177\ mV\ (\text{Answer})$$

Representing sinusoidal waveforms with phasors

You are probably already familiar with the concept of **vectors**. Quantities such as **force** or **velocity** can be represented by means of a **vector** – where the *length* of the vector represents the *magnitude* of the force or velocity, and the *direction* in which the vector points represents the *direction* in which the force or velocity acts. Two or more forces may be added or subtracted, by adding or subtracting their corresponding vectors – this can either be done graphically, to scale, or by applying the rules of simple geometry and trigonometry.

'**Phasors**' are, to electrical engineering, what **vectors** are to mechanical engineering. However, while the *length* of a phasor represents the *magnitude* (normally, expressed in terms of r.m.s. values) of an alternating voltage or alternating current, its '*direction*' (or, more accurately, its 'angle') doesn't represent direction but, rather, the *time displacement* of that voltage or current –

expressed in terms of *displacement angle measured in a counterclockwise direction*. For this reason, phasors have been described as '**rotating vectors**'.

Although, in Figure 18.24, the length of the phasor is equal to the peak value of voltage, *in practice, they usually represent the root-mean-square of that voltage.*

The phasor's arrow head serves two functions: (1) it indicates the 'rotating' end of the phasor, and (2) it helps distinguish phasors from each other, when different phasors lie at the same angle.

Phasors that represent the *same* quantities (voltages *or* currents) may be added or subtracted in exactly the same way as vectors are added or subtracted, and by using exactly the same techniques – graphically or mathematically.

Phasors that represent *different* quantities (voltages *and* currents) *cannot* be added or subtracted, but the angle between them represents the phase relationship between those quantities.

Fortunately, as you will see, *phasors are much easier to use than they are to explain!*

A **phasor** is an arrowed line whose *length* represents the *magnitude* of an alternating voltage or current and whose *angle*, measured counterclockwise from the real positive axis, represents the *displacement angle* of that voltage or current.

A phasor's arrow head represents the 'rotating' end of that phasor and *not* its direction.

The rules of **vector addition** and **subtraction** apply to phasors representing like quantities (i.e. two or more voltage phasors, or two or more current phasors).

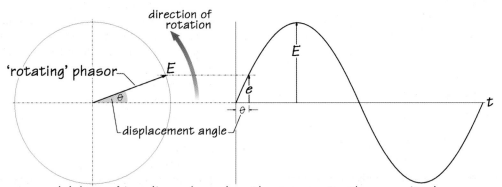

A 'phasor' is a line, whose length represents the magnitude of an a.c. voltage or current, and whose angle, measured counterclockwise from the real positive axis, represents its displacement angle.

Figure 18.24

As you will discover in the following chapters, it's very much easier to *use* phasors than it is to describe what they are!

All you need to know to understand alternating current

The chapters that follow explain how phasors can be used to understand (and solve problems on) the behaviour of alternating-current circuits. If you

- are already able to add or subtract vectors
- understand how to apply Pythagoras's Theorem
- can apply the trigonometric ratios, sine, cosine and tangent, to simple right-angled triangles,

... then you are already well on your way to being able to solve most problems relating to behaviour of series, parallel and series-parallel single-phase a.c. circuits, as well as three-phase a.c. circuits!

And you will be able to do all of this *without having to remember more than just **two** of the many equations associated with a.c. theory!*

Let's expand on the promise made in the last sentence of the previous paragraph.

To understand the behaviour of a.c. circuits, both single-phase and three-phase, you will need to:

- recall, and be able to apply, Pythagoras's Theorem.
- recall, and be able to apply to right-angled triangles, each of the following trigonometric ratios:

$$\sin\phi = \frac{\text{opposite}}{\text{hypotenuse}}; \quad \cos\phi = \frac{\text{adjacent}}{\text{hypotenuse}};$$

$$\tan\phi = \frac{\text{opposite}}{\text{adjacent}}$$

- remember, and be able to apply, the following two equations:

$$X_L = 2\pi f L \quad \text{and} \quad X_C = \frac{1}{2\pi f C}$$

- remember, and be able to apply, the mnemonic **CIVIL** – but more of this, later.

Really! That's *all* there is to it! Well . . . not quite! Of course, there really *is* more to a.c. theory than just that – but armed with only the knowledge described above, you will soon be well on your way to gaining a tremendous understanding of the behaviour of single-phase and three-phase a.c. circuits, and you will quickly be able to solve a great many of the problems that you will be faced with both in the classroom and on the job.

Tip! As you progress through the remaining chapters of this book, you'll come across an increasing number of equations. *Don't bother to try to remember them!* You don't actually *need* to remember them, because *you are going to learn how to derive them.* Once you know how to derive them, you'll *never* need to remember any individual equations. However, after a while, you'll probably find yourself remembering these equations without even realising it!

Finally . . .

Now that you have completed this chapter, are you able to achieve the objectives or learning outcomes listed at the beginning of this chapter?

Ask yourself, 'Can I . . .'

1. explain how the voltage induced into a conductor varies according to the angle at which the conductor moves through a magnetic field.
2. apply Fleming's Right-Hand Rule to determine the direction of the voltage induced into a conductor moving through a magnetic field.
3. explain why a conductor or loop, rotating within a magnetic field, generates a sinusoidal voltage.
4. explain each of the following terms specifying, where applicable, their SI units of measurement:
   a amplitude
   b instantaneous value
   c period
   d wavelength
   e cycle
   f frequency.
5. given the peak value of an a.c. voltage or current, calculate its r.m.s. (or 'effective') value, and *vice versa.*

Online resources
The companion website to this book contains further resources relating to this chapter. The website can be accessed via the following link:
www.routledge.com/cw/waygood

Chapter 19

Series alternating-current circuits

On completion of this chapter, you should be able to:

1 sketch a waveform for each of the following, showing the phase relationship between the current and supply voltage:
   a purely resistive circuit
   b purely inductive circuit
   c purely capacitive circuit
   d series *R-L* circuit
   e series *R-C* circuit.
2 state the phase relationship between the current and supply voltage for
   a a purely resistive circuit
   b a purely inductive circuit
   c a purely capacitive circuit.
3 sketch the phasor diagram representing a
   a purely resistive circuit
   b purely inductive circuit
   c purely capacitive circuit
   d series *R-L* circuit
   e series *R-C* circuit
   f series *R-L-C* circuit.
4 develop an impedance diagram for a
   a series *R-L* circuit
   b series *R-C* circuit
   c series *R-L-C* circuit.
5 from impedance diagrams, derive equations for resistance, inductive reactance, capacitive reactance and impedance, in terms of voltages and currents.
6 state the equation for inductive reactance, in terms of inductance and frequency.
7 state the equation for capacitive reactance, in terms of capacitance and frequency.

8 explain what is meant by the term 'series resonance'.
9 list the effects of series resonance.
10 solve problems on series a.c. circuits, including series-resonant circuits.

Introduction

Important! The key to understanding and solving a.c. circuits, whether they are series circuits, parallel circuits or even three-phase circuits, is your ability to sketch a phasor diagram which represents that circuit and, then, use this phasor diagram to generate all the equations you need, simply by treating it as a simple exercise in geometry or trigonometry.

All 'real' alternating-current circuits exhibit combinations of **resistance** (symbol: *R*), **inductance** (symbol: *L*) and **capacitance** (symbol: *C*). The amount of each of these quantities appearing in any particular circuit is determined by the *configuration* and *design characteristics* of that particular circuit.

For example, *all* conductors (by virtue of their length, cross-sectional area and resistivity) exhibit *natural* amounts of resistance. Overhead power lines, due to the configuration of their individual conductors, exhibit relatively high *natural* values of inductance, as well as some capacitance and resistance. Underground cables, because of the closeness of their individual conductors, exhibit relatively high *natural*

An Introduction to Electrical Science, Waygood, ISBN 9780415810029, 2013, © Taylor & Francis

values of capacitance as, well as some inductance and resistance.

'Real' a.c. circuits, then, are relatively complicated because they contain a *combination* of resistance, inductance and capacitance. So, in order to understand the behaviour of an a.c. circuit, it is necessary to start by considering how an 'ideal' circuit would behave. In this context, an 'ideal' circuit is one which is '**purely resistive**', '**purely inductive**' or '**purely capacitive**'.

'Ideal' circuits *only exist theoretically*. But if we are able to understand how these relatively simple, ideal circuits would behave if they *did* exist, then we will be able to move on to combine these behaviours in order to understand how *real* and *more complicated* circuits behave.

> **Important**! In the circuit diagrams that follow, it is important to understand that the symbols represent quantities *not* components. That is, **resistance** *not* resistors; **inductance** *not* inductors; **capacitance** *not* capacitors.

Throughout the rest of this chapter, the voltages and currents we will be referring to are 'phasor' quantities. In order to remind ourselves of this, the symbols for voltage and current will be shown with small 'bars' above them (i.e. $\bar{E}, \bar{U}_R, \bar{U}_L, \bar{U}_C$ and $\bar{I}$). This is one way of indicating that these are phasor quantities. It's unncessary to do this when labelling phasor diagrams as the quantities involved are obviously phasors.

Resistance, inductive reactance, capacitive reactance and impedance are *not* phasor quantities and, so, will not have bars placed above their symbols.

Purely resistive circuit

So let's start with the simplest of all 'ideal' circuit: the **purely resistive circuit**.

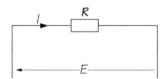

Figure 19.1

In a purely resistive circuit, the current $(\bar{I})$ and the potential difference $(\bar{E})$ across that resistance are said to be **in phase** with each other – i.e. the peak and zero

points of their two separate waveforms correspond exactly throughout each complete cycle. This is exactly how we would instinctively expect a circuit to behave but, as we shall learn, this is *not* the case with other types of circuits.

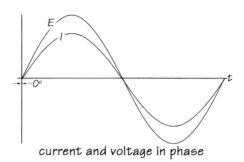

current and voltage in phase

Figure 19.2

The waveforms shown in Figure 19.2 can be represented by means of what is known as a '**phasor diagram**', with the current and voltage phasors each lying alongside each other. In this case, it's usual to draw them both along the horizontal positive axis (i.e. horizontally, pointing to the right). In Figure 19.3, and in those that follow, the small curved arrow to the right is used simply to remind ourselves that, by common consent, phasors 'rotate' in a counter-clockwise direction.

> Whenever we measure the angles between phasor quantities, such as currents and voltages, counterclockwise is *always* considered to be the positive direction, and clockwise is *always* considered to be the negative direction.

Figure 19.3

The *length* of the voltage phasor $(\bar{E})$ represents the *r.m.s. value of the supply voltage*, and the *length* of the current phasor $(\bar{I})$ represents the *r.m.s. value of the current*. There is absolutely *no* relationship between the scales of the two phasors, because they each represent different quantities (for example, the voltage phasor could be drawn to a scale of 10 volts per millimetre, while the current phasor is drawn to a scale of 2 amperes per millimetre). What *is* important, however, is the **angle** (or, in this case, the *lack* of any angle) between the two phasors, which indicates that the two quantities are 'in phase' with each other.

As is the case for *any* circuit, the ratio of voltage to current represents the *opposition* to current. In a **purely resistive circuit**, this 'opposition' is, of course, the **resistance** (symbol: **R**) of the circuit, measured in ohms:

$$R = \frac{\overline{E}}{\overline{I}}$$

However, as we have already learnt, this equation tells us what the resistance happens to be for that particular ratio of voltage to current – the resistance itself being determined by the load's physical factors (length, cross-sectional area and resistivity).

Purely inductive circuit

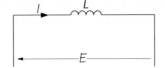

Figure 19.4

You will recall, from the section on *inductance* that, whenever the current through an inductor varies, a voltage is induced into that inductor through the process of *self-induction*. By Lenz's Law, the direction of this induced voltage *(u)* always acts to *oppose any change in current*. And, in accordance with Faraday's

Law, this induced voltage is *directly proportional to the rate of change of current*, as expressed below:

$$u \propto -\frac{\Delta i}{\Delta t}$$

The Greek letter 'delta' (Δ) simply means 'change in', so the expression $\Delta i / \Delta t$ means 'rate of change of current'.

The greatest rate of change in current occurs when the current waveform is at its steepest. As can be seen in Figure 19.5, this occurs whenever the current waveform passes through the zero axis. So, at point **A**, for example, the current is *increasing* at its greatest rate of change, so the maximum self-induced voltage (point **B**) will occur at the same time but, in accordance with Lenz's Law, must act in the negative sense (i.e. opposing the increase in current). This induced voltage waveform is shown as a broken line in Figure 19.5.

By Kirchhoff's Voltage Law, the induced voltage must be equal but *opposite* to the supply voltage (point **C**). The supply-voltage waveform is shown as a solid line. So, the current is clearly one-quarter of a wavelength behind the supply voltage, $(\overline{E})$. We say, therefore, that '*the current **lags** the supply voltage by 90°*'.

In a **purely inductive** circuit, then, the current $(\overline{I})$ is said to **lag** the supply voltage $(\overline{E})$ by 90° (and it is equally true to say that the *voltage leads the current by 90°*). If we now redraw the waveform, ignoring the self-induced voltage, it will look like Figure 19.6.

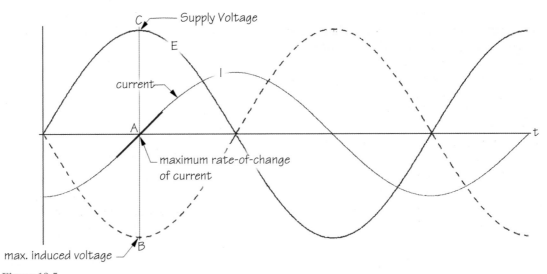

Figure 19.5

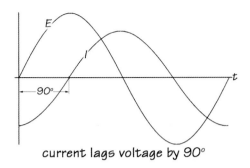

current lags voltage by 90°

Figure 19.6

Again, the waveforms can be represented by means of a **phasor diagram**, with the current and voltage phasors *lying at right angles to each other*. By convention, phasors are considered to 'rotate' counterclockwise, so the voltage phasor *(Ē)* is drawn 90° counterclockwise from the current phasor *(Ī)*, as shown below. It would be equally correct to place the voltage phasor horizontally, with the current phasor drawn 90° clockwise but, for consistency throughout this chapter, we'll draw the current phasor horizontally.

Figure 19.7

As before, the *length* of the voltage phasor represents the *r.m.s. value of voltage*, and the *length* of the current phasor represents the *r.m.s. value of the current*. Again, there is absolutely *no* relationship between the lengths of the two phasors, as one represents voltage and the other current. What *is* important, however, is that *the voltage phasor is drawn 90° counterclockwise relative to the current phasor*.

Once again, the ratio of voltage to current represents the *opposition* to current. In a **purely inductive circuit**, of course, *there is no resistance*, so we call this 'opposition' to current **inductive reactance** (symbol: X_L) measured in ohms, with the term, 'reactance', meaning 'reacting against' the passage of current.

$$X_L = \frac{\bar{E}}{\bar{I}}$$

Once again, it's important to understand that the ratio of voltage to current tells us what the inductive reactance happens to be for that particular ratio of voltage to current. The **inductive reactance** itself is determined by the inductance of the load, and the frequency of the supply, as shown in the following equation:

$$X_L = 2\pi f L$$

where:

X_L = inductive reactance (ohms)
f = supply frequency (hertz)
L = inductance (henrys)

Unfortunately, in order to understand how the above equation is derived, it is necessary to have some knowledge of calculus, which is beyond the scope of this text. *So it is necessary to commit this equation to memory.*

Purely capacitive circuit

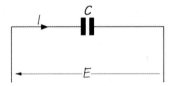

Figure 19.8

You will recall, from the section on *capacitance* that, for a capacitor, the current *(i)* is *directly proportional to the rate of change of voltage*, as expressed below:

$$i \propto \frac{\Delta e}{\Delta t}$$

Where $\Delta e/\Delta t$ simply means 'rate of change of voltage.'

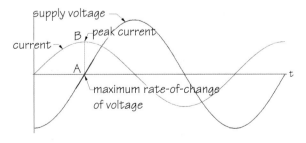

Figure 19.9

In Figure 19.9, the greatest rate of change of voltage occurs when the supply voltage waveform passes through the zero axis and is at its *steepest* – for example, at point **A**. This is the point at which the maximum current occurs (point **B**). So, the current is clearly one-quarter of a waveform *ahead* of the supply voltage. We say that '*the current **leads** the supply voltage by 90°*'.

In a **purely capacitive** circuit, then, the current $(\overline{I})$ is said to **lead** the voltage drop $(\overline{E})$ by 90° (and it is equally true to say the *voltage lags the current by 90°*).

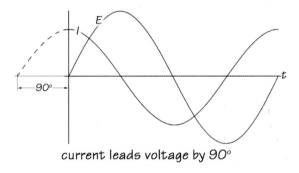

current leads voltage by 90°

Figure 19.10

Again, the waveforms in Figure 19.10 can be represented by means of a **phasor diagram**, with the current phasor and voltage phasor each lying at right angles to each other. The voltage phasor $(\overline{E})$ is drawn 90° clockwise relative to the current phasor $(\overline{I})$, as illustrated in Figure 19.11. Again, it would be equally correct to draw the voltage phasor horizontally, with the current phasor 90° counterclockwise.

As with all phasor diagrams, the *length* of the voltage phasor represents the *r.m.s. value of voltage*, and the *length* of the current phasor represents the *r.m.s. value of the current*, but there is absolutely *no* relationship between the scales of the two phasors. What *is* important, however, is that *the current phasor is drawn 90° counterclockwise from the voltage phasor.*

As always, the ratio of voltage to current determines the opposition to current. In a **purely capacitive circuit**,

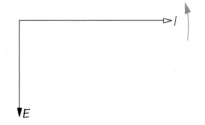

Figure 19.11

there is no resistance, so we call the 'opposition' to current the **capacitive reactance** (symbol: X_C) of the circuit, measured in ohms:

$$X_C = \frac{\overline{E}}{\overline{I}}$$

The ratio of voltage to current tells us what the capacitive reactance happens to be for that particular ratio of voltage to current. The **capacitive reactance** itself being inversely proportional to the capacitance of the load, and the frequency of the supply, as specified in the following equation:

$$X_C = \frac{1}{2\pi fC}$$

where:

X_C = capacitive reactance (ohms)

f = supply frequency (hertz)

C = capacitance (farads)

Once again, in order to understand how the above equation is derived, it is necessary to have some knowledge of calculus, which is beyond the scope of this text. So, again, *it is necessary to commit this equation to memory.*

Don't worry, though, as this equation, together with that for inductive reactance, are the *only two equations you will be asked to commit to memory* in these chapters on alternating current. From now on, **all other equations can be derived from phasor diagrams!**

Does alternating current flow through a capacitor?

Not really, although it certainly *appears* to be doing so from the behaviour of the current in the external circuit. However, if you refer back to the chapter on *capacitors and capacitance*, you will recall that there can be no conduction current within a capacitor's dielectric due to the lack of any free electrons. However, the changing voltage applied across the plates causes a **displacement current** to take place.

The term 'displacement current' describes the distortion and polarisation of the electron orbits around the fixed nuclei of the dielectric's atoms. As the applied voltage increases, the orbits 'stretch' more and more, causing an increase in the displacement current; as the voltage decreases and reverses direction, so too does the direction of the displacement current. You will also recall that the direction of the field set up by these polarised atoms is such that it always opposes and, therefore, reduces the electric field set up within the dielectric by the applied voltage – 'reacting' against any change in the external voltage.

To summarise, for a capacitive circuit, a 'conduction' current (electron flow) takes place *around the external circuit*, whereas a 'displacement' current (distortion of electron orbits) takes place *within the dielectric*.

CIVIL

It is **absolutely essential** to remember the phase relationships between currents and voltages for purely resistive, purely inductive and purely capacitive circits. *Unless you do so, you will* **not** *be able to construct phasor diagrams!*

And if you cannot construct phasor diagrams, you will never understand the behaviour of alternating current!

To help you, you should learn the mnemonic 'CIVIL', in which 'C' stands for 'capacitive circuit', and 'L' stands for 'inductive circuit' (see Figure 19.12).

capacitive (**C**) circuits: inductive (**L**) circuits:
I *before* ('leads') Voltage   I *after* ('lags') Voltage

Figure 19.12

An alternative mnemonic you might wish to consider is '**ELI the ICEman**', which (not surprisingly!) is popular with American and Canadian students, where '*ELI*' indicates that, in an inductive (**L**) circuit, voltage (**E**) is before (leads) current (**I**); and where '*ICE*' indicates that, in a capacitive (**C**) circuit, current (**I**) is before (leads) voltage (**E**).

'Real' circuits

Now that we have learned how these three 'ideal' (theoretical) circuits behave, let's now turn our attention to 'real' circuits.

'Real' circuits exhibit *combinations* of resistance, inductance and capacitance.

Since *all* circuits exhibit resistance, we will look at **series Resistive-Inductive (series *R-L*)** circuits, then at **series Resistive-Capacitive (series *R-C*)** circuits and, finally, at **series Resistive-Inductive-Capacitive (series *R-L-C*)** circuits.

Again, it's worth reminding ourselves that the circuit symbols used throughout this chapter represent the *quantities* resist*ance*, induct*ance* and capacit*ance* – *not* resistors, inductors and capacitors.

Series R-L circuits

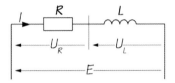

Figure 19.13

The circuit diagram shown in Figure 19.13 represents a **series resistive-inductive** circuit. It does *not* necessarily represent a resistor in series with an inductor, but a load (such as a coil) which exhibits both resistance and inductance. We have, if you like, 'separated out' the resistance and the inductance from the coil or other inductive component, so that they can be considered separately.

We now know that, in a purely resistive circuit, *the current and voltage are in phase* with each other and, in a purely inductive circuit, *the current lags the voltage by 90°*. So, what happens in a series *R-L* circuit? Well, clearly, our instincts tell us that *the current is likely to lag the voltage by some angle between 0° and 90°* – this angle being called the circuit's **phase angle** (symbol: φ, pronounced '*phi*').

The general definition of **phase angle** is '*the angle by which the* **current** *leads or lags the supply voltage*'. Note, we *always* measure phase angles in terms of what the load *current* is doing, relative to the supply voltage, *never the other way around*.

So, for an *R-L circuit*, because the current *always* lags the supply voltage, the phase angle is *always* described as **lagging**. Whenever a phase angle is quoted, *it is usual to specify whether it's leading or lagging*.

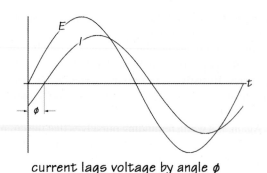

current lags voltage by angle φ

Figure 19.14

Of course, whenever a current $(\bar{I})$ flows through a **series R-L circuit**, a voltage drop, $\bar{U}_R$, will appear across the resistive component of the circuit, and a voltage drop, $\bar{U}_L$, will appear across the inductive component of the circuit – as shown in the schematic diagram in Figure 19.13.

> In the following step-by-step construction of a phasor diagram, the step being described is illustrated in **blue**, while the previous steps are shown in black.

Drawing the phasor diagram

Step 1

In a series R-L circuit, the **current** is common to both the resistive and the inductive components and, so, current is **always** chosen as the **reference phasor**. The reference phasor is *always drawn first, and always along the horizontal positive axis*. It's also drawn fairly long in order to distinguish it from the other phasors. In the following diagrams, we will further distinguish it by using an outline, rather than solid, arrow head (although this is not absolutely necessary) (Figure 19.15).

Figure 19.15

Step 2

Since the voltage drop, $\bar{U}_R$, across the resistive component is *in phase with the current,* it is also drawn along the horizontal positive axis. Some textbooks

show the phasor $\bar{U}_R$ *superimposed* over the reference phasor; others show it drawn *very close and parallel* with the reference phasor – the method preferred in this text (Figure 19.16).

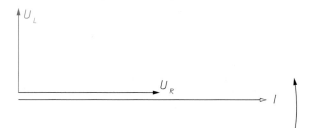

Figure 19.16

Step 3

The voltage drop, $\bar{U}_L$, across the inductive component *leads* the current by 90° (remember C**IVIL**), so is drawn 90° counterclockwise ('leading') from the reference phasor. Both $\bar{U}_R$ and $\bar{U}_L$, of course, represent voltage drops and, so, are both drawn to the same scale as each other (Figure 19.17).

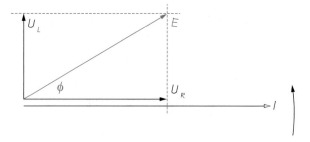

Figure 19.17

Step 4

We know, from Kirchhoff's Voltage Law, that in a series circuit, the supply voltage is the sum of the individual voltage drops. However, because, in this case, the two voltage drops, $\bar{U}_R$ and $\bar{U}_L$, lie at right angles to each other, we have to add them *vectorially* (Figure 19.18).

Figure 19.18

From the completed phasor diagram (Figure 19.18), we can see that the supply voltage, $\bar{E}$, is the **phasor** sum (or **vector sum**) of $\bar{U}_R$ and $\bar{U}_L$, which can be determined using Pythagoras's Theorem:

$$\bar{E} = \sqrt{\bar{U}_R^2 + \bar{U}_L^2}$$

It's completely unnecessary to commit this equation to memory, because it has been derived from the phasor diagram, using Pythagoras's Theorem. If you can draw the phasor diagram, and know Pythagoras's Theorem, then you don't need to bother to remember this equation!

Worked example 1   The voltage drop across the resistive component of a series *R-L* circuit is 30 V, and the voltage drop across the inductive component is 40 V. What is the value of the supply voltage?

Solution   *Always* start by sketching the circuit diagram (Figure 19.19), and inserting all the values given to you in the question.

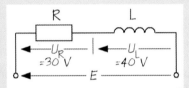

Figure 19.19

Next, sketch the phasor diagram, following the steps described above. You *don't* have to draw the phasor diagram to scale (Figure 19.20).

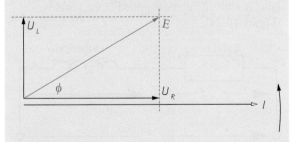

Figure 19.20

Now, you can apply Pythagoras's Theorem to solve the problem:

$$\bar{E} = \sqrt{\bar{U}_R^2 + \bar{U}_L^2}$$
$$= \sqrt{30^2 + 40^2}$$
$$= \sqrt{2500} = 50 \text{ V (Answer)}$$

Impedance diagram

The current flowing through a series *R-L* circuit will be opposed by *both* resistance (*R*) *and* inductive reactance (*X_L*). The combination of these two 'oppositions' is called the **impedance** (symbol: **Z**) of the circuit, also measured in ohms. 'Impedance' is yet another word, meaning to 'oppose' or 'impede' the passage of current.

But we *cannot* simply add the resistance and inductive reactance – so how can we find the impedance? The answer is by means of an **impedance diagram.** As before, in the following series of diagrams, the step being described is shown in **blue**; those previously described are shown in black.

Drawing the impedance diagram

Step 1

We start by drawing the circuit's phasor diagram, following the steps already explained (Figure 19.21).

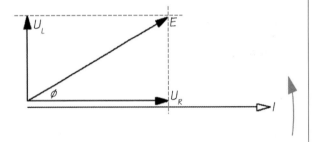

Figure 19.21

Step 2

Next, we *divide each of the voltage phasors by the reference phasor (**I**)* (Figure 19.22).

The resulting diagram is known as an '**impedance diagram**' (sometimes called an *'impedance triangle'*), and is useful because it generates the following important equations:

$$R = \frac{\bar{U}_R}{\bar{I}} \quad X_L = \frac{\bar{U}_L}{\bar{I}} \quad Z = \frac{\bar{E}}{\bar{I}}$$

Again, you don't have to commit these equations to memory, because they are derived when you convert a voltage phasor diagram into an impedance diagram!

Also from the impedance diagram, you can also see that the impedance is also the *vector sum of resistance and inductive reactance*, which can be calculated by simply applying Pythagoras's Theorem (Figure 19.23).

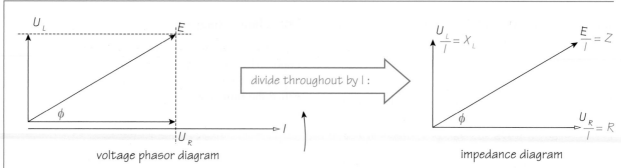

voltage phasor diagram   impedance diagram

Figure 19.22

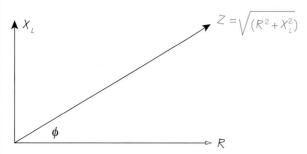

$Z = \sqrt{(R^2 + X_L^2)}$

applying Pythagoras's Theorem

Figure 19.23

$$Z = \sqrt{(R^2 + X_L^2)}$$

If required, you can also write similar equations for resistance and inductive reactance, by applying Pythagoras's Theorem. That is:

$$R = \sqrt{(Z^2 - X_L^2)} \qquad X_L = \sqrt{(Z^2 - R^2)}$$

Once again, you don't need to commit any of these equations to memory, *providing* you can draw a phasor diagram, convert it to an impedance diagram, and apply Pythagoras's Theorem!

We can also find the circuit's **phase angle**, using basic trigonometry, utilising either the *sine*, *cosine* or *tangent* ratios. In practice, for a reason we'll see later in this text, the best choice is always to use the *cosine*:

$$\cos\phi = \frac{\text{adjacent}}{\text{hypotenuse}} = \frac{R}{Z}$$

$$\angle\phi = \cos^{-1}\frac{R}{Z}$$

Important! Dividing a voltage phasor diagram by current produces an impedance diagram which generates each of the equations shown above. *So you don't have to learn any of these*

equations – they can all be generated provided you learn how to draw the phasor and impedance diagrams!

Worked example 2   An inductor, of resistance 5 Ω and inductance 0.02 H, is connected across a 230-V, 50 Hz, a.c. supply. Calculate each of the following:

a   inductive reactance
b   impedance
c   current
d   voltage drop across the resistive component of the circuit
e   voltage drop across the inductive component of the circuit
f   phase angle of the circuit.

Solution   The first set in solving any a.c. circuit problem is to sketch the circuit diagram, and label it with all values supplied in the problem (Figure 19.24).

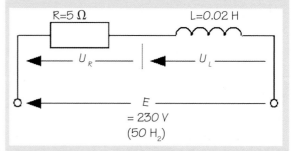

Figure 19.24

The next step is to draw the voltage phasor diagram, following the steps described earlier (Figure 19.25).

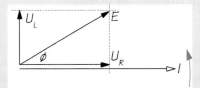

Figure 19.25

As the problem relates to inductive reactance, impedance, etc., the next step is to convert the voltage phasor diagram into an impedance diagram by dividing throughout by the reference quantity – i.e. by the current. This generates all the equations that we need to solve the problem (Figure 19.26).

a   To find the **inductive reactance** (X_L) of the circuit, we start by looking at the equations generated when we constructed the impedance diagram. There's only one equation for inductive reactance, $X_L = \dfrac{\bar{U}_L}{\bar{I}}$. Unfortunately, we don't know the value of $\bar{U}_L$, so we *cannot* use this formula. What about applying Pythagoras's Theorem $\left(X_L = \sqrt{Z^2 - R^2}\right)$?

Could we use this equation to find X_L? Unfortunately, no, because we don't know the value of Z. If we *can't* use any of the equations generated by the impedance diagram, then we must fall back on the basic equation for inductive reactance, as follows:

$$X_L = 2\pi f L = 2\pi \times 50 \times 0.02 = 6.28\ \Omega \text{ (Answer a.)}$$

b   To find the impedance, we *can* use an equation generated by the impedance diagram:

$$Z = \sqrt{R^2 + X_L^2}$$
$$= \sqrt{5^2 + 6.28^2}$$
$$= \sqrt{25 + 39.44}$$
$$= \sqrt{64.44} = 8.03\ \Omega \text{ (Answer b.)}$$

c   To find the current, we use the following equation that was generated by the impedance diagram:

$$\bar{I} = \frac{\bar{E}}{Z} = \frac{230}{8.03} = 28.64\ \text{A (Answer c.)}$$

d   Again, using the equation generated by the impedance diagram:

$$\bar{U}_R = \bar{I}R = 28.64 \times 5 = 143.20\ \text{V (Answer d.)}$$

e   Again, using the equation generated by the impedance diagram:

$$\bar{U}_L = \bar{I}X_L = 28.64 \times 6.28 = 179.86\ \text{V (Answer e.)}$$

f   Using the cosine function:

$$\angle\phi = \cos^{-1}\frac{R}{Z} = \cos^{-1}\frac{5}{8.03} = \cos^{-1}0.6227$$
$$= 51.48° \text{ lagging (Answer f.)}$$

(*'Lagging'*, because the supply current lags the supply voltage in an inductive circuit.)

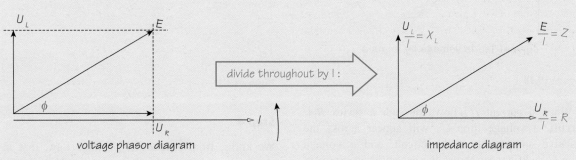

voltage phasor diagram   divide throughout by I :   impedance diagram

Figure 19.26

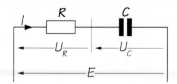

Series R-C circuits

Figure 19.27

Again, it's important to realise that the circuit shown in Figure 19.27 does not necessarily represent a resistor and a capacitor; rather, it simply represents resistance in series with capacitance. For example, it could represent the resistance and capacitance of a very long underground cable.

We know that in a purely resistive circuit, the current and voltage are in phase with each other; and, in a purely capacitive circuit, the current leads the voltage by 90°. So, what happens in a *series R-C* circuit? Well, clearly this time, we instinctively know that the *current must lead the voltage by some angle between 0° and 90°* – this angle is called the circuit's **phase angle** (symbol: ϕ, pronounced 'phi').

Remember, the general definition of **phase angle** is *the angle by which the current leads or lags the supply voltage*. So for *resistive-capacitive circuits*, because the current *always* leads the supply voltage, the phase angle is *always* described as **leading**.

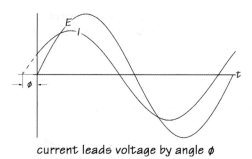

current leads voltage by angle ϕ

Figure 19.28

When a current $(\bar{I})$ flows through a **series R-C circuit**, a voltage drop $\bar{U}_R$ will appear across the resistive component of the circuit, and a voltage drop $\bar{U}_C$ will appear across the capacitive component of the circuit.

Drawing the phasor diagram

Step 1

In a series circuit, the **current** is common to each component and, so, current is *always* chosen as the **reference phasor**. The reference phasor is *always drawn along the horizontal positive axis*, and it's also normally drawn fairly long in order to distinguish it from the others (Figure 19.29).

Figure 19.29

Step 2

The voltage drop, $\bar{U}_R$, across the resistive component is *in phase with the current* and, so, is also drawn along the horizontal positive axis (Figure 19.30).

Figure 19.30

Step 3

The voltage drop, $\bar{U}_C$, across the capacitive component *lags the current by 90°* (remember **CIVIL**), so is drawn 90° clockwise from the reference phasor (Figure 19.31).

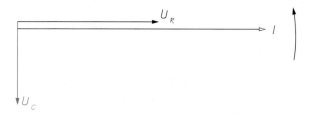

Figure 19.31

Step 4

We know, from Kirchhoff's Voltage Law, that in a series circuit, the total voltage drop is the sum of the individual voltage drops. Because, in this case, the two voltage drops, $\bar{U}_R$ and $\bar{U}_C$, lie at right angles to each other, we have to add them *vectorially*.

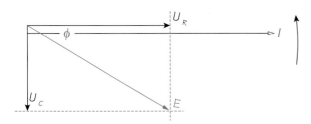

Figure 19.32

From the completed phasor diagram shown in Figure 19.32, we can see that the supply voltage, $\bar{E}$, is the **phasor sum** (or **vector sum**) of $\bar{U}_R$ and $\bar{U}_C$, which can be found using Pythagoras's Theorem:

$$\bar{E} = \sqrt{\bar{U}_R^2 + \bar{U}_C^2}$$

Worked example 3   The voltage drop across the resistive component of a series R-C circuit is 40 V, and the voltage drop across the capacitive component is 30 V. What is the value of the total voltage drop?

Solution   *Always* start by sketching the circuit diagram (Figure 19.33), and inserting all the values given to you in the question.

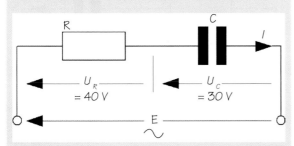

Figure 19.33

Next, sketch the phasor diagram, following the steps described above. You *don't* have to draw the phasor diagram to scale (Figure 19.34).

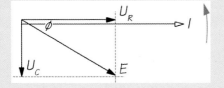

Figure 19.34

Now, you can apply Kirchhoff's Voltage Law, and use Pythagoras's Theorem to solve the problem:

$$\begin{aligned}\bar{E} &= \sqrt{\bar{U}_R^2 + \bar{U}_C^2} \\ &= \sqrt{40^2 + 30^2} \\ &= \sqrt{2500} = 50 \text{ V (Answer)}\end{aligned}$$

Impedance diagram

The current flowing through a series *R-C* circuit will be opposed by both its resistance (*R*) *and* by its capacitive reactance (X_C). The combination of these two 'oppositions' is called the **impedance** (symbol: Z) of the circuit, and is measured in ohms. Again, we *cannot* simply add the resistance and capacitive reactance – so how can we find the impedance? Again, the answer is by means of an **impedance diagram**.

Drawing the impedance diagram

Step 1

We start by drawing the circuit's phasor diagram, following the steps already explained (Figure 19.35).

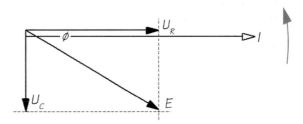

Figure 19.35

Step 2

Next, we *divide each of the voltage phasors by the reference phasor ($\bar{I}$)*.

The resulting diagram (Figure 19.36) is an **impedance diagram** (or *'impedance triangle'*), and is useful because it generates the following equations:

$$R = \frac{\bar{U}_R}{\bar{I}} \qquad X_C = \frac{\bar{U}_C}{\bar{I}} \qquad Z = \frac{\bar{E}}{\bar{I}}$$

Also from the impedance diagram, you can see that the impedance is the *vector sum of resistance and capacitive reactance*, which can be calculated by applying Pythagoras's Theorem (Figure 19.37).

$$Z = \sqrt{R^2 + X_C^2}$$

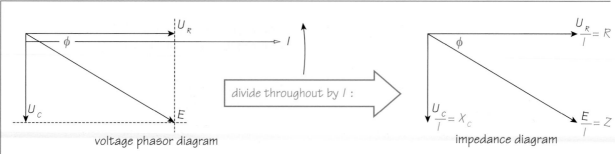

voltage phasor diagram

impedance diagram

Figure 19.36

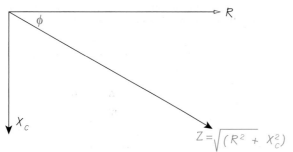

applying Pythagoras's Theorem

Figure 19.37

You can, if necessary, write similar equations for the resistance and capacitive reactance, by applying Pythagoras's Theorem. That is:

$$R = \sqrt{(Z^2 - X_C^2)} \quad X_C = \sqrt{(Z^2 - R^2)}$$

We can also find the circuit's **phase angle**, using basic trigonometry, utilising either the *sine*, *cosine* or *tangent* ratios – as before, the best choice is to use the *cosine*:

$$\cos\phi = \frac{\text{adjacent}}{\text{hypotenuse}} = \frac{R}{Z}$$

$$\angle\phi = \cos^{-1}\frac{R}{Z}$$

> **Important!** Dividing a voltage phasor diagram by current produces an impedance diagram which generates each of the equations shown above. *So you don't have to learn any of these equations – they can all be generated provided you learn how to draw the phasor and impedance diagrams!*

Worked example 4 A capacitor, of resistance 40 Ω and capacitance 50 μF is connected across a 110-V, 50 Hz, a.c. supply. Calculate each of the following:

a capacitive reactance
b impedance
c current
d voltage drop across the resistive component of the circuit
e voltage drop across the capacitive component of the circuit
f phase angle of the circuit.

Solution The first step in solving any a.c. circuit problem is to sketch the circuit diagram, and label it with all values supplied in the problem (Figure 19.38).

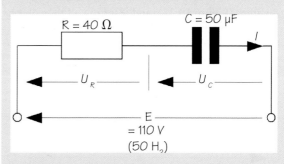

Figure 19.38

The next step is to draw the voltage phasor diagram, following the steps described earlier (Figure 19.39).

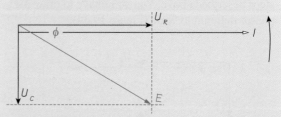

Figure 19.39

As the problem relates to inductive reactance, impedance, etc., the next step is to convert the voltage phasor diagram into an impedance diagram, by dividing throughout by the reference quantity – i.e. current. This generates the equations that we need to solve the problem (Figure 19.40).

a To find the **capacitive reactance** (X_c) of the circuit, we start by looking at the equations generated when we constructed the impedance diagram. There's only one equation for capacitive reactance, $X_c = \dfrac{\bar{U}_c}{\bar{I}}$. Unfortunately, we don't know the value of $\bar{U}_c$ so we can't use this formula. What about applying Pythagoras's Theorem $(X_c = \sqrt{Z^2 - R^2})$? Could we use this to find X_c? Unfortunately, no, because we don't know the value of Z. Clearly, then, we must fall back on the basic equation for capacitive reactance, as follows:

$$X_c = \frac{1}{2\pi f C} = \frac{1}{2\pi \times 50 \times (50 \times 10^{-6})}$$
$$= 63.67\ \Omega \text{ (Answer a.)}$$

b To find the impedance, we can now use the equation generated by the impedance diagram:

$$Z = \sqrt{R^2 + X_L^2}$$
$$= \sqrt{40^2 + 63.67^2}$$
$$= \sqrt{1600 + 4054}$$
$$= \sqrt{5654} = 75.2\ \Omega \text{ (Answer b.)}$$

c To find the current, we use the following equation generated by the impedance diagram:

$$\bar{I} = \frac{\bar{E}}{Z} = \frac{110}{75.2} = 1.46\ \text{A (Answer c.)}$$

d Again, using the equation generated by the impedance diagram:

$$\bar{U}_R = \bar{I} R = 1.46 \times 40 = 58.53\ \text{V (Answer d.)}$$

e Again, using the equation generated by the impedance diagram:

$$\bar{U}_C = \bar{I} X_C = 1.46 \times 63.67 = 93.14\ \text{V (Answer e.)}$$

f Using the cosine function:

$$\angle\phi = \cos^{-1}\frac{R}{Z} = \cos^{-1}\frac{40}{75.2} = \cos^{-1} 0.5319$$
$$= 57.87° \text{ leading (Answer f.)}$$

(*'Leading'*, because the supply current leads the supply voltage in a capacitive circuit.)

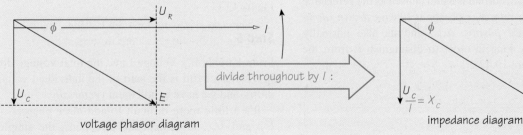

voltage phasor diagram   impedance diagram

divide throughout by I :

Figure 19.40

Series R-L-C circuits

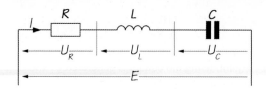

Figure 19.41

We have learnt that:

- in a series *R-L* circuit, *the current lags the supply voltage by some angle* and,
- in a series *R-C* circuit, *the current leads the supply voltage by some angle*.

So, what happens in a *series R-L-C* circuit? Well, clearly the *current could either lag **or** lead the voltage* – depending on the values of the inductive reactance and the capacitive reactance! Whatever value this angle happens to be, it will be the circuit's **phase angle** (symbol: φ, pronounced 'phi').

When a current $(\bar{I})$ flows through a series *R-L-C* circuit, a voltage drop $\bar{U}_R$ will appear across the resistive component of the circuit, a voltage drop $\bar{U}_L$ will appear across the inductive component, and a voltage drop $\bar{U}_C$ will appear across the capacitive component of the circuit.

Drawing the phasor diagram

Step 1

In a series circuit, the **current** is common to each component and, so, current is again chosen as the **reference phasor**. The reference phasor is *always drawn along the horizontal positive axis*, and its also normally drawn fairly long in order to distinguish it from the others (Figure 19.42).

Figure 19.42

Step 2

The voltage drop, $\bar{U}_R$, across the resistive component is *in phase with the current* and, so, is also drawn along the horizontal positive axis (Figure 19.43).

Figure 19.43

Step 3

The voltage drop, $\bar{U}_L$, across the inductive component *leads the current by 90°* (remember **CIVIL**), so is drawn 90° counterclockwise from the reference phasor (Figure 19.44).

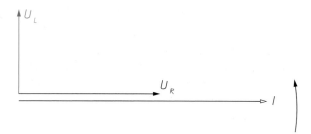

Figure 19.44

Step 4

The voltage drop, $\bar{U}_C$, across the capacitive component *lags the current by 90°* (remember **CIVIL**), so is drawn 90° clockwise from the reference phasor (Figure 19.45).

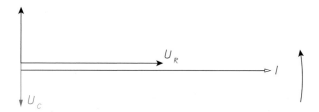

Figure 19.45

Step 5

From Kirchhoff's Voltage Law, the total voltage drop in a series circuit is the sum of the individual voltage drops, and we have to add them *vectorially*.

It's a little more difficult to add *three* phasors. As $\bar{U}_L$ and $\bar{U}_C$ lie in *opposite* directions, the simplest thing to do is to start by subtracting them and, *then*, add the difference to phasor $\bar{U}_R$.

The snag is, of course, that we might not know whether $\bar{U}_L$ is bigger than $\bar{U}_C$, or vice versa! Fortunately, *it doesn't matter!* The purpose of the phasor diagram is simply to *generate equations*, not to accurately represent the actual conditions in the circuit to scale! And the phasor diagram will *always* generate the correct equations whether $\bar{U}_L$ is actually bigger than $\bar{U}_C$, or vice versa!

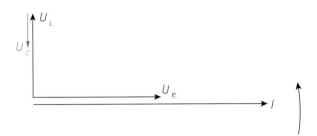

Figure 19.46

So, the simplest solution is to get into the habit of **always** *drawing $\bar{U}_L$ longer than $\bar{U}_C$* – or the other way around, if you prefer! But, for a reason that will be revealed later, whatever you do, *never* **ever** *draw them the same length!!*

Figure 19.47 shows what the finished phasor diagram will look like.

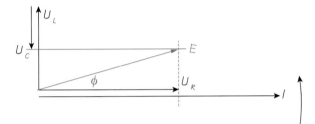

Figure 19.47

From the completed phasor diagram, we can see that $\bar{E}$ is the **phasor sum** (or **vector sum**) of $\bar{U}_R$, $\bar{U}_L$ and $\bar{U}_C$, which can be found using Pythagoras's Theorem:

$$\bar{E} = \sqrt{\bar{U}_R^2 + (\bar{U}_L - \bar{U}_C)^2}$$

If, when we draw the phasor diagram, we make $\bar{U}_L$ bigger than $\bar{U}_C$ when, really, it's the other way around, *it doesn't really matter.* When we subtract the two, we'll end up with a negative quantity inside the brackets which, when squared, will result in a positive value.

Worked example 5 In a series *R-L-C* circuit, the voltage drop across the resistive component is 4 V, the voltage drop across the inductive component is 10 V, and the voltage drop across the capacitive component is 7 V. What is the value of the total voltage drop?

Solution *Always* start by sketching the circuit diagram, and inserting all the values given to you in the question (Figure 19.48).

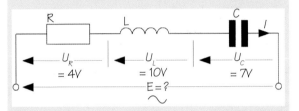

Figure 19.48

Next, sketch the phasor diagram, following the steps described above. You *don't* have to draw the phasor diagram to scale (Figure 19.49).

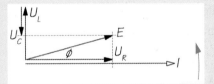

Figure 19.49

Now, you can apply Kirchhoff's Voltage Law, and use Pythagoras's Theorem to solve the problem:

$$\bar{E} = \sqrt{\bar{U}_R^2 + (\bar{U}_L - \bar{U}_C)^2}$$
$$= \sqrt{4^2 + (10 - 7)^2}$$
$$= \sqrt{4^2 + 3^2} = \sqrt{25} = 5 \text{ V (Answer)}$$

Impedance diagram

The current flowing through a series *R-L-C* circuit will be opposed by resistance (*R*) *and* by inductive reactance (X_L), *and* by capacitive reactance (X_C). The combination of these three 'oppositions' is called the **impedance** (symbol: **Z**) of the circuit, measured in ohms. But, again, we *cannot* simply add the resistance, inductive reactance and capacitive reactance – so how can we find the impedance? Again, the answer is by means of an **impedance diagram**.

Drawing the impedance diagram

Step 1

We start by drawing the circuit's phasor diagram, following the steps already explained (Figure 19.50).

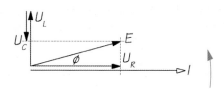

Figure 19.50

Step 2

Next, we *divide each of the voltage phasors by the reference phasor ($\overline{I}$)* (Figure 19.51).

The resulting diagram is an **impedance diagram** (or *'impedance triangle'*), and is useful because it generates the following equations:

$$R = \frac{\overline{U}_R}{\overline{I}} \qquad X_L = \frac{\overline{U}_L}{\overline{I}} \qquad X_C = \frac{\overline{U}_C}{\overline{I}} \qquad Z = \frac{\overline{E}}{\overline{I}}$$

Also from the impedence diagram, you can see that the impedance is also the *vector sum of resistance, inductive reactance* and *capacitive reactance*, which can be calculated by applying Pythagoras's Theorem (Figure 19.52).

$$Z = \sqrt{R^2 + (X_L - X_C)^2}$$

Once again, we can also find the circuit's **phase angle**, using basic trigonometry:

$$\cos\phi = \frac{\text{adjacent}}{\text{hypotenuse}} = \frac{R}{Z}$$

$$\angle\phi = \cos^{-1}\frac{R}{Z}$$

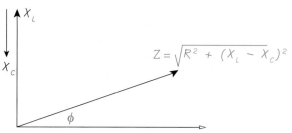

applying Pythagoras's Theorem

Figure 19.52

> **Important!** Dividing a voltage phasor diagram by current produces an impedance diagram which generates each of the equations shown above. *So you don't have to learn any of these equations – they can all be generated* **provided you learn how to draw the phasor and impedance diagrams!**

> **Worked example 6**   A circuit of resistance of 1.5 Ω, inductance of 16 mH, and capacitance 500 μF, is connected across a 230-V, 50 Hz, a.c. supply. Calculate each of the following:
>
> a   inductive reactance
> b   capacitive reactance

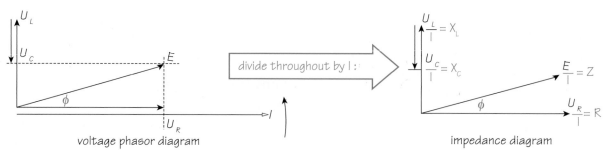

voltage phasor diagram   impedance diagram

Figure 19.51

c impedance
d current
e voltage drop across the resistive component of the circuit
f voltage drop across the inductive component of the circuit
g voltage drop across the capacitive component of the circuit
h phase angle of the circuit.

Solution The first step in solving *any* a.c. circuit problem is to sketch the circuit diagram, and label it with all values supplied in the problem (Figure 19.53).

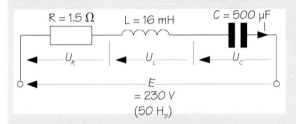

Figure 19.53

The next step is to draw the voltage phasor diagram, following the steps described earlier (Figure 19.54).

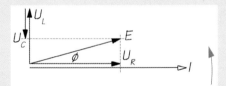

Figure 19.54

As the problem relates to inductive reactance, capacitive reactance, impedance, etc., the next set is to convert the voltage phasor diagram into an

impedance diagram (Figure 19.55), by dividing throughout by the reference quantity – i.e. current. This generates the equations that we need to solve the problem.

a To find the **inductive reactance** (X_L) of the circuit, we start by looking at the equations generated when we constructed the impedance diagram. There's only one equation for inductive reactance, $X_L = \bar{U}_L / \bar{I}$. Unfortunately, we don't know the value of $\bar{U}_L$ so we can't use this formula. What about applying Pythagoras's Theorem? Could we use this to find X_L? Unfortunately, no, because we don't know the value of Z. Clearly, then, we must fall back on the basic equation for inductive reactance, as follows:

$$X_L = 2\pi f L = 2\pi \times 50 \times 16 \times 10^{-3}$$
$$= 5.03\ \Omega\ \text{(Answer a.)}$$

b To find the **capacitive reactance** (X_C) of the circuit, we start by looking at the equations generated when we constructed the impedance diagram. There's only one equation for capacitive reactance, $X_C = \bar{U}_C / \bar{I}$. Unfortunately, we don't know the value of $\bar{U}_C$ so we can't use this formula. What about applying Pythagoras's Theorem? Could we use this to find X_C? Unfortunately, no, because we don't know the value of Z. Clearly, then, we must fall back on the basic equation for capacitive reactance, as follows:

$$X_C = \frac{1}{2\pi f C} = \frac{1}{2\pi \times 50 \times (500 \times 10^{-6})}$$
$$= 6.37\ \Omega\ \text{(Answer b.)}$$

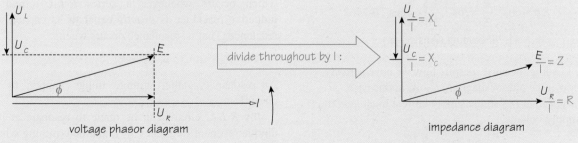

voltage phasor diagram   divide throughout by I: impedance diagram

Figure 19.55

c   To find the impedance, we can now use the equation generated by the impedance diagram:

$$Z = \sqrt{R^2 + (X_L - X_C)^2}$$
$$= \sqrt{1.5^2 + (5.03 - 6.37)^2}$$
$$= \sqrt{1.5^2 + (-1.34)^2}$$
$$= \sqrt{2.25 + 1.8}$$
$$= \sqrt{4.05} = 2.01\ \Omega\ \text{(Answer c.)}$$

Note! Despite X_C being larger than X_L, the above equation still delivers the correct answer, as the negative sign inside the bracket becomes positive when the bracket is squared!

d   To find the current, we use the following equation generated by the impedance diagram:

$$\overline{I} = \frac{\overline{E}}{Z} = \frac{230}{2.01} = 114.43\ \text{A (Answer d.)}$$

e   Again, using the equation generated by the impedance diagram:

$$\overline{U}_R = \overline{I}R = 114.43 \times 1.5 = 171.65\ \text{V (Answer e.)}$$

f   Again, using the equation generated by the impedance diagram:

$$\overline{U}_L = \overline{I}X_L = 114.43 \times 5.03 = 575.38\ \text{V (Answer f.)}$$

g   Again, using the equation generated by the impedance diagram:

$$\overline{U}_C = \overline{I}X_C = 114.43 \times 6.37 = 728.92\ \text{V (Answer g.)}$$

h   Using the cosine function:

$$\angle\phi = \cos^{-1}\frac{R}{Z} = \cos^{-1}\frac{1.5}{2.01} = \cos^{-1}0.7462$$
$$= 41.74°\ \text{leading (Answer h.)}$$

('Leading' because, despite how we have actually drawn the phasor diagram, $\overline{U}_C$ is, in this case, *larger* than $\overline{U}_L$, so the supply current must *lead* the supply voltage.)

Series resonance

From:

$$X_L = 2\pi fL \quad \text{and} \quad X_C = \frac{1}{2\pi fC}$$

... we know that the inductive reactance of a circuit is directly proportional to the supply frequency, whereas the capacitive reactance is inversely proportional to the supply frequency. So, if we gradually increase the supply frequency, the inductive reactance will gradually *increase*, while the capacitive reactance will *decrease*.

Eventually, a point will be reached for any *R-L-C* circuit, when the values of the inductive reactance and capacitive reactance will equal one another – as illustrated in Figure 19.56.

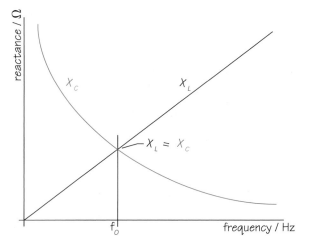

Figure 19.56

The frequency at which this occurs is called the '**resonant frequency**' (f_o) of the circuit, and the circuit is said to be at '**series resonance**'.

'**Series resonance**', then, is a unique condition which occurs whenever a series *R-L-C* circuit's inductive reactance is *exactly* equal to its capacitive reactance. That is, resonance occurs when:

$$X_L = X_C$$

At resonance, rather strange things happen to a circuit!

Any R-L-C circuit can be made to resonate at its unique resonant frequency (f_o), and determining what that frequency is for any circuit is straightforward. Starting with the basic condition for resonance:

$$X_L = X_C$$

and expanding:

$$2\pi f_o L = \frac{1}{2\pi f_o C}$$

$$f_o^2 = \frac{1}{(2\pi)^2 LC}$$

$$f_o = \frac{1}{2\pi \sqrt{LC}}$$

Worked example 7   What is the resonant frequency for a circuit having an inductance of 16 mH and a capacitance of 100 μF?

Solution

$$f_o = \frac{1}{2\pi \sqrt{LC}} = \frac{1}{2\pi \sqrt{(16 \times 10^{-3}) \times (100 \times 10^{-6})}}$$

$$= \frac{1}{2\pi \sqrt{1600 \times 10^{-9}}} = \frac{1}{2\pi \times (1.26 \times 10^{-3})}$$

$$= 126 \text{ Hz (Answer)}$$

Impedance diagram for series resonance

If we were to draw an **impedance diagram** for a series R-L-C circuit at resonance, it would look like Figure 19.57.

As you can see, the inductive reactance and the capacitive reactance are identical, but *act in opposite directions*. So, their vector sum must be zero.

This leaves **resistance** as the circuit's *only opposition to the flow of current*, from which we can say:

> At **resonance**, a circuit's impedance is equal to its resistance.

So if, at resonance, the vector sum of a circuit's inductive reactance and capacitive reactance is zero, leaving only its resistance to oppose current, then it follows that the circuit's current will reach its maximum value at resonance.

> A circuit's current reaches its maximum value when resonance occurs.

If we were to conduct a simple experiment, by adjusting the supply frequency of a series R-L-C circuit, until resonance occurs, while measuring its current, the resulting graph would look something like Figure 19.58.

As the frequency approaches the resonant frequency, the reactance (i.e. the combined inductive and capacitive reactance) falls towards zero ohms, and the corresponding current increases. The value which the current reaches, at resonance, is limited by the resistance of the circuit. If the resistance is low, then the resulting current will be high; if the resistance is high, then the resulting current will be low. As the applied frequency passes beyond the resonant frequency, the reactance starts to increase again, and the current starts to fall again.

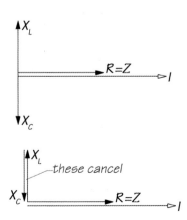

Figure 19.57

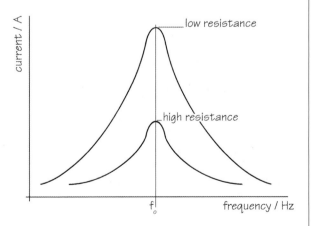

Figure 19.58

Now, let's look at the **voltage phasor diagram** for an *R-L-C* circuit at resonance, shown as Figure 19.59.

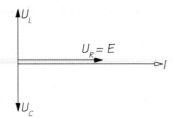

Figure 19.59

As the voltage drops, $\bar{U}_L$ and $\bar{U}_C$, are equal and opposite, then the phasor sum of $\bar{U}_L$, $\bar{U}_C$ and $\bar{U}_R$ will simply be $\bar{U}_R$. In other words, $\bar{U}_R = \bar{E}$ or, to put in another way: the *entire supply voltage will appear across the resistive component of the circuit*.

> You will recall that when we learnt how to draw general voltage phasor diagrams for *R-L-C* circuits, we were warned *always* to make the $\bar{U}_L$ phasor larger than the $\bar{U}_C$ phasor *(or vice versa)*. Well, now you know why! Making the two phasors the same lengths results in resonance, which is a *special* condition, not the normal condition!

It's very important to understand that we are *not* saying that the voltages $\bar{U}_L$ and $\bar{U}_C$ don't exist. They most certainly *do* exist! What we *are* saying is that, if we tried to measure the voltage drop across *both* the inductive *and* capacitive components, then a voltmeter would read zero because the two voltage drops are in *antiphase* – that is, they are equal in magnitude, but 180° out of phase with each other, as illustrated in Figure 19.60.

So, at any instant, the sum of the instantaneous voltages, $u_L + (-u_C)$, must be zero. Or, in general:

$$\bar{U}_L + (-\bar{U}_C) = 0$$

This can be summarised as shown in Table 19.1.

The voltage phasor diagram also tells us another important thing about resonance. At resonance, *the circuit's phase angle is zero*.

> At resonance the circuit's phase angle is zero.

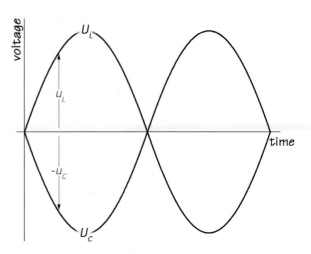

Figure 19.60

Table 19.1

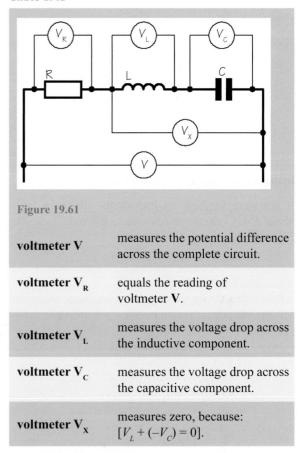

Figure 19.61

voltmeter V	measures the potential difference across the complete circuit.
voltmeter V_R	equals the reading of voltmeter **V**.
voltmeter V_L	measures the voltage drop across the inductive component.
voltmeter V_C	measures the voltage drop across the capacitive component.
voltmeter V_X	measures zero, because: $[V_L + (-V_C) = 0]$.

The final 'strange' thing which occurs, or rather, which *could* occur, at resonance is revealed in the following worked example.

Worked example 8 A circuit has a resistance of 10 Ω. At resonance, the values of the circuit's inductive reactance and capacitive reactance are each 100 Ω. If the supply voltage is 200 V, calculate (a) the current, (b) the voltage drops across the circuit's resistive component and (c) across its inductive and capacitive components.

Solution At resonance, the *only* opposition to current is the resistance of the circuit. So, the resulting current must be:

$$\bar{I} = \frac{\bar{E}}{R} = \frac{200}{10} = 20 \text{ A (Answer a.)}$$

At resonance, the voltage drop across the resistive component will be equal to the supply voltage, i.e. 200 V (Answer b.)

At resonance, the voltage drop across the inductive component is as follows:

$$\bar{U}_L = \bar{I} \, X_L = 20 \times 100 = 2000 \text{ V (Answer c.)}$$

At resonance, the voltage drop across the capacitive component is as follows:

$$\bar{U}_C = \bar{I} \, X_C = 20 \times 100 = 2000 \text{ V (Answer c.)}$$

Yes! The above answers *are* correct! Even though the supply voltage to the circuit is just 200 V, the voltage drop across each of the reactive components is indeed 2000 V!

If the circuit's resistance was just 1 Ω, instead of 10 Ω, then the current would rise to 200 A, and the voltage drops across the reactive components would become 20 000V!

Don't forget, these very large voltages are what will appear across the inductive component and the capacitive component, if they are measured *separately*. If we measure the voltage drop across *both* components, then the result would be zero, because the two voltage drops are in antiphase with each other!

So the final thing we can say about resonance is that, *if the resistance of the circuit is low compared with the inductive and capacitive reactances*, then very large voltages can appear across those components – *voltages that are many times the value of the supply voltage!*

At resonance, if the resistance of the circuit is low compared to the inductive and capacitive reactances, then the individual values of voltage drops, $\bar{U}_L$ and $\bar{U}_C$ can be *many* times larger than the supply voltage.

There are *two* ways of looking at this strange phenomenon. The first is from the electronics engineer's point of view; the second is from the transmission/distribution engineer's point of view.

Electronics engineers have to deal with signals that are frequently in the microvolt range. Resonance can boost these tiny voltages hundreds of times, providing what is, essentially, 'free amplification' of those signals.

Electricity transmission/distribution engineers, on the other hand, are already dealing with very high, and very dangerous, voltages. Unintentional resonance could boost these voltages far higher, with disastrous results to system insulation. For example, resonance could occur when a highly capacitive underground cable feeds a highly inductive transformer. If the resulting values of capacitive reactance and inductive reactance happen to be such that resonance, or even near-resonance, conditions result, then the resulting voltages could well exceed the dielectric strengths of the cable/transformer winding insulation, leading to failure.

Summary

To summarise what we have learnt about **series resonance**, we can say:

- series resonance occurs when: $X_L = X_C$
- a series circuit will resonate at a frequency, called its 'resonant frequency' (f_o), which is determined from:

$$f_o = \frac{1}{2\pi\sqrt{LC}}$$

- at resonance, a circuit's impedance will equal its resistance: $Z = R$.
- a circuit's current will reach its maximum value at resonance, and it will be in phase with the supply voltage.
- at resonance, the voltage drop across a circuit's resistive component will equal its supply voltage.
- at resonance, the sum of the voltage drops across a circuit's inductive and capacitive components will be zero.

- at resonance, the individual voltage drops appearing across the inductive or capacitive components can be very much higher than the supply voltage.

Finally . . .

Now that you have completed this chapter, are you able to achieve the objectives or learning outcomes listed at the beginning of this chapter?

Ask yourself, 'Can I . . .'

1 sketch a waveform for each of the following, showing the phase relationship between the current and supply voltage:
   a purely resistive circuit
   b purely inductive circuit
   c purely capacitive circuit
   d series *R-L* circuit
   e series *R-C* circuit.
2 state the phase relationship between the current and supply voltage for
   a a purely resistive circuit
   b a purely inductive circuit
   c a purely capacitive circuit.

3 sketch the phasor diagram representing a
   a purely resistive circuit
   b purely inductive circuit
   c purely capacitive circuit
   d series *R-L* circuit
   e series *R-C* circuit
   f series *R-L-C* circuit.
4 develop an impedance diagram for a
   a series *R-L* circuit
   b series *R-C* circuit
   c series *R-L-C* circuit.
5 from impedance diagrams, derive equations for resistance, inductive reactance, capacitive reactance and impedance, in terms of voltages and currents.
6 state the equation for inductive reactance, in terms of inductance and frequency.
7 state the equation for capacitive reactance, in terms of capacitance and frequency.
8 explain what is meant by the term 'series resonance'.
9 list the effects of series resonance.
10 solve problems on series a.c. circuits, including series-resonant circuits.

Online resources
The companion website to this book contains further resources relating to this chapter. The website can be accessed via the following link:
www.routledge.com/cw/waygood

Chapter 20

Power in alternating-current circuits

On completion of this chapter, you should be able to:

1 define the terms energy, work, heat and power, and specify their SI units of measurement.
2 explain the behaviour of power in
   a a purely resistive a.c. circuit
   b a purely inductive a.c. circuit
   c a purely capacitive a.c. circuit
   d an *R-L* a.c. circuit
   e an *R-C* a.c. circuit.
3 state the relationship between true (or 'active') power, reactive power and apparent power.
4 state the units of measurement for
   a true (or 'active') power
   b reactive power
   c apparent power.
5 change the voltage phasor diagram into a power diagram, and derive equations for true, reactive and apparent power for *R-L*, *R-C* and *R-L-C* circuits.
6 define the term 'power factor'.
7 solve problems on power in a.c. circuits.

Introduction

In this chapter, we are going to examine the behaviour of **energy** and **power** in alternating-current circuits.

Before we do so, though, we need to remind ourselves of how energy can be manipulated, so you may want to revise the earlier chapter on *energy, work, heat and power*.

As we have learnt, energy can be 'manipulated' in either of *two* ways: it can be *changed from one form into another*, or it can be *transferred from one body to another*.

When energy is *changed from one form into another*, we say that '**work**' is being done. For example, when an electric motor converts electrical energy into kinetic energy, the motor is doing work.

Energy is *transferred between objects* whenever those objects are at different temperatures. We call this process '**heat transfer**'. You will recall that the term 'heat' describes *'the transfer of energy from a warmer body to a cooler body'*.

Both work *and* heat are expressed in joules.

> **Work** is defined as *'the conversion of energy from one form into another'*.

> **Heat** is defined as *'the transfer of energy from a body at a higher temperature to one at a lower temperature'*.

Finally, we should also remind ourselves that '**power**' is defined as *'the rate of doing work'* or 'the rate of heat transfer', expressed in watts.

> **Power** is defined as *'the rate of doing work, or the rate of heat transfer'*.

An Introduction to Electrical Science, Waygood, ISBN 9780415810029, 2013. © Taylor & Francis

Behaviour of energy in a.c. circuits

Behaviour of energy in purely resistive circuits

Whenever an electric current overcomes the resistance of a conductor, it does **work** on that conductor $(W = I^2Rt)$, causing its internal energy (the vibration of its atoms) to increase. An increase in internal energy is always accompanied by an increase in temperature and, if the temperature of the conductor exceeds that of its surroundings, then energy will be transferred *away* from the conductor into its surroundings through **heat transfer**.

As you may recall, from the chapter on *resistance*, the *consequence of resistance is heat.*

Heat transfer away from a conductor is *completely irreversible*. That is, you cannot transfer that energy back into the conductor as electrical energy, and send it back to the supply. Once it's gone, it's gone for good!

The *rate* of heat transfer away from a conductor to its surroundings is termed 'power', and is expressed in watts. For reasons that will shortly become clear, in a purely resistive circuit, it is traditional to refer to this as either '**true power**', '**real power**' or '**active power**' (symbol: P). Throughout this rest of this book, we will stick with the term 'true power'.

Behaviour of energy in a purely inductive circuit

You will recall that a purely inductive circuit is an 'ideal' circuit, in which there is no resistance.

If a circuit has no resistance, then no heating can take place. So there can be no expenditure of energy, through heat transfer, away from the circuit.

However, due to the circuit's inductance, the presence of a current creates a *magnetic field*. Because the a.c. current is continuously changing in both magnitude and direction, the resulting magnetic field also varies in magnitude and direction and induces a voltage into the circuit which always acts to oppose the change in current.

During the first quarter-cycle, then, the *increasing* current is *opposed* by this induced voltage. Energy is required to overcome this opposition, and this energy is drawn from the supply and then stored within the magnetic field. During the second quarter-cycle, as the current *falls*, the direction of the induced voltage reverses, and acts to oppose the reduction of the current – that is, it tries to *sustain* the current. In other words, the energy that was previously *stored* within the magnetic field, is now being *returned* to the supply.

This process repeats itself during subsequent quarter-cycles, with energy being alternately *stored* in the magnetic field and, then, *released* back to the supply. So, although there *is* energy conversion taking place *within* the circuit, there is no net loss of energy away *from* the circuit.

As always, the *rate* at which this energy conversion is taking place, is power. But to distinguish it from 'true power', we call it '**reactive power**' (symbol: Q). It's also traditional to measure reactive power in **reactive volt amperes** (symbol: **var**), rather than in watts.

Some textbooks describe reactive power as the rate at which energy is continually *'sloshing back and forth, between the supply and the load'.*

> Some textbooks refer to reactive power as *'wattless power'* or *'imaginary power'*. The term *'imaginary power'*, though, *doesn't* mean it exists only in the mind! In this sense, the word 'imaginary' is used by mathematicians to mean 'quadrature' (i.e. 'at right angles'), in other words it is created by a load current that lags (or, as we shall see, leads) the supply voltage by 90°.

Behaviour of energy in a purely capacitive circuit

A purely capacitive circuit also has no resistance so, again, no heating effect can take place to cause any transfer of energy away from the circuit.

The behaviour of energy in a purely capacitive circuit is almost identical to its behaviour within a purely inductive circuit, except that energy from the circuit is being alternately stored in, and returned from, an *electric* field rather than a magnetic field.

Once again, although there *is* energy conversion taking place within the circuit, there is no net loss of energy away *from* the circuit.

As with a purely inductive circuit, the rate at which this energy conversion is taking place is also called 'reactive power', which is also measured in reactive volt amperes to distinguish it from true power.

Behaviour of energy in R-L, R-C and R-L-C circuits

In **R-L**, **R-C** and **R-L-C** circuits, the behaviour of energy is a *combination* of the behaviour of energy in purely resistive and purely reactive circuits. That is *some* energy is being irreversibly lost from the circuit through heat transfer or by the work done by the load, while *some* energy is being alternately stored, and returned to the supply from, the magnetic or electric fields (or *both* magnetic *and* electric fields, in the case of an *R-L-C* circuit).

So resistive-reactive circuits have *both* 'true power' *and* 'reactive power' occuring at the same time. As we shall see, shortly, we *cannot* simply add these two quantities together, but the *combination* of the two we call '**apparent power**' (symbol: *S*) and, to distinguish apparent power from true power and reactive power, it is traditional to measure it in **volt amperes*** (symbol: **V·A**).

> *It's worth pointing out that the units, **reactive volt amperes (var)** and **volt amperes (V·A)**, are *traditional*, in order to easily distinguish reactive power, apparent power and true power from each other. SI doesn't recognise either, and uses the watt to measure all 'forms' of power.

Further explanation of true power

For a *purely resistive* circuit, we learnt that **true power** is the rate at which energy is irreversibly lost from the circuit through heat transfer to its surroundings. However, in some circuits, it can be a little more complicated that this.

Motors, for example, do work by converting electrical energy into kinetic energy, in order to drive their mechanical loads. As the output power of a motor is expressed in **watts**, the rate at which a motor does work driving its load must represent its '**true power**'. Since motors are obviously not 'purely resistive' loads that dissipate energy through heat transfer, how do we account for this?

Because of its windings, a motor is an *R-L* circuit and, as we have learnt ('**CIVIL**'), the

motor's supply current will lag the supply voltage by some angle, φ.

As can be seen in Figure 20.1, the motor's (lagging) supply current can actually be resolved into two 'components', called an *'in-phase'* (or *'active'*) component, and a *'quadrature'* (meaning *'at 90°'*) component.

The in-phase component represents the machine's **load current** (i.e. used, in this sense, to describe the component of current responsible for the work done by the motor in driving its **load**), whereas the quadrature component represents its **magnetising current**.

These two components of the motor's current don't *actually* exist as two individual currents, but the circuit *behaves as if they do*. In the equivalent circuit for a motor, shown in Figures 20.2 and 20.3, we show the winding and its equivalent resistance as two separate branches of the same circuit although, in reality, they are one and the same.

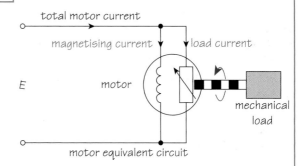

Figure 20.2

The quadrature (**magnetising current**) component of current (which accounts for the reactive power providing the motor's magnetic field) is constant, but the in-phase component (**load current**) *varies with load* – if the motor's

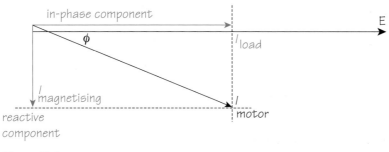

Figure 20.1

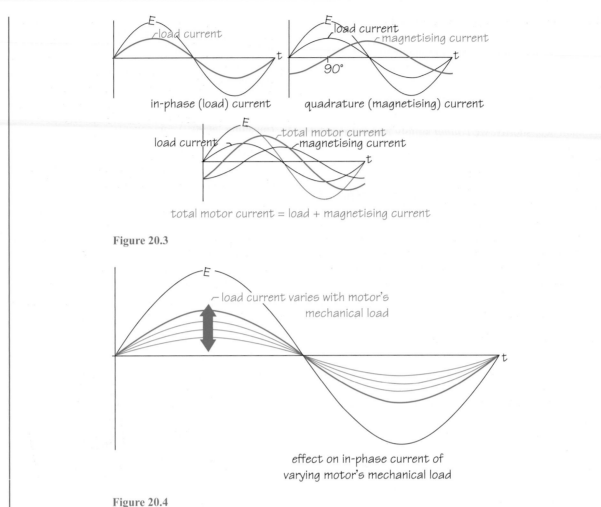

Figure 20.3

Figure 20.4

mechanical load *increases*, then the in-phase component of the current will also *increase* in order to supply more 'true power' to drive that load. So, in a sense, the varying mechanical load on the motor behaves like a varying resistive load.

So, the 'equivalent resistance' of the motor is *not* the same as its *actual* resistance, and it *varies* (which is why we have shown it as a variable resistor in the schematic diagram, Figure 20.4) as the load varies – accounting for why the in-phase component of the current varies with load.

Summary

To summarise our explanations for **true power**, **reactive power** and **apparent power**, we can say that

- the *in-phase* or *active* component of a lagging or leading load current is responsible for **true power**, which represents the rate at which energy is expended by the supply – i.e. the rate at which energy is either permanently lost through heat transfer, or the rate at which work is done by a motor, driving its mechanical load (e.g. when a motor converts electrical energy into kinetic energy), or a combination of the two.
- the *reactive* or *quadrature* component of the lagging or leading load current is responsible for **reactive power**, which is the rate at which energy is alternately stored in, and returned from, a magnetic or electric field.
- the load current is responsible for **apparent power**, which is the combination (but *not* the sum) of true power and reactive power.

As we shall learn, the apparent power is the vector sum of true power and reactive power.

A.C. power waveforms

Another approach to understanding the behaviour of power in a.c. circuits is to examine the **power waveform**, which is derived from the corresponding voltage and current waveforms.

Throughout the following, it is conventional to describe the rate at which energy is delivered *from the supply to the load* as being '**positive**' **power**, while the rate at which energy is *returned from the load to the supply* as being '**negative**' **power**. In this context, the terms 'positive' and 'negative' describe *flow directions*, and *not* polarities.

Purely resistive circuits

For a purely resistive circuit, the load current and supply voltage are *in phase* with each other. In Figure 20.5, the power waveform is constructed by multiplying those values of instantaneous voltage and current, that occur at the same instant in time, over a complete cycle. That is: $p = e\,i$.

To clarify this process, we'll consider just two instants, occurring at points **A** and **B**.

At point **A**, then, the resulting point on the power waveform is the product of the instantaneous voltage, *e*, and the instantaneous current, *i*. Both of these are positive, so the corresponding point on the power

curve is, therefore, also positive. The power waveform is thus plotted by repeating this process for numerous instantaneous values of current and voltage throughout one complete cycle of voltage/current.

At point **B**, the point on the power waveform is, again, the product of the instantaneous voltage and current at that particular point along the axis. However, this time, both the instantaneous voltage *and* the instantaneous current are *negative*. The resulting point on the power curve is, therefore, positive, because the product of two negatives is a positive – that is:

$$-e\,(-i) = +p$$

Throughout the second half-cycle, of course, we will continue to multiply *negative* instantaneous voltages by *negative* instantaneous currents, which result in *positive* instantaneous power.

The resulting power waveform is illustrated in Figure 20.6.

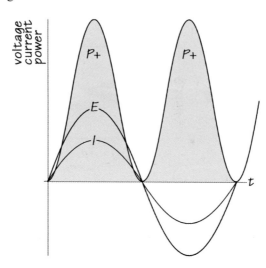

Figure 20.6

The resulting power waveform is called a 'sine-squared' (sine^2) waveform, meaning that it is a sine wave that varies entirely *above* the horizontal axis (it is also double the frequency of the voltage frequency), so that it is entirely *positive* – indicating, by convention, that the energy flow is *from* the supply *towards* the load. The *amount* of power is represented by the grey areas enclosed between the power waveform and the horizontal axis.

So, for a purely resistive load, energy is being *expended* by the supply and is being entirely *consumed* by the load or *lost* to the surroundings, through heat transfer, and the rate at which this is happening is termed the **true power** of the circuit, expressed in **watts**.

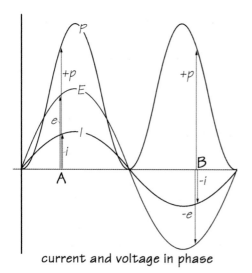

current and voltage in phase

Figure 20.5

Some students are confused as to why, when a.c. current reverses direction every half-cycle, the direction of power doesn't *also* reverse every half-cycle. Hopefully, the 'sense arrows' (described in the chapter on *series, parallel and series-parallel circuits*) will provide the answer.

During the first half-cycle (top illustration in Figure 20.7), the instantaneous potential differences and currents are both acting in the same *directions* as the assumed sense arrows and, so, their product represents 'positive power': $e \times i = +p$.

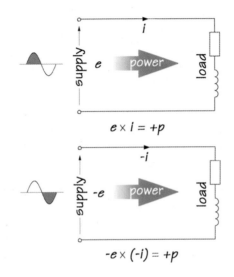

Figure 20.7

During the second half-cycle (bottom illustration), the instantaneous potential differences and currents are both acting in the *opposite* directions to the assumed sense arrows and, so, are each considered to be negative in direction. Their product of two negatives, of course, is a positive – so, once again, we have 'positive power': $-e \times (-i) = +p$.

Purely inductive circuits

For a purely inductive circuit, the load current lags the supply voltage by 90°. In Figure 20.8, the power waveform is, again, constructed by multiplying those values of instantaneous voltage and current, that occur at the same instant of time, throughout a complete cycle. Again, to clarify this process, we'll consider just two instants, at points **A** and **B**.

At point **A**, the point on the power waveform is the product of the instantaneous voltage, e, and the instantaneous current, $-i$, because at this point along the axis, the instantaneous voltage is positive, whereas the corresponding instantaneous current is negative. The corresponding point on the power curve, therefore, is *negative*:

$$e\,(-i) = -p$$

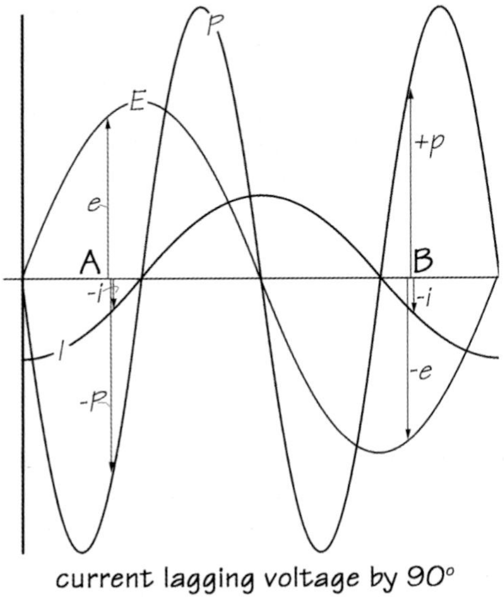

current lagging voltage by 90°

Figure 20.8

At point **B**, the point on the power waveform is, again, the product of the instantaneous voltage and current at that particular point along the axis. However, this time, both the instantaneous voltage *and* the instantaneous current are *positive*. The resulting point on the power curve is, therefore, positive, because the product of two negatives is a positive – that is:

$$+e\,(+i) = +p$$

This process is repeated for numerous values of instantaneous voltages and currents throughout a complete cycle of voltage/current, and the resulting power waveform, which is sinusoidal with twice the frequency of the voltage, is illustrated in Figure 20.9. You will notice, the grey areas enclosed by the power waveform *above* the horizontal axis ('positive' power) are exactly equal to the grey areas enclosed by the power waveform *below* that axis

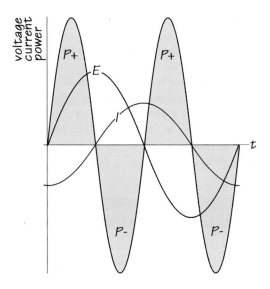

Figure 20.9

('negative' power). So, the rate at which energy is *delivered to* the circuit (the 'positive' power) during each quarter-cycle, and stored in a magnetic field, is *balanced exactly* by the rate at which energy is *returned to* the supply from the collapsing magnetic field (the 'negative' power) during the following quarter-cycle.

So, as already explained, the energy in this circuit is being alternately *stored in*, and *returned from*, a magnetic field. No energy is being consumed by the load or lost to the surroundings through heat transfer, so the 'true power' is zero. On the other hand, the rate at which this energy is being transferred back and forth represents the '**reactive power**' of the circuit, expressed in **reactive volt amperes**.

Purely capacitive circuits

If we were to draw the power waveform for a purely capacitive circuit, then we would end up with a similar situation to that of a purely inductive circuit. That is, the amount of 'positive' power would be exactly balanced by the amount of 'negative' power. So, once again, the rate at which energy is being *delivered* every quarter-cycle to the load and stored in an electric field is balanced exactly by the rate at which energy is being *returned* to the supply when that field collapses during the following quarter-cycle.

So, as for a purely capacitive circuit, with no energy being consumed by the load or lost through heat transfer, the 'true power' is zero. But the rate at which

energy is being transferred back and forth represents the '**reactive power**' of the circuit.

Power in resistive-reactive circuits

In the chapter on *a.c. series circuits*, we learnt that 'real' circuits are a *combination* of resistance, inductance and/or capacitance. So, in a 'real circuit' – i.e. resistive-reactive circuits – then:

- some energy is being *permanently lost* due to the resistive component of the circuit, through heat transfer and/or the work done by loads such as motors, while
- some energy is being alternately stored, and returned to the load from the magnetic and/or electric fields associated with the circuit's *reactive (inductive and/or capacitive) components*.

In 'real' circuits, therefore, there exists both **true power** (the rate at which energy is *permanently* lost) *and* **reactive power** (the rate at which energy is continually stored in, and returned from, magnetic or electric fields).

In the power waveform diagram for an *R-L* circuit, Figure 20.10, the current lags the supply voltage by some phase angle, φ. You will notice that the amount of 'positive' power *above* the horizontal axis, is larger than the amount of 'negative' power, *below* that axis. This means that the rate of energy transfer *to* the load is *greater* than the rate of energy transfer (temporarily stored in the magnetic field) from the load back to the supply.

To summarise, we can say that **true power** is the rate at which energy is lost through heat transfer or

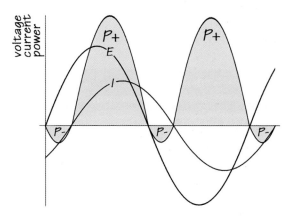

Figure 20.10

by the work done by a load such as a motor, whereas **reactive power** is the rate at which energy must be supplied in order to *sustain the magnetic and/or electric fields*.

> We should not assume that reactive power is unimportant; in fact, reactive power is *essential* to the operation of an electrical transmission and distribution system, in order to maintain magnetic/electric fields, and to maintain the system voltages required to 'push' the power demanded by loads along the power lines.

Apparent power, true power and reactive power

Now let's move on to examine the mathematical relationship between **apparent power**, **true power** and **reactive power**.

As we have already learnt, apparent power is the *combination* of a circuit's true power and reactive power. However, the relationship between apparent power, true power and reactive power is *not* a simple algebraic relationship but, rather, a *vectorial* relationship, as we will see shortly.

Units of measurement

Whether we are discussing **apparent power**, **true power** *or* **reactive power**, we should bear in mind that 'power' is *always* the rate at which energy is being transferred – *regardless* of whether that work is reversible or irreversible, useful or useless! So, there is absolutely no technical reason, therefore, why each of these quantities shouldn't be measured using the *same* unit of measurement – the **watt** (symbol: **W**). In fact,

Table 20.1

Quantity	Symbol	Unit of measurement	Symbol
apparent power	S	volt ampere	V·A
true power	P	watt	W
reactive power	Q	reactive volt ampere	var

that is precisely what SI does! SI doesn't acknowledge the following units.

Traditionally, however, *in order to clearly differentiate between each of these quantities*, different units of measurement have been allocated to them, and it is unlikely we will ever see them being replaced by the watt! In fact, using these distinctive units of measurement is very useful, as there can then be absolutely no doubt as to which 'form' of power is then being referred to. This is summarised in Table 20.1.

Power in a series R-L circuit

You will recall that to convert a phasor diagram into an *impedance diagram*, we divided throughout by the reference phasor, $\overline{I}$. Doing this generated various useful equations for *resistance*, *inductive reactance* and *impedance*, as well as for the *phase angle* (Figure 20.11).

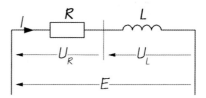

Figure 20.11

Well, by *multiplying* the phasor diagram by the reference phasor, $\overline{I}$, we can produce a power **diagram** and, at the same time, generate various equations that will allow us to determine the circuit's **true power**, **reactive power**, **apparent power** and **power factor** (more on this, later).

Step 1: We start by drawing the voltage phasor diagram for the circuit, as we did in the chapter on *a.c. series circuits* (Figure 20.12).

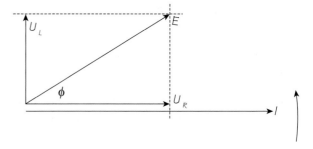

Figure 20.12

Step 2: *Multiply* throughout by the reference phasor, $\overline{I}$ (Figure 20.13).

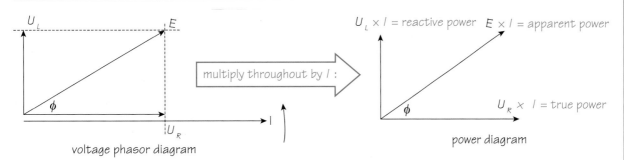

voltage phasor diagram   power diagram

multiply throughout by I :

Figure 20.13

So, by simply *multiplying the voltage phasor diagram by $\overline{I}$*, we have created a **power diagram**, from which the following equations have been derived:

$$\text{apparent power} = \overline{I}\,\overline{E} \quad \text{true power} = \overline{I}\,\overline{U}_R$$
$$\text{reactive power} = \overline{I}\,\overline{U}_L$$

If we know *any two* out of these three, then *we can find the third*, by applying *Pythagoras's Theorem*. For example, to find apparent power in terms of true and reactive power (Figure 20.14).

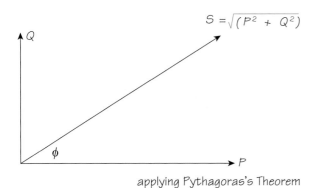

$$S = \sqrt{(P^2 + Q^2)}$$

applying Pythagoras's Theorem

Figure 20.14

$$(\text{apparent power})^2 = (\text{true power})^2$$
$$+ (\text{reactive power})^2$$

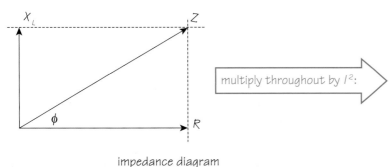

impedance diagram

multiply throughout by I^2:

Figure 20.15

$$S = \sqrt{P^2 + Q^2}$$

To demonstrate further how useful phasor diagrams are in generating equations for power, let's look at another example of using this technique.

We know that power in a resistive circuit can be obtained from $P = \overline{I}^2 R$, so if we redraw the impedance diagram, and multiply throughout by $\overline{I}^2$, we'll end up with another version of the power diagram, as shown in Figure 20.15.

Thus generating the following additional equations for power!

$$\text{apparent power} = \overline{I}^2 Z \quad \text{true power} = \overline{I}^2 R$$
$$\text{reactive power} = \overline{I}^2 X_L$$

If a load is 'static' such as, say, an **inductor** having a resistance R and inductive reactance, X_L, then the above equation for **true power** may be manipulated to determine the *actual* **resistance** of that circuit, if the true power is known. That is,

$$\text{since true power} = \overline{I}^2 R \text{ then } R = \frac{\text{true power}}{\overline{I}^2}$$

However, if the load is 'dynamic', such as a **motor**, then its true power will vary as its mechanical load varies. So if you use the same equation to determine the

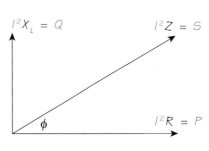

'resistance' of the motor circuit, what you will obtain is the '**equivalent resistance**' of the machine, *not* its true resistance. As we learnt, earlier in this chapter, the equivalent resistance of a motor will vary as the mechanical load varies.

> **Note**! Hopefully, you are now beginning to realise how important it is to be able to construct a voltage phasor diagram and to be able to change this into an impedance diagram and into a power diagram. These diagrams will generate all the equations that you will ever need to solve a.c. problems, without the need to commit any of these equations to memory!

Power factor

In a d.c. circuit, *regardless of the type of load*, we can determine the power simply by multiplying together the readings of a voltmeter and an ammeter.

In a resisitive-reactive a.c. circuit, however, the product of the supply voltage and the load current gives us the **apparent power** of the load, *not* its true power.

To determine its **true power**, in practice, we must use a *wattmeter*, which is designed specifically to measure true power, by monitoring the supply voltage together with the *in-phase (resistive) component* of the load current (Figure 20.16).

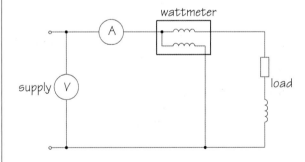

Figure 20.16

The *ratio of a load's true power to its apparent power* is *very* important, and is called the '**power factor**' of the load:

$$\text{power factor} = \frac{\text{true power}}{\text{apparent power}}$$

If we re-examine the power diagram, derived from the voltage phasor diagram, it should be obvious that the ratio of true power to apparent power (adjacent over hypotenuse) is the cosine of the phase angle:

$$\text{power factor} = \frac{\text{true power}}{\text{apparent power}} = \frac{\text{adjacent}}{\text{hypotenuse}} = \cos\phi$$

So an alternative definition for power factor is that it is '*the cosine of the load's phase angle*'.

Power factor is usually expressed as a *per-unit* value (e.g. 0.85), although it may still (usually in older textbooks) occasionally be seen expressed as a *percentage* value (e.g. 85%).

A circuit's power factor *must* also be specified as '**leading**' or '**lagging**' (e.g '0.85 lagging'). These terms refer to *where the* **load current** *phasor lies in relation to the supply voltage phasor* (never the other way around!). Resistive-inductive circuits, therefore, always have 'lagging' power factors, while resistive-capacitive circuits always have 'leading' power factors.

The reason that power factor is so important is that it may be thought of as an indication of the 'efficiency' of the rate of energy conversion, and it might prove helpful to think of it in terms of the mechanical analogy, shown in Figure 20.17.

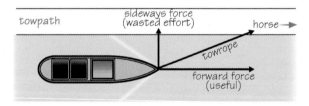

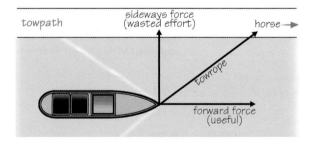

Figure 20.17

The two diagrams in Figure 20.17 represent a plan view of a barge being pulled along a canal by a horse, walking along the canal's towpath.

The force in the tow-rope can be resolved into two (imaginary) forces: one *in the direction of the barge*, and the other *towards the canal's bank*.

Of these two resolved forces, it is the one acting in the direction of the barge's motion that contributes the *useful* force, as it's acting to pull the barge forward. On the other hand, the resolved force acting towards the bank (quadrature force) contributes nothing to the forward motion to the barge. Yet neither force can exist without the other!

In the second illustration, the canal is wider, and the barge is further from the bank, so the tow-rope assumes a somewhat greater angle than before.

If we, again, resolve the force in the tow-rope (assuming that the same forward resolved force is required to keep the barge moving at the same velocity), we notice that the resolved force towards the canal bank is *greater* than it was before.

In this new situation, *in order to provide the same amount of forward force*, a greater force *has to be made available in the direction of the bank*.

This situation may be likened to the power supplied to an resistive-reactive circuit. The **true power** is equivalent to *the forward force acting on the barge*; the **reactive power** is equivalent to the *'wasted' force acting towards the canal bank* (which, despite being *apparently* 'wasted', is still *essential* to the behaviour of the barge). The greater the phase angle (equivalent to the angle between the forward force on the barge and the force in the tow-rope), the more 'apparently wasted' (but, nonetheless essential!) reactive power that must be provided!

By expressing the angle between the tow-rope and the forward-resolved force in terms of its cosine, the 'efficiency' of the arrangement is indicated. The maximum efficiency (×1 or 100%) occurs when the tow-rope acts in the *same direction as the necessary forward force* (cos 0° = 1); and the minimum efficiency (×0 or 0%) occurs when the tow-rope is at 90° (cos 90° = 0) to the forward direction of the barge – i.e. no forward motion whatsoever!

In terms of **power factor**, 'efficiency' isn't really the appropriate term to describe what it represents. It would be more accurate to say that

> **Power factor** is *'the percentage of apparent power that represents the rate of doing real work'*.

In other words,

- when the power factor is unity (1), the apparent power is equal to the true power.
- when the power factor is zero (0), the apparent power is equal to the reactive power.

For most circuits, the power factor lies somewhere in between these two extremes.

Power in a series R-C circuit

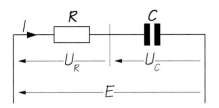

Figure 20.18

As with the *R-L* circuit, described above, to convert a phasor diagram into a *power diagram*, we simply multiply the phasor diagram by the reference phasor, $\bar{I}$, and generate equations that will allow us to determine the circuit's true power, reactive power, apparent power and power factor (Figure 20.18).

Step 1: Draw the voltage phasor diagram for the circuit, as we did in the chapter on *a.c. series circuits* (Figure 20.19).

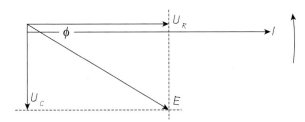

Figure 20.19

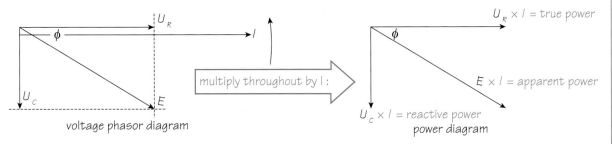

Figure 20.20

Step 2: *Multiply* throughout by the reference phasor, $\overline{I}$ (Figure 20.20).

So, by simply *multiplying the voltage phasor diagram by* $\overline{I}$, we have created a **power diagram**, from which the following equations have been generated:

$$\text{apparent power} = \overline{I}\,\overline{E} \quad \text{true power} = \overline{I}\,\overline{U}_R$$

$$\text{reactive power} = \overline{I}\,\overline{U}_C$$

If we know *any two* out of these three, then *we can find the third*, by applying *Pythagoras's Theorem* (Figure 20.21). For example:

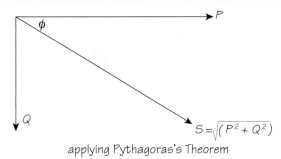

applying Pythagoras's Theorem

Figure 20.21

$$(\text{apparent power})^2 = (\text{true power})^2$$
$$+ (\text{reactive power})^2$$
$$S = \sqrt{P^2 + Q^2}$$

So, once again, the ability to draw a power diagram *saves you having to remember numerous equations*, as they can all be generated from the power diagram, by applying simple geometry (Pythagoras's Theorem) or simple trigonometry (the cosine ratio)! Let's look at what else we can derive from the power diagram.

For example, if we know the *apparent power* and the *true power*, then we could rearrange Pythagoras' Theorem, as follows:

reactive power

$$= \sqrt{(\text{apparent power})^2 - (\text{true power})^2}$$

To find the **power factor**, we simply need to find the cosine of the phase angle:

$$\cos\phi = \frac{\text{adjacent}}{\text{hypotenuse}} = \frac{\text{true power}}{\text{apparent power}}$$

This time, we *always* describe the power factor as **leading**, because, by definition, the *current leads the supply voltage*.

Again, we can manipulate this equation, to find another equation for the true power of a circuit:

$$\text{since } \cos\phi = \frac{\text{true power}}{\text{apparent power}}$$

$$\text{then true power} = \text{apparent power} \times \cos\phi$$

This gives us a very important equation:

$$\text{true power} = (\overline{E}\,\overline{I})\cos\phi$$

As was the case for the series *R-L* circuit, we can also convert an *R-C* impedance diagram into a power diagram, by simply multiplying throughout by I^2 (Figure 20.22).

Thus generating the following additional equations for power:

$$\text{apparent power} = \overline{I}^2 Z \quad \text{true power} = \overline{I}^2 R$$

$$\text{reactive power} = \overline{I}^2 X_C$$

Note! If you learn *nothing else* from this chapter, you *must* learn **the importance of being able to draw a voltage phasor diagram and to be able to convert it into a power diagram,** for this will generate *all* the equations you will *ever* need to know in order to solve a.c. power problems.

This will save you from *ever* needing to memorise these equations!

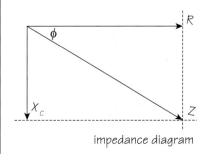

impedance diagram

multiply throughout by I^2:

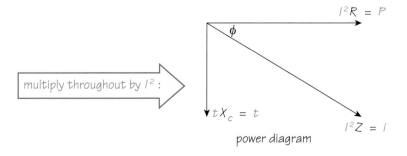

power diagram

Figure 20.22

Worked example 1 An inductor, of inductance 16 mH and resistance 2 Ω, is connected across a 24-V, 50 Hz, supply. Calculate each of the following:

a inductive reactance
b impedance
c current.
d apparent power.
e true power.
f reactive power.
g power factor.

Solution As with any problem, always start by sketching the circuit and inserting all the values given in the problem (Figure 20.23). You notice that this circuit consists of an inductor; you are given its inductance and resistance, so you can assume that it is equivalent to a *series R-L circuit*.

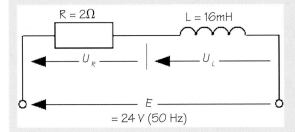

Figure 20.23

Next, sketch the corresponding voltage phasor diagram (Figure 20.24).

Next, convert this into an impedance diagram, by dividing throughout by $\overline{I}$ (Figure 20.25).

a None of the generated equations allow us to determine the inductive reactance, so we must

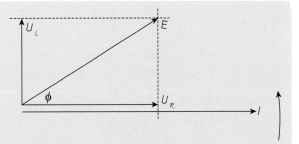

Figure 20.24

fall back on the basic equation that we have committed to memory:

$$X_L = 2\pi f = 2\pi \times 50 \times (16 \times 10^{-3}) = 5.03 \; \Omega$$

(Answer a.)

b $Z = \sqrt{R^2 + X_L^2} = \sqrt{2^2 \times 5.03^2} = \sqrt{29.30} = 5.41 \; \Omega$

(Answer b.)

c To find the current, we can use the following equation generated by the impedance diagram:

$$\overline{I} = \frac{E}{Z} = \frac{24}{5.41} = 4.44 \; A \; \text{(Answer c.)}$$

For the rest of this problem, redraw the impedance diagram, and multiply by $\overline{I}^2$ to create a power diagram. Then, we can utilise the equations generated by the power diagram to solve the remaining parts of the problem (Figure 20.26).

d. apparent power $= \overline{I}^2 Z$

$$= 4.44^2 \times 5.41$$

$$= 106.65 \; V \cdot A \; \text{(Answer d.)}$$

e. true power $= \overline{I}^2 R = 4.44^2 \times 2 = 39.43 \; W$

(Answer e.)

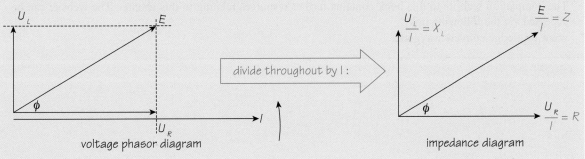

Figure 20.25

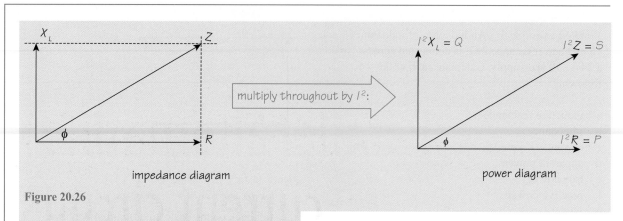

Figure 20.26

f. reactive power $= \bar{I}^2 X_L$

$\qquad = 4.44^2 \times 5.02$

$\qquad = 98.96$ var (Answer f.)

g. power factor $= \dfrac{\text{true power}}{\text{apparent power}}$

$\qquad = \dfrac{39.43}{106.65}$

$\qquad = 0.3697$ lagging (Answer g.)

('Lagging', because it is an **inductive** circuit, and current always lags the supply voltage in an inductive circuit.)

Finally . . .

Now that you have completed this chapter, are you able to achieve the objectives or learning outcomes listed at the beginning of this chapter?

Ask yourself, 'Can I . . .'

1   define the terms energy, work, heat and power, and specify their SI units of measurement.
2   explain the behaviour of power in
    a   a purely resistive a.c. circuit
    b   a purely inductive a.c. circuit
    c   a purely capacitive a.c. circuit
    d   an *R-L* a.c. circuit
    e   an *R-C* a.c. circuit.
3   state the relationship between true (or 'active') power, reactive power and apparent power.
4   state the units of measurement for
    a   true (or 'active') power
    b   reactive power
    c   apparent power.
5   change the voltage phasor diagram into a power diagram, and derive equations for true, reactive and apparent power for *R-L*, *R-C* and *R-L-C* circuits.
6   define the term 'power factor'.
7   solve problems on power in a.c. circuits.

Online resources

The companion website to this book contains further resources relating to this chapter. The website can be accessed via the following link:

www.routledge.com/cw/waygood

Chapter 21

Parallel alternating-current circuits

On completion of this chapter, you should be able to

1   draw a current phasor diagram for a parallel:
   a   resistive-inductive *(R-L)* circuit
   b   resistive-capacitive *(R-C)* circuit
   c   resistive-inductive-capacitive *(R-L-C)* circuit.
2   derive an admittance diagram, and derive expressions for conductance, susceptance, admittance and phase angle for a parallel:
   a   resistive-inductive *(R-L)* circuit
   b   resistive-capacitive *(R-C)* circuit
   c   resistive-inductive-capacitive *(R-L-C)* circuit.
3   derive a power diagram, and derive expressions for true power, reactive power, apparent power and power factor for a parallel:
   a   resistive-inductive *(R-L)* circuit
   b   resistive-capacitive *(R-C)* circuit
   c   resistive-inductive-capacitive *(R-L-C)* circuit.
4   solve problems on parallel *R-L*, *R-C* and *R-L-C* circuits.

Introduction

As it was with the earlier chapters on *series a.c. circuits* and *power in a.c. circuits*, the key to understanding and solving **parallel a.c. circuits** is the ability to be able to draw phasor diagrams from which *practically every equation you will ever need can be generated – providing you know how to use Pythagoras's Theorem and basic trigonometry*. So you are urged *not* to waste your time trying to memorise all the various equations!

We have learnt how all 'real' a.c. circuits exhibit combinations of **resistance** (symbol: R), **inductance** (symbol: L) and **capacitance** (symbol: C).

We also learnt that 'real' circuits are relatively complicated because they contain *combinations of resistance, inductance and capacitance* and, in order to understand the behaviour of a.c. circuits, it is necessary to start by first considering how 'ideal' circuits would behave. We described these 'ideal' circuits as **purely resistive**, **purely inductive** and **purely capacitive**, and we discovered that

a   in a **purely resistive** circuit, *the current and voltage are in phase* with each other.
b   in a **purely inductive** circuit, *the current lags the voltage by 90°.*
c   in a **purely capacitive** circuit, *the current leads the voltage by 90°.*

To help us remember these *very* important relationships, we can use the mnemonic, '**CIVIL**', in which '**C**' stands for 'capacitive circuit', and '**L**' stands for '**inductive** circuit'.

Finally, we discovered that the opposition to current in a purely resistive circuit is called **resistance** (R), the opposition to current in a purely inductive circuit is called **inductive reactance** (X_L) and that the opposition to current in a purely capacitive circuit is called **capacitive reactance** (X_C), where:

$$X_L = 2\pi f L \quad X_C = \frac{1}{2\pi f C}$$

An Introduction to Electrical Science, Waygood, ISBN 9780415810029, 2013. © Taylor & Francis

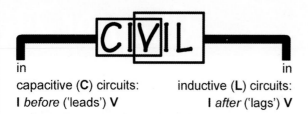

in   in

capacitive (**C**) circuits: inductive (**L**) circuits:

I *before* ('leads') **V**   **I** *after* ('lags') **V**

Figure 21.1

where:

$$X_L = \text{inductive reactance (ohms)}$$
$$X_C = \text{capacitive reactance (ohms)}$$
$$f = \text{supply frequency (hertz)}$$
$$L = \text{inductance (henrys)}$$
$$C = \text{capacitance (farads)}$$

Unlike *all* the other equations that we will need to use for solving a.c. circuits, these two equations have to be committed to memory, as their derivation requires a knowledge of calculus, which is beyond the scope of this text.

All other equations can be derived by learning to represent circuits using phasor diagrams and their derivatives (e.g. impedance diagrams and power diagrams). So if we learn how to draw these diagrams, then we will have absolutely no need to remember all the other various equations.

In this chapter we will be examining **parallel a.c. circuits**, specifically, **parallel *R-L*, parallel *R-C*** and **parallel *R-L-C*** circuits.

Again, a reminder that the circuit symbols used throughout this chapter represent the *quantities* resist*ance*, induct*ance*, and capacit*ance* – **not** resist*ors*, induct*ors* and capacit*ors*.

Parallel R-L circuits

When a potential difference $(\bar{E})$ is applied across a parallel circuit, *that potential difference is common to each branch*. For an *R-L* parallel circuit, the resistive branch will then draw a current $\bar{I}_R$, and the inductive branch will draw a current $\bar{I}_L$. In accordance with Kirchhoff's Current Law, the supply current $(\bar{I})$ will then be the sum of the two branch currents. However, as we learnt in the chapter on *a.c. series circuits*, because these currents are not in phase with each other, *we must add them vectorially*.

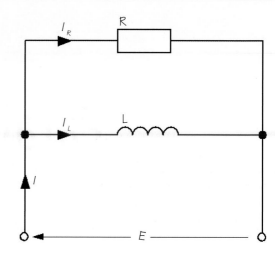

Figure 21.2

So, let's examine this a little more closely, by drawing the phasor diagram for the circuit. In the following series of diagrams, the step being described is shown in **blue**, while previous steps are shown in black.

Drawing the phasor diagram

Step 1

In a parallel circuit, the **supply voltage** $(\bar{E})$ is common to each branch and, so, *voltage is always chosen as the* **reference phasor**. The reference phasor is *always drawn first, and always along the horizontal positive axis* (i.e. horizontally and to the right), and it's also normally drawn fairly long in order to distinguish it from the others. In Figure 21.3, we have also given the reference phasor an outline, rather than a solid, arrow head although this is not really necessary. The small, curved, arrow head is there to remind us that phasors 'rotate' counterclockwise.

Figure 21.3

Step 2

The current, $\bar{I}_R$, flowing in the resistive branch is *in phase with the supply voltage* and, so, is also drawn along the horizontal positive axis (Figure 21.4).

Figure 21.4

Step 3

The current, $\overline{I}_L$, flowing in the inductive branch *lags* the voltage by 90° (remember C**IVIL**) so, as phasors 'rotate' counterclockwise, it is drawn 90° clockwise from the reference phasor (Figure 21.5).

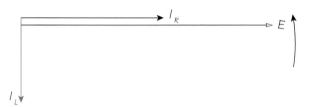

Figure 21.5

Step 4

We know, from Kirchhoff's Current Law that, in a parallel circuit, the supply current $(\overline{I})$ is the sum of the individual branch currents. However, because, in this case, the two currents, $\overline{I}_R$ and $\overline{I}_L$, lie at right angles to each other, we have to add them *vectorially* (Figure 21.6).

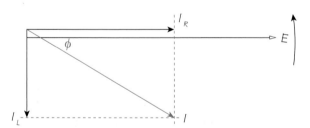

Figure 21.6

From the completed phasor diagram,, we can see that the supply current, $\overline{I}$, is the **phasor sum** (or **vector sum**) of $\overline{I}_R$ and $\overline{I}_L$, which can be found by simply applying Pythagoras's Theorem:

$$\overline{I} = \sqrt{\overline{I}_R^2 + \overline{I}_L^2}$$

To find the phase angle, we can use the *cosine ratio*:

$$\cos\phi = \frac{adjacent}{hypotenuse} = \frac{\overline{I}_R}{\overline{I}}$$

so

$$\angle\phi = \cos^{-1}\frac{\overline{I}_R}{\overline{I}}$$

and, of course, because this is an inductive circuit, it is **lagging**.

We *could* have used the sine or tangent ratios, too, but by using the cosine ratio, we are able to work out the circuit's **power factor** (which is the cosine of the phase angle) at the same time! That is, you are 'killing two birds with one stone'!

Worked example 1   The current drawn by the resistive branch of a parallel *R-L* circuit is 15 A, and the current drawn by the inductive branch is 20 A. What is the value of the supply current? Also, what is the circuit's phase angle?

Solution   *Always* start by sketching the circuit diagram, and inserting all the values given to you in the question (Figure 21.7).

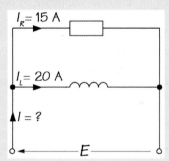

Figure 21.7

Next, sketch the phasor diagram, following the steps described above. You *don't* have to draw the phasor diagram to scale (Figure 21.8).

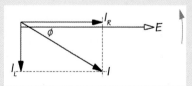

Figure 21.8

Now, you can apply Kirchhoff's Current Law, using Pythagoras's Theorem to solve the problem:

$$\bar{I} = \sqrt{\bar{I}_R^2 + \bar{I}_L^2}$$
$$= \sqrt{15^2 + 20^2}$$
$$= \sqrt{625} = 25 \text{ A (Answer)}$$

Since this is an inductive circuit, and the current *lags* the supply voltage, then the phase angle will be *lagging*:

$$\angle\phi = \cos^{-1}\frac{\bar{I}_R}{\bar{I}} = \cos^{-1}\frac{15}{25} = \cos^{-1}0.6$$
$$= 53.13° \text{ lagging (Answer)}$$

Admittance diagram

You will recall that, in a **series R-L** diagram, we changed the voltage phasor diagram into an *impedance diagram* simply by dividing throughout by the reference phasor which, in that case, happened to be the supply current, $\bar{I}$.

Well, let's follow exactly the same rule and, again: *divide the phasor diagram by the reference phasor*. For our parallel circuit, however, the reference phasor is the supply voltage, $\bar{E}$, so let's see what happens this time.

Drawing the admittance diagram

Step 1

We start by drawing the circuit's phasor diagram, following the steps already explained (Figure 21.9).

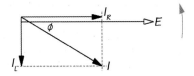

Figure 21.9

Step 2

Next, we *divide each of the current phasors by the reference phasor* $(\bar{E})$ (Figure 21.10).

As you can see, what we now have are expressions for the *reciprocals* of resistance, inductive reactance and impedance. We call these conductance, inductive susceptance and admittance – hence the term, '**admittance diagram**'.

Figure 21.11 shows exactly the same admittance diagram, but expressed directly in terms of **conductance (G)**, **inductive susceptance (B_L)** and **admittance (Y)**, each expressed in siemens (S).

As you can see, we have ended up with the *reciprocals* of resistance, inductive reactance and impedance. The resulting diagram is called an **admittance diagram** (sometimes called an *'admittance triangle'*), and is useful because it generates the following equations:

$$\frac{\bar{I}_R}{\bar{E}} = \frac{1}{R} \qquad \frac{\bar{I}_L}{\bar{E}} = \frac{1}{X_L} \qquad \frac{\bar{I}}{\bar{E}} = \frac{1}{Z}$$

No doubt you will have noticed that each of these equations can be turned upside down to give us the more familiar equations:

$$\frac{\bar{E}}{\bar{I}_R} = R, \quad \frac{\bar{E}}{\bar{I}_L} = X_L, \quad \text{and} \quad \frac{\bar{E}}{\bar{I}} = Z$$

... so you might be wondering why bother to express them in the way we have (i.e. as reciprocals)! Well, by applying Pythagoras's Theorem to them, as reciprocals, you can also use the following relationship:

$$\left(\frac{1}{Z}\right)^2 = \left(\frac{1}{R}\right)^2 + \left(\frac{1}{X_L}\right)^2$$

... which is exactly equivalent to the following equation that we are already familiar with for calculating the total resistance of a parallel d.c. circuit:

$$\frac{1}{R} = \frac{1}{R_1} + \frac{1}{R_2}$$

Manipulating the 'admittance' equation, then gives us:

$$\frac{1}{Z} = \sqrt{\frac{1}{R^2} + \frac{1}{X_L^2}}$$

... from which we can find the **impedance** of the parallel circuit.

We can also find the circuit's **phase angle**, using basic trigonometry, utilising either the *sine, cosine*

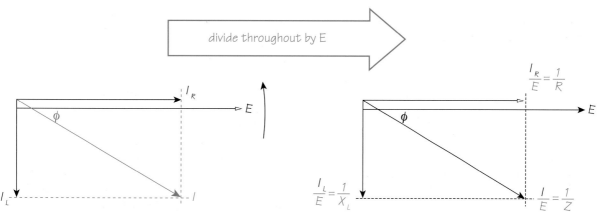

divide throughout by E

Figure 21.10

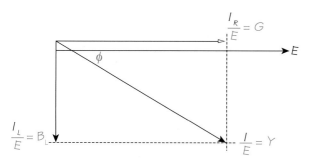

Figure 21.11

or *tangent* ratios. In practice, for a reason we'll see shortly, the best choice is always to use the *cosine*:

$$\cos\phi = \frac{\text{adjacent}}{\text{hypotenuse}} = \frac{\left(\dfrac{1}{R}\right)}{\left(\dfrac{1}{Z}\right)} = \frac{Z}{R}$$

$$\angle\phi = \cos^{-1}\frac{Z}{R}$$

You will notice that this is rather different from the corresponding equation for a series *R-L* circuit! This is because the effective impedance is smaller than either the resistance or the inductive reactance (in just the same way that the effective resistance of a d.c. parallel circuit is less than either of the branch resistances).

Important! Dividing a current phasor diagram by voltage produces an admittance diagram

which generates each of the equations shown above. So you don't have to learn *any* of these equations – they can all be generated *provided you learn how to draw the phasor and impedance diagrams!*

However, *if you prefer*, you can rewrite the above equations directly in terms conductance, inductive susceptance and conductance, as follows:

$$\frac{\overline{I_R}}{\overline{E}} = G \quad \frac{\overline{I_L}}{\overline{E}} = B_L \quad \frac{\overline{I}}{\overline{E}} = Y$$

By applying Pythagoras's Theorem, you can also find the following relationship:

$$Y^2 = G^2 + B_L^2$$
$$Y = \sqrt{G^2 + B_C^2}$$

We can also find the circuit's **phase angle**, using basic trigonometry, utilising either the *sine, cosine* or *tangent* ratios. As usual, the best choice is always to use the *cosine*, because it also tells you what the circuit's power factor happens to be, should you need to know:

$$\cos\phi = \frac{\text{adjacent}}{\text{hypotenuse}} = \frac{G}{Y}$$

$$\angle\phi = \cos^{-1}\frac{G}{Y}$$

In many respects, it's far more convenient to work in terms of conductance, inductive susceptance and admittance, as it avoids the complications of having to work with reciprocals (fractions). But it's entirely up to you!

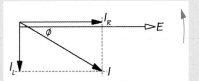

Figure 21.13

Worked example 2 A parallel circuit of resistance 5 Ω in parallel with an inductance 0.02 H is connected across a 230-V, 50 Hz, a.c. supply. Calculate each of the following:

a inductive reactance
b impedance
c current through the resistive branch
d current through the inductive branch
e supply current
f phase angle of the circuit.

Solution As always, the first set in solving *any* circuit problem is to sketch the circuit diagram, and label it with all values supplied in the problem (Figure 21.12).

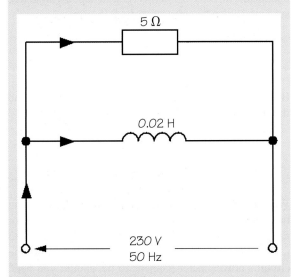

$5\ \Omega$

$0.02\ \text{H}$

$230\ \text{V}$
$50\ \text{Hz}$

Figure 21.12

The next step is to sketch the current phasor diagram, following the steps described earlier (Figure 21.13).

Next, we must convert the current phasor diagram into an admittance diagram (Figure 21.14), by dividing throughout by the reference quantity – i.e. the supply voltage. This generates all the equations that we need to solve the problem (without you having to remember them!).

a To find the **inductive reactance (X_L)** of the circuit, we start by looking at the equations generated when we constructed the admittance diagram. There's only one equation with inductive reactance shown, $\dfrac{\overline{I_L}}{\overline{E}} = \dfrac{1}{X_L}$.

Unfortunately, we don't know the value of $\overline{I_L}$ so we can't use this formula. So we must fall back on the basic equation for inductive reactance, as follows:

$$X_L = 2\pi f L = 2\pi \times 50 \times 0.02 = 6.28\ \Omega\ \text{(Answer a.)}$$

b To find the impedance, we can now use the equation generated by the admittance diagram:

$$\frac{1}{Z} = \sqrt{\frac{1}{R^2} + \frac{1}{X_L^2}}$$

$$\frac{1}{Z} = \sqrt{\frac{1}{5^2} + \frac{1}{6.28^2}} = \sqrt{\frac{1}{25} + \frac{1}{39.44}} = \sqrt{\frac{39.44 + 25}{25 \times 39.44}}$$

$$\frac{1}{Z} = \sqrt{\frac{64.44}{986}} = \sqrt{0.0654} = 0.2557$$

$$Z = \frac{1}{0.2557} \approx 3.91\ \Omega\ \text{(Answer b.)}$$

This answer makes sense, because, in a parallel circuit, the impedance must be lower than either the resistance or the reactance (in exactly the same way that the equivalent resistance of a d.c. parallel circuit must be less than the resistance of any branch).

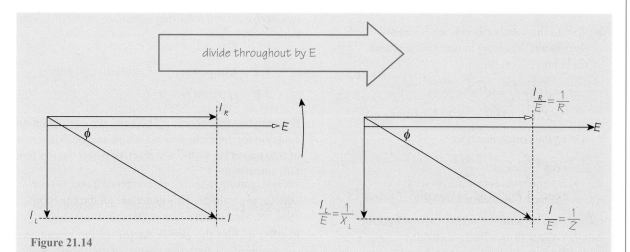

Figure 21.14

The alternative, and somewhat simpler, way of solving this part of the question, is to work directly with *conductance*, *inductive susceptance* and *admittance*.

First, find the **conductance** and the **inductive susceptance**:

$$G = \frac{1}{R} = \frac{1}{5} = 0.20 \text{ S} \quad \text{and} \quad B_L = \frac{1}{X_L} = \frac{1}{6.28} = 0.16 \text{ S}$$

Next, apply the Pythagoras's Theorem equation for **admittance**:

$$Y = \sqrt{G^2 + B_L^2} = \sqrt{0.2^2 + 0.16^2} = \sqrt{0.0656} = 0.256 \text{ S}$$

Finally, the **impedance** is the inverse of the admittance:

$$Z = \frac{1}{Y} = \frac{1}{0.256} \approx 3.91 \, \Omega \text{ (Answer b.)}$$

c   To find the current through the resistive branch, we use the following equation generated by the admittance diagram:

$$\frac{1}{R} = \frac{\overline{I}_R}{\overline{E}}$$

rearranging, $\overline{I}_R = \frac{\overline{E}}{R} = \frac{230}{5} = 46 \text{ A (Answer c.)}$

Again, if you prefer, you can use *conductance*, rather than resistance, to work our the current through the resistive branch:

$$\frac{\overline{I}_R}{\overline{E}} = G$$

rearranging, $\overline{I}_R = \overline{E} \, G = 230 \times 0.2 = 46 \text{ A (Answer c.)}$

d   Again, using the equation generated by the admittance diagram:

$$\frac{1}{X_L} = \frac{\overline{I}_L}{\overline{E}}$$

rearranging, $\overline{I}_L = \frac{\overline{E}}{X_L} = \frac{230}{6.28} \approx 36.62 \text{ A (Answer d.)}$

Again, if you prefer, you can use *conductance*, rather than resistance, to work our the current through the resistive branch:

$$\frac{\overline{I}_L}{\overline{E}} = B_L$$

rearranging, $\overline{I}_R = \overline{E} \, B_L = 230 \times 0.16$
$$\approx 36.8 \text{ A (Answer d.)}$$

(*minor difference in answer due to rounding up/down*)

e To find the supply current, we can apply Pythagoras' Theorem to the current phasor diagram,

$$\overline{I} = \sqrt{\overline{I}_R^2 + \overline{I}_L^2} = \sqrt{46^2 + 36.62^2}$$
$$= \sqrt{2116 + 1341} \approx 58.78 \text{ A (Answer e.)}$$

f Using the cosine function:

$$\angle\phi = \cos^{-1}\frac{Z}{R} = \cos^{-1}\frac{3.911}{5}$$
$$= \cos^{-1}0.7822 \approx 38.54° \text{ lagging (Answer f.)}$$

('*Lagging*', because the *supply current lags the supply voltage* in an inductive circuit.)

Once again, if you prefer to work in terms of *admittance* and *conductance*, then:

$$\angle\phi = \cos^{-1}\frac{G}{Y} = \cos^{-1}\frac{0.2}{0.256} = \cos^{-1}0.7813$$
$$\approx 38.62° \text{ lagging (Answer f.)}$$

(minor difference in answer, due to rounding up/down)

Power diagram

To create a **power diagram** (or '**power triangle**') for a parallel *R-L* circuit we simply *multiply* throughout by the reference phasor (Figure 21.15).

This action generates the exact same power equations as we have already seen for a *series R-L* circuit. As always, if you can draw the circuit's current phasor diagram, and convert it into a power diagram, *then you*

can **derive** all the following equations, without having to remember them:

$$\overline{I}_R\overline{E} = \text{true power} \quad \overline{I}_L\overline{E} = \text{reactive power}$$
$$\overline{I}\ \overline{E} = \text{apparent power}$$

By applying Pythagoras's Theorem, we can obtain an equation for apparent power (in volt amperes) in terms of true power (in watts) and reactive power (in reactive volt amperes):

$$(\text{apparent power})^2 = (\text{true power})^2 + (\text{reactive power})^2$$

Finally, to find the **power factor** of the circuit, we simply need to find the cosine of the phase angle, which is:

$$\cos\phi = \frac{\text{adjacent}}{\text{hypotenuse}} = \frac{\text{true power}}{\text{apparent power}}$$

Parallel R-C circuits

We know that in a purely resistive circuit, the current and voltage are in phase with each other; and, in a purely capacitive circuit, the current *leads* the voltage by 90°. So, in a *parallel R-C* circuit, the *current must lead the voltage by some angle between 0° and 90°* – this angle is called the circuit's **phase angle** (symbol: ϕ, pronounced 'phi').

When a potential difference $(\overline{E})$ is applied across a **parallel R-C circuit**, a current, $\overline{I}_R$, will flow through the resistive branch, and a current, $\overline{I}_C$, will flow through the capacitive branch.

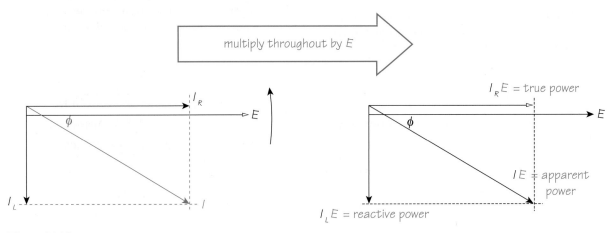

Figure 21.15

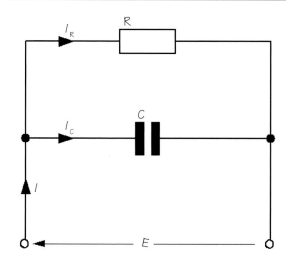

Figure 21.16

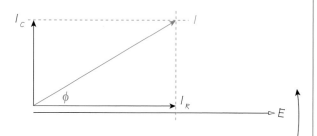

Figure 21.17

To find the phase angle, we can use the *cosine ratio*:

$$\cos\phi = \frac{\text{adjacent}}{\text{hypotenuse}} = \frac{\overline{I}_R}{\overline{I}}$$

so

$$\angle\phi = \cos^{-1}\frac{\overline{I}_R}{\overline{I}}$$

and, of course, because this is an inductive circuit, it is **leading**.

Admittance diagram

Again, by dividing *the phasor diagram by the reference phasor*, $\overline{E}$, we can change the phasor diagram into an admittance diagram (Figure 21.18). Remember, 'admittance' is the *reciprocal* of impedance.

The same admittance diagram, expressed directly in terms of **conductance (G)**, **capacitive susceptance (B_C)** and **admittance (Y)** is shown in Figure 21.19.

We needn't go through the step-by-step process of creating the current phasor diagram for this circuit, as it is practically the same as the procedure we have already gone through to create the phasor diagram for an *R-L* circuit. The important difference, of course, is that, for an *R-C* circuit, the current through the capacitive branch, $\overline{I}_C$, leads the supply voltage.

So, the finished current phasor diagram for a parallel *R-C* circuit will look like Figure 21.17.

From the completed phasor diagram, we can see that the supply current, $\overline{I}$, is the **phasor sum** (or **vector sum**) of $\overline{I}_R$ and $\overline{I}_C$, which can be found using Pythagoras's Theorem:

$$\overline{I} = \sqrt{\overline{I}_R^2 + \overline{I}_C^2}$$

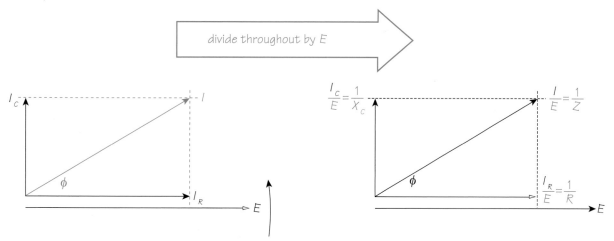

divide throughout by *E*

Figure 21.18

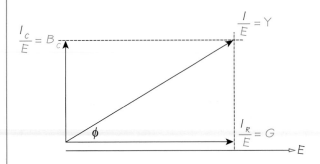

Figure 21.19

The resulting diagram is an **admittance diagram** (or *'admittance triangle'*), and is useful because it generates the following equations:

$$\frac{\overline{I}_R}{\overline{E}} = \frac{1}{R} \qquad \frac{\overline{I}_C}{\overline{E}} = \frac{1}{X_C} \qquad \frac{\overline{I}}{\overline{E}} = \frac{1}{Z}$$

Also from the admittance diagram, you can also see that the admittance can be calculated by applying Pythagoras's Theorem:

$$\frac{1}{Z} = \sqrt{\left(\frac{1}{R}\right)^2 + \left(\frac{1}{X_C}\right)^2}$$

We can also find the circuit's **phase angle**, using basic trigonometry, utilising the *cosine* ratio.

$$\cos\phi = \frac{\text{adjacent}}{\text{hypotenuse}} = \frac{\left(\dfrac{1}{R}\right)}{\left(\dfrac{1}{Z}\right)} = \frac{Z}{R}$$

$$\angle\phi = \cos^{-1}\frac{Z}{R}$$

> **Important**! Dividing a current phasor diagram by voltage produces an admittance diagram which generates each of the equations shown above. So you don't have to learn *any* of these equations – they can all be generated *provided you learn how to draw the phasor and admittance diagrams!*

Again, you may prefer to work in terms of conductance, capacitive susceptance and admittance. In which case,

$$\frac{\overline{I}_R}{\overline{E}} = G \qquad \frac{\overline{I}_C}{\overline{E}} = B_C \qquad \frac{\overline{I}}{\overline{E}} = Y$$

Also from the admittance diagram, you can also see that the admittance can be calculated by applying Pythagoras's Theorem:

$$Y = \sqrt{G^2 + B_C{}^2}$$

We can also find the circuit's **phase angle**, using basic trigonometry, utilising the *cosine* ratio.

$$\cos\phi = \frac{\text{adjacent}}{\text{hypotenuse}} = \frac{G}{Y}$$

$$\angle\phi = \cos^{-1}\frac{G}{Y}$$

Once again, in many respects, it's more convenient to work in terms of conductance, inductive susceptance and admittance, as it avoids the complications of having to work with fractions. But it's entirely up to you!

Power diagram

To create a **power diagram** (or '**power triangle**') for a parallel *R-C* circuit, we simply multiply throughout by the reference phasor (Figure 21.20).

This action generates the same power equations as we have already seen for a *series R-L* circuit. As always, if you can draw the circuit's current phasor diagram, and convert it into a power diagram, *then you can **derive** all the following equations, without having to remember them:*

$$\overline{I}_R\overline{E} = \text{true power} \qquad \overline{I}_C\overline{E} = \text{reactive power}$$

$$\overline{I}\,\overline{E} = \text{apparent power}$$

By applying Pythagoras's Theorem, we can obtain an equation for apparent power (in volt amperes) in terms of true power (in watts) and reactive power (in reactive volt amperes):

$$(\text{apparent power})^2 = (\text{true power})^2 + (\text{reactive power})^2$$

Finally, to find the **power factor** of the circuit, we simply need to find the cosine of the phase angle, which is:

$$\cos\phi = \frac{\text{adjacent}}{\text{hypotenuse}} = \frac{\text{true power}}{\text{apparent power}}$$

multiply throughout by E

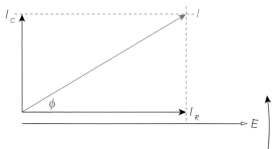

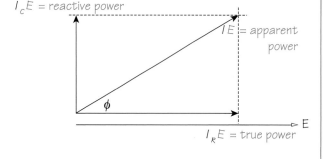

Figure 21.20

Worked example 3   A parallel *R-C* circuit comprises a 30-Ω resistor in parallel with a capacitor of capacitive reactance 20 Ω, connected across a 120-V supply. Calculate the circuit's (a) true power, (b) reactive power, (c) apparent power and (d) power factor.

Solution   As always, the first step is to sketch a circuit diagram, with all the supplied information inserted (Figure 21.21).

Before we can work out the true, reactive and apparent power, *we need to work out the current flowing in each branch*. So we must sketch the current phasor diagram for the circuit, and divide throughout by the reference phasor $(\overline{E})$ to produce the **admittance diagram** (Figure 21.22). *All the equations you will need are then generated without you having to remember them.*

To find the current through the resistor, $\overline{I}_R$:

$$\frac{\overline{I}_R}{\overline{E}} = \frac{1}{R} \quad \text{rearranging:} \quad \overline{I}_R = \frac{\overline{E}}{R} = \frac{120}{30} = 4 \text{ A}$$

To find the current through the capacitor, $\overline{I}_C$:

$$\frac{\overline{I}_C}{\overline{E}} = \frac{1}{X_C} \quad \text{rearranging:} \quad \overline{I}_C = \frac{\overline{E}}{X_C} = \frac{120}{20} = 6 \text{ A}$$

To find the supply current, since we don't know the circuit's impedance, we apply Pythagoras's Theorem:

$$I = \sqrt{I_R^2 + I_C^2} = \sqrt{4^2 + 6^2} = 7.21 \text{ A}$$

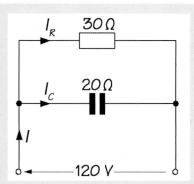

Figure 21.21

Now we can convert the current phasor diagram into a power diagram (Figure 21.23), by multiplying throughout by the reference phasor $(\overline{E})$.

To find the true power:

true power $= \overline{I}_R\overline{E} = 4 \times 120 = 480$ W (Answer a.)

To find the reactive power:

reactive power $= \overline{I}_C\overline{E} = 6 \times 120 = 720$ var (Answer b.)

To find the apparent power:

apparent power $= \overline{I}\overline{E} = 7.21 \times 120 \approx 865$ V · A (Answer c.)

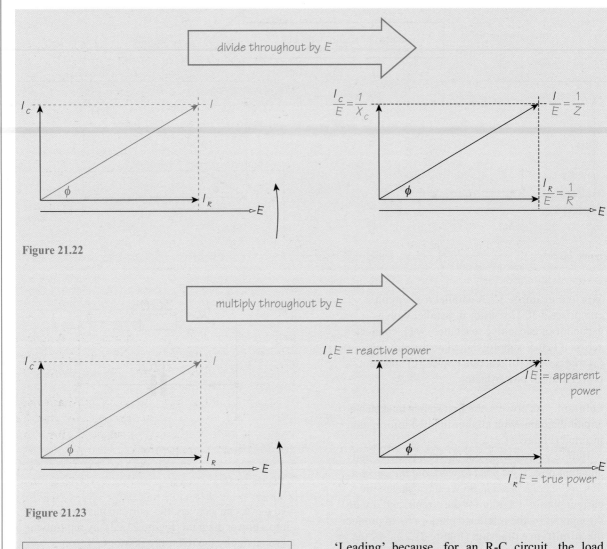

Figure 21.22

Figure 21.23

An alternative method, would be to apply Pythagoras's Theorem, as follows:

$$\text{apparent power} = \sqrt{\text{true power}^2 + \text{reactive power}^2}$$

$$= \sqrt{480^2 + 720^2} \approx 865 \text{V} \cdot \text{A}$$

Finally, to find the power factor, we use the cosine ratio:

$$\text{power factor} = \cos\phi = \frac{\text{adjacent}}{\text{hypotenuse}}$$

$$= \frac{\text{true power}}{\text{apparent power}}$$

$$= \frac{480}{865} = 0.555 \text{ leading (Answer d.)}$$

'Leading' because, for an R-C circuit, the load current leads the supply voltage.

Parallel R-L-C circuits

When a potential difference $(\bar{E})$ is applied across a **parallel R-L-C circuit**, a current, $\bar{I}_R$, will flow through the resistive branch of the circuit, a current, $\bar{I}_L$, will flow through the inductive branch, and a current, $\bar{I}_C$, will flow through the capacitive branch.

Drawing the phasor diagram

Step 1

In a parallel circuit, the **supply voltage** $(\bar{E})$ is common to each branch and, so, this is chosen as the **reference**

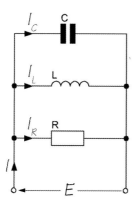

Figure 21.24

phasor. As we have learnt, the reference phasor is *always drawn along the horizontal positive axis*, and its also normally drawn fairly long in order to distinguish it from the others (Figure 21.25).

Figure 21.25

Step 2

The current, $\overline{I}_R$, flowing through the resistive branch, is *in phase with the supply voltage* and, so, is also drawn along the horizontal positive axis (Figure 21.26).

Figure 21.26

Step 3

The current, $\overline{I}_L$, flowing through the inductive branch *lags the supply voltage by 90°* (remember CI**V**IL), so is drawn 90° clockwise from the reference phasor (Figure 21.27).

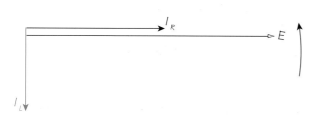

Figure 21.27

Step 4

The current, $\overline{I}_C$, flowing through the capacitive branch *leads the supply voltage by 90°* (remember **CIV**IL), so is drawn 90° counterclockwise from the reference phasor (Figure 21.28). ***Always*** draw $\overline{I}_C$ shorter than $\overline{I}_L$ (or *vice versa*), or we will end up with a unique condition called **resonance** – just as we did in the *series R-L-C* circuit!

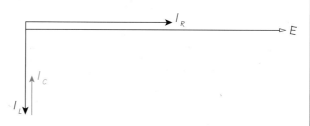

Figure 21.28

Step 5

We know from Kirchhoff's Current Law that, in a parallel circuit, the supply current is the sum of the individual branch currents and, for a.c. circuits, we have to add them *vectorially*.

As always, it's a little more difficult to add *three* phasors. As $\overline{I}_L$ and $\overline{I}_C$ lie in opposite directions, the simplest thing to do is to start by subtracting them and, then, add the difference to phasor $\overline{I}_R$.

The snag is, of course, that we might not know whether $\overline{I}_L$ is bigger than $\overline{I}_C$, or *vice versa*! Fortunately, *it doesn't matter!* Once again, the purpose of the phasor diagram is to *generate equations*, not to accurately represent the *actual* conditions in the circuit to scale! And the phasor diagram will *always* generate the correct equations, whether or not $\overline{I}_L$ is bigger than $\overline{I}_C$!

So, the simplest solution is to get into the habit of *always drawing $\overline{I}_L$ longer than $\overline{I}_C$ – but whatever you do*, **never draw them the same length***!!*

So, as explained, start by subtracting $\overline{I}_C$ from $\overline{I}_L$, and *then* vectorially add the resultant to $\overline{I}_R$ to give the completed phasor diagram, as shown in Figure 21.29.

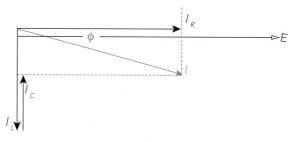

Figure 21.29

From the completed phasor diagram, we can see that the supply current, $\overline{I}$, is the **phasor sum** (or **vector sum**) of $\overline{I}_R$, $\overline{I}_L$ and $\overline{I}_C$, which can then be found using Pythagoras's Theorem:

$$\overline{I} = \sqrt{\overline{I}_R^2 + (\overline{I}_L - \overline{I}_C)^2}$$

To find the phase angle, we can use the *cosine ratio*:

$$\cos\phi = \frac{\text{adjacent}}{\text{hypotenuse}} = \frac{\overline{I}_R}{\overline{I}}$$

If $\overline{I}_L$ really is larger than $\overline{I}_C$, then the circuit is predominantly inductive and, so, the resulting phase angle will be **lagging**. On the other hand, if $\overline{I}_C$ happens to be larger than $\overline{I}_L$, then the circuit will be predominantly capacitive and, so, the resulting phase angle will be **leading**.

Admittance diagram

How can we now proceed to find out further equations for solving a parallel *R-L-C* circuit?

Again, the answer is by means of an **impedance diagram**.

Drawing the admittance diagram

Step 1

We start by drawing the circuit's phasor diagram, following the steps already explained (Figure 21.30).

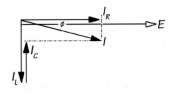

Figure 21.30

Step 2

Next, we *divide each of the voltage phasors by the reference phasor ($\overline{E}$)* (Figure 21.31).

The resulting diagram is an **admittance diagram** (or *'admittance triangle'*), and is useful because it generates the following equations:

$$\frac{\overline{I}_R}{\overline{E}} = \frac{1}{R} \qquad \frac{\overline{I}_L}{\overline{E}} = \frac{1}{X_L} \qquad \frac{\overline{I}_C}{\overline{E}} = \frac{1}{X_C}$$

$$\frac{\overline{I}}{\overline{E}} = \frac{1}{Z}$$

Also from the admittance diagram, you can see that the admittance is the *vector sum of conductance (1/R) and capacitive susceptance (1/X_C)*, which can be calculated by applying Pythagoras' Theorem:

$$\left(\frac{1}{Z}\right)^2 = \left(\frac{1}{R}\right)^2 + \left(\frac{1}{X_L} - \frac{1}{X_C}\right)^2$$

$$\frac{1}{Z} = \sqrt{\left(\frac{1}{R}\right)^2 + \left(\frac{1}{X_L} - \frac{1}{X_C}\right)^2}$$

We can also find the circuit's **phase angle**, using basic trigonometry, utilising the *cosine* ratio:

$$\cos\phi = \frac{\text{adjacent}}{\text{hypotenuse}} = \frac{\left(\dfrac{1}{R}\right)}{\left(\dfrac{1}{Z}\right)}$$

$$\angle\phi = \cos^{-1}\frac{Z}{R}$$

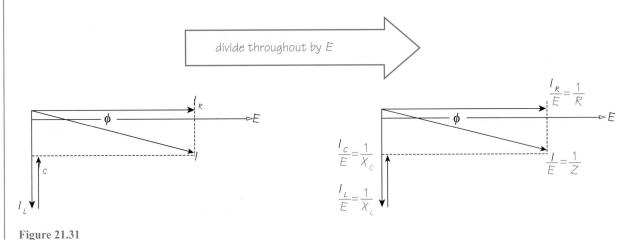

Figure 21.31

Important! Dividing a current phasor diagram by voltage produces an admittance diagram which generates each of the equations shown above. So you don't have to learn *any* of these equations – they can all be generated *provided you learn how to draw the phasor and impedance diagrams!*

Once again, you may prefer to work in terms of conductance, inductive susceptance, capacitive susceptance and admittance (after all, it is easier!), in which case the following equations apply.

$$\frac{\overline{I}_R}{E} = G \qquad \frac{\overline{I}_L}{E} = B_L \qquad \frac{\overline{I}_C}{E} = B_C \qquad \frac{\overline{I}}{E} = Y$$

Also from the admittance diagram, you can see that the admittance is the *vector sum of conductance (G), inductive susceptance (B$_L$)* and *capacitive susceptance (B$_C$)*, which can be calculated by applying Pythagoras's Theorem:

$$Y^2 = G^2 + \left(B_L - B_C\right)^2$$

$$Y = \sqrt{G^2 + \left(B_L - B_C\right)^2}$$

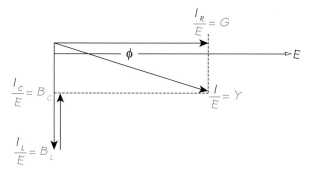

Figure 21.32

We can also find the circuit's **phase angle**, using basic trigonometry, utilising the *cosine* ratio:

$$\cos\phi = \frac{\text{adjacent}}{\text{hypotenuse}} = \frac{G}{Y}$$

$$\angle\phi = \cos^{-1}\frac{G}{Y}$$

Power diagram

To create a **power diagram** (or '**power triangle**') for a parallel *R-C* circuit, we simply multiply throughout by the reference phasor (Figure 21.33).

This action generates the same power equations as we have already seen for a *series R-L-C* circuit. As always, if you can draw the circuit's current phasor diagram, and convert it into a power diagram, *then you can **derive** all the following equations, without having to remember them:*

$$\overline{I}_R\overline{E} = \text{true power} \qquad \overline{I}\,\overline{E} = \text{apparent power}$$

$$(\overline{I}_L - \overline{I}_C)\overline{E} = \text{reactive power}$$

By applying Pythagoras's Theorem, we can obtain an equation for apparent power (in volt amperes) in terms of true power (in watts) and reactive power (in reactive volt amperes):

$$(\text{apparent power})^2 = (\text{true power})^2 + (\text{reactive power})^2$$

Finally, to find the **power factor** of the circuit, we simply need to find the cosine of the phase angle, which is:

$$\cos\phi = \frac{\text{adjacent}}{\text{hypotenuse}} = \frac{\text{true power}}{\text{apparent power}}$$

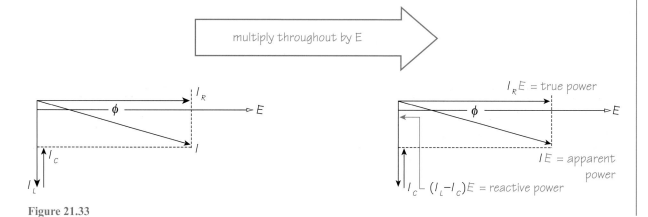

Figure 21.33

Worked example 4 A parallel *R-L-C* circuit comprises a 6-Ω purely resistive branch, a 19-mH purely inductive branch and a 66-μF purely capacitive branch, supplied by a 48-V, 100-Hz supply. Calculate:

a each of the branch currents
b the supply current
c the impedance
d true power of the complete circuit.

Solution As always, the first step is to sketch the circuit, with all the supplied information inserted (Figure 21.34).

Before proceeding any further, we should determine the inductive reactance and capacitive reactance of the inductive and capacitive branches, as we will need to use these values:

$$X_L = 2\pi f L = 2 \times \pi \times 100 \times (19 \times 10^{-3}) \simeq 12 \ \Omega$$

$$X_C = \frac{1}{2\pi f C} = \frac{1}{2\pi \times 100 \times (66 \times 10^{-6})} \simeq 24 \ \Omega$$

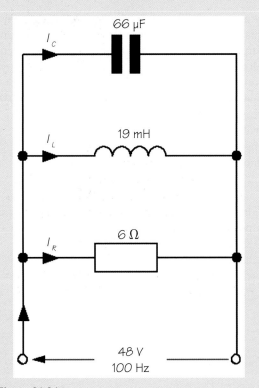

Figure 21.34

The next step is to sketch the current phasor diagram, and convert this into an admittance diagram, by dividing throughout by the reference phasor $(\bar{E})$ (Figure 21.35).

To find the current through the **resistive** branch:

$$\frac{\bar{I}_R}{\bar{E}} = \frac{1}{R} \quad \text{from which:} \quad \bar{I}_R$$

$$= \frac{\bar{E}}{R} = \frac{48}{6} = 8 \text{ A (Answer a.)}$$

To find the current through the **inductive** branch:

$$\frac{\bar{I}_L}{\bar{E}} = \frac{1}{X_L} \quad \text{from which:}$$

$$\bar{I}_L = \frac{\bar{E}}{X_L} = \frac{48}{12} = 4 \text{ A (Answer a.)}$$

To find the current through the **capacitive** branch:

$$\frac{\bar{I}_C}{\bar{E}} = \frac{1}{X_C} \quad \text{from which:}$$

$$\bar{I}_C = \frac{\bar{E}}{X_C} = \frac{48}{24} = 2 \text{ A (Answer a.)}$$

We now need to refer back to the current phasor diagram, and apply Pythagoras's Theorem, to determine the supply current:

$$I = \sqrt{I_R^2 + (I_L - I_C)^2} = \sqrt{8^2 + (4-2)^2}$$

$$= \sqrt{68} = 8.25 \text{ A (Answer b.)}$$

To find the impedance of the circuit, we can use the appropriate equation generated by the admittance diagram:

$$\frac{\bar{I}}{\bar{E}} = \frac{1}{Z} \quad \text{from which:}$$

$$Z = \frac{\bar{E}}{\bar{I}} = \frac{48}{8.25} = 5.82 \ \Omega \text{ (Answer c.)}$$

To find the true power of the circuit, we must convert the current phasor diagram into a power diagram, by multiplying throughout by the reference phasor $(\bar{E})$ (Figure 21.36).

From the various equations generated, we can use the following:

$$\text{true power} = I_R E = 8 \times 48 = 384 \text{ W (Answer d.)}$$

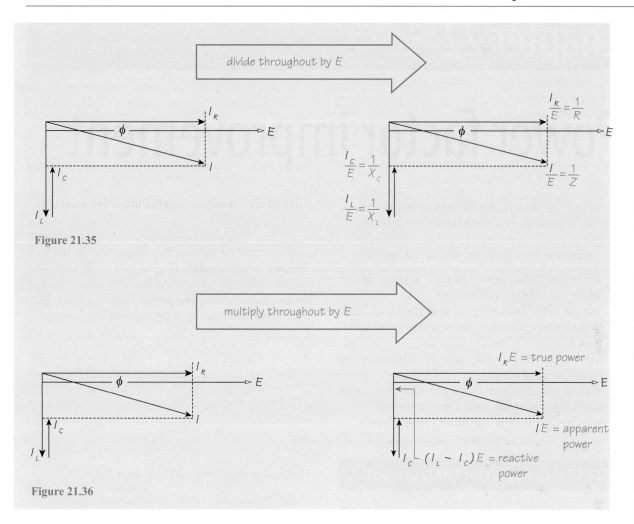

Figure 21.35

Figure 21.36

Finally . . .

Now that you have completed this chapter, are you able to achieve the objectives or learning outcomes listed at the beginning of this chapter?

Ask yourself, 'Can I . . .'

1   draw a current phasor diagram for a parallel:
    a   resistive-inductive *(R-L)* circuit
    b   resistive-capacitive *(R-C)* circuit
    c   resistive-inductive-capacitive *(R-L-C)* circuit.
2   derive an admittance diagram, and derive expressions for conductance, susceptance, admittance and phase angle for a parallel:

a   resistive-inductive *(R-L)* circuit
b   resistive-capacitive *(R-C)* circuit
c   resistive-inductive-capacitive *(R-L-C)* circuit.
3   derive a power diagram, and derive expressions for true power, reactive power, apparent power and power factor for a parallel:
    a   resistive-inductive *(R-L)* circuit
    b   resistive-capacitive *(R-C)* circuit
    c   resistive-inductive-capacitive *(R-L-C)* circuit.
4   solve problems on parallel *R-L*, *R-C* and *R-L-C* circuits.

Online resources
The companion website to this book contains further resources relating to this chapter. The website can be accessed via the following link:
www.routledge.com/cw/waygood

Chapter 22

Power factor improvement

On completion of this chapter, you should be able to

1 define the terms, 'power factor', 'lagging' power factor and 'leading' power factor.
2 explain why increasing a load's power factor towards unity will reduce the current supplied to that load.
3 outline the main advantages of improving the power factor of a load.
4 explain why a power factor capacitor is rated in reactive volt amperes, rather than in farads.
5 solve problems on power factor improvement.

Need for power factor improvement

As we have learned, '**power factor**' is defined as *the ratio of true power to apparent power of a load*, and is expressed as the *cosine of a circuit's phase angle –* i.e. the cosine of the angle by which the load's *supply current* lags or leads its supply voltage. We *always* express phase angle in terms of the position of the load current, relative to the supply voltage, not the other way around. So,

$$\text{power factor} = \cos\phi = \frac{\text{true power}}{\text{apparent power}}$$

Power factor can be expressed as 'per unit', e.g. '0.75 lagging', or as a 'percentage', e.g. '75% lagging'. The 'per unit' method is more widely used, and expressing it as a percentage is now considered rather old fashioned.

It's important to specify whether a power factor is 'lagging' or 'leading'. For inductive circuits, where the load current *lags* the supply voltage, we use the term '**lagging power factor**' (Figure 22.1); for capacitive circuits, where the load current *leads* the supply voltage, we use the term '**leading power factor**' (Figure 22.2) (remember that phasors 'rotate' counterclockwise).

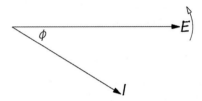

Figure 22.1 circuit with a lagging power factor.

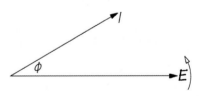

Figure 22.2 circuit with a leading power factor.

Most domestic, commercial, and industrial loads are **resistive-inductive** (**R-L**) comprising resistive loads (i.e. lighting and heating loads) and inductive loads (i.e. motors, relays, etc.) loads. The majority of loads, therefore, have **lagging power factors**.

The circuit diagram in Figure 22.3 represents an equivalent circuit for an *R-L* load, with the resistive and inductive branches taking currents, $\overline{I}_R$ and $\overline{I}_L$ respectively.

An Introduction to Electrical Science, Waygood, ISBN 9780415810029, 2013. © Taylor & Francis

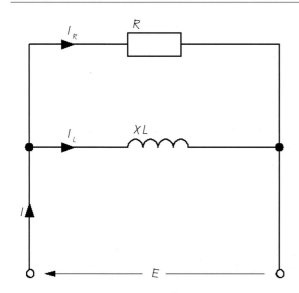

Figure 22.3

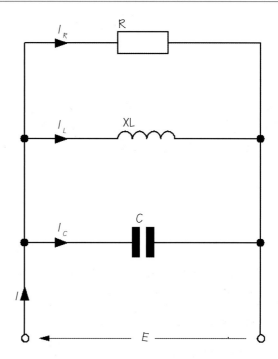

Figure 22.5

The phasor diagram for this circuit is shown in Figure 22.4. As you see, the supply current is the phasor sum of the two branch currents.

$$\overline{I} = \overline{I}_R + \overline{I}_L$$

And, of course, this current is *lagging* the supply voltage by some angle, ϕ, and, so, the circuit is said to have a 'lagging power factor'.

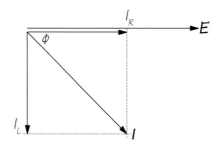

Figure 22.4

Now, let's see what happens if we add a capacitor in parallel with the circuit's existing branches.

As we already know, with a parallel circuit, *adding another branch will have absolutely no effect whatsoever on the behaviour or operation of the existing branches – they will continue to operate normally and draw exactly the same amounts of current regardless of how many additional branches are added.*

However, adding an additional branch *will*, of course, affect the value of the load current the circuit draws from the supply.

By applying Kirchhoff's Current Law, we can express the supply current in terms of the branch currents, as follows:

$$\overline{I} = \overline{I}_R + \overline{I}_L + \overline{I}_C$$

You would, perhaps, be forgiven for assuming that, by adding another branch current $\left(\overline{I}_C\right)$, then the load current is *bound* to increase due to the additional branch current.

However, this is *not* the case!

Let's redraw the phasor diagram, to see what effect the capacitive-branch current will have on the supply current (Figure 22.6).

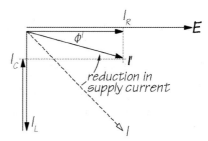

Figure 22.6

As you can see, because $\overline{I}_C$ leads the supply voltage, it is in *antiphase* (acts in the opposite sense) to $\overline{I}_L$ and therefore acts to reduce the circuit's overall reactive current (i.e. $\overline{I}_L - \overline{I}_C$) which, of course, reduces the phase angle and, therefore, reduces the value of the supply current – in this case, from $\overline{I}$ to $\overline{I}'$.

If we use a capacitor that draws an even bigger leading current, then the phase angle will reduce even further – reducing the supply current from $\overline{I}'$ to $\overline{I}''$ (Figure 22.7).

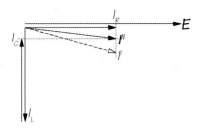

Figure 22.7

So, while adding a capacitive branch has no effect whatsoever on the behaviour of the currents in the other two branches, it acts *to reduce the circuit's phase angle which, in turn, reduces the value of the supply current*. The more leading current the capacitive branch draws, the more it acts to reduce the phase angle and the corresponding supply current. Eventually, the capacitive branch will draw enough current to cause the load current to become 'leading', rather than 'lagging', and start to increase in value again. However, it is both unnecessary and, in fact, undesirable to reduce the phase angle to zero, for reasons we will discuss shortly.

> The practice of adding capacitors in parallel with a load in order to adjust the lagging power factor towards unity is termed **power factor improvement**, or **power factor correction**.

The result of **power factor improvement**, then, is to reduce the size of the load current and, thus, allowing the sizes of supply equipment, such as cables, switchgear, transformers, etc., to be smaller and, therefore, less expensive than would be the case if the power factor had *not* been improved. A lower supply current also means a reduction in the voltage drop along the supply cables and, therefore, a higher terminal voltage at the load.

To summarise, then. From the electricity network company's point of view, an improved power factor results in

- reduced system losses
- improved system voltages
- improved system capacity.

But power factor improvement also offers advantages to the industrial or large commercial users too. Unlike residential consumers, these larger consumers not only have to pay for the **energy** they consume, but they must also pay an *additional fee* based on their installation's **demand** (i.e. its apparent power, expressed in kilovolt amperes) usually together with a **power factor surcharge** if the power factor of their load is too low.

To minimise these additional charges, it would seem sensible to improve the installation's power factor to unity (i.e. reduce the phase angle to zero), by adding as much capacitance in parallel with a load as is necessary, until this is achieved. Perhaps, surprisingly, however, this is *not* the case, because the savings achieved by doing so are offset by the costs incurred in purchasing, installing and maintaining the required capacitors and their associated control equipment! *So, it is rarely financially worthwhile to try to achieve a unity power factor.* It is possible to determine the optimum power factor for any given load, however, this is well beyond the scope of this chapter.

To further complicate matters, practical loads vary continuously: lighting and heating are always being switched on and off, and motors are being started, their loads adjusted, and switched off. So, even if it were financially sensible to achieve a unity power factor, any temporary or instantaneous reduction in load would then result in a *leading* power factor – i.e. an increasingly leading load current would result! For reasons beyond the scope of this chapter, *leading power factors are highly undesirable,* as they lead to system instability and, so, must be avoided at all cost.

> In practice, for financial reasons, it is desirable for an installation's lagging power factor to approach, *but to never achieve,* unity.

It should be understood that the advantages of power factor improvement *do* **not** *apply to residential loads*. Residential consumers (you and me, in other words) are charged simply for the **energy** that they consume. Residential energy meters monitor the supply voltage and in-phase component of the load current, so their readings are based on the true power of the load. And, of course, there is no surcharge imposed on residential consumers for 'poor' power factor.

There are a number of online companies selling 'energy-reduction capacitors' which, they promise, 'will substantially reduce your electricity bill'! Some of these are even demonstrated on *YouTube*, showing how they 'reduce the current' drawn by a residential load. And they do, indeed, reduce the load current. But, of course, they fail to mention that residences are *not* charged for the amount of current they draw from the supply, but for the energy they consume! These devices are a complete and utter waste of money, and the whole business is a scam!

Power factor improvement in practice

We have learnt that **power factor improvement** is a method of adjusting a load's power factor towards unity as a means of reducing a commercial or industrial installation's (but *not* a residential's) supply current in order to minimise the cross-sectional area requirements of the supply cables, switchgear and transformers in order to reduce copper and, therefore, their costs, and to reduce voltage drops and improve system capacity.

The greatest beneficiary of any power factor improvement is the electricity network company, whose cables, switchgear and transformers supply these installations. So, encouraging its commercial and industrial consumers to improve the power factor of their installations means that the distribution company's line losses and installation costs are reduced, and its system capacity and earning potential is increased.

For these reasons, as already explained, network companies offer their commercial and industrial consumers tariffs (pricing systems) which encourage those consumers who implement power factor improvement to their installations – essentially, by penalising those consumers who don't, through the use of surcharges!

At the beginning of this chapter, we said that power factor is defined as *the ratio of **true power** to **apparent power***. Let's remind ourselves what a power diagram for an inductive load looks like (Figure 22.8).

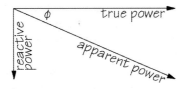

Figure 22.8

In order to reduce ('improve') the power factor, we must reduce the phase angle. We can do this by *reducing the circuit's overall reactive power*. How can we do this in practice? Well, there are a number of ways including through the use of synchronous machines but, in general, the least expensive way is, as already explained, through the use of **individal capacitors** or **capacitor banks** (sometimes called 'static capacitors').

There are *three* approaches to power factor correction using capacitors:

1 Motors in excess of around 35 kW can have their individual power factor improved by connecting capacitors directly across their terminals.
2 Installing capacitor banks at motor control-centre bus will improve the power factor of several motors.
3 Installing capacitor banks at the installation's service entrance.

Of these three approaches, the third is usually the least expensive. However, while this method will decrease the supply current in the distribution company's feeder, it will have absolutely no effect on the current between the service entrance (where the network company's cables are terminated) and the bulk of the load, whereas the other two methods will.

Capacitors intended for power factor improvement must be capable of withstanding the peak voltages of the installation's supply voltage – this information, together with their reactive power ratings is shown on their information nameplates. Power factor capacitors are normally oil-impregnated paper dielectric types, sealed within oil-filled metal containers.

Overcorrecting power factor can be avoided through the use of automatic switching, in which a number of individual capacitors are added or removed, to meet changes in the load's power factor.

So, power factor improvement can be achieved by adding a capacitor bank with an appropriate value of leading reactive power which, of course, acts in the *opposite sense* to the load's lagging reactive power (Figure 22.9).

To achieve any desired value of power factor, we need to know what size of capacitor to place in parallel with the load. In this case, a power factor improvement capacitor's 'size' is expressed in **reactive volt amperes (var)**, and *not* in terms of its capacitance.

Knowing the *capacitance* of a power factor improvement capacitor is of no practical use, and is of academic interest only.

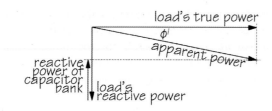

Figure 22.9

Power factor capacitors are usually oil-impregnated paper dielectric capacitors, with their plates enclosed in oil-filled metal tanks. These capacitors are always rated in **reactive volt amperes (var)**, *never* in farads. Banks of capacitors enable individual capacitors to be brought on line, automatically or manually, to ensure optimum power factor improvement for variations in the load's reactive power.

Let's illustrate this with a worked example.

Worked example 1 A single-phase a.c. supply of 400 V at 50 Hz, supplies a 50-kW load having a power factor of 0.5 lagging. Find the reactive power of a capacitor bank necessary to bring the power factor to 0.85.

Solution The first step is to determine the load's phase angle, given its power factor:

$$\cos\phi = 0.5$$
$$\phi = \cos^{-1}0.5 = 60°$$

The next step is to find out what the load's *existing reactive power* is. We can do that by applying basic trigonometry to its power diagram (Figure 22.10).

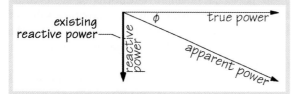

Figure 22.10

$$\tan\phi = \frac{\text{opposite}}{\text{adjacent}} = \frac{\text{reactive power}}{\text{true power}}$$

so, reactive power = true power × tan φ

$$= 50 \times \tan 60°$$
$$= 50 \times 1.732$$
$$= 86.60 \text{ kvar}$$

Now, we need to find out *what value of reactive power would be required* to achieve a power factor of 0.85:

$$\phi' = \cos^{-1}0.85 = 31.79°$$

reactive power = true power × tan φ

$$= 50 \times \tan 31.79°$$
$$= 50 \times 0.62$$
$$= 31.00 \text{ kvar}$$

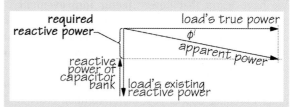

Figure 22.11

To reduce the load's existing reactive power of 86.60 kvar down to 31.00 kvar, we must supply the circuit with:

$$86.60 - 31.00 = 55.60 \text{ kvar}$$

... which must act in *the opposite sense*. In other words, the necessary capacitor bank must be rated at **55.60 kvar** (Answer).

Finally . . .

Now that you have completed this chapter, are you able to achieve the objectives or learning outcomes listed at the beginning of this chapter?

Ask yourself, 'Can I . . .'

1 define the terms, 'power factor', 'lagging' power factor and 'leading' power factor.

2 explain why increasing a load's power factor towards unity will reduce the current supplied to that load.

3 outline the main advantages of improving the power factor of a load.

4 explain why a power factor capacitor is rated in reactive volt amperes, rather than in farads.

5 solve problems on power factor improvement.

Online resources

The companion website to this book contains further resources relating to this chapter. The website can be accessed via the following link:

www.routledge.com/cw/waygood

Chapter 23

Balanced three-phase a.c. systems

On completion of this chapter, you should be able to

1 briefly explain the major advantages of three-phase a.c. systems, compared with single-phase a.c. systems.
2 identify delta- and star-connected systems.
3 distinguish between 'phases' and 'lines'.
4 briefly explain how lines are identified.
5 sketch a balanced, three-phase, delta-connected load, showing all voltage and current 'sense arrows'
6 construct a three-phase phasor diagram for a balanced star- and delta-connected load, showing all line and phase voltages and currents.
7 for a star-connected balanced load, define the relationships between its line and phase voltages and currents.
8 for a delta-connected balanced load, define the relationships between its line and phase voltages and currents.
9 express the total power of a balanced star- or delta-connected three-phase load, in terms of both their phase values and their line values.
10 Solve three-phase problems for balanced loads.

Introduction

Almost without exception, alternating current is generated, transmitted and distributed using **three-phase** systems. The main reason for this is **economy**, because it can be shown that, for a given load, the total volume of copper required by a three-phase system is approximately 75 per cent of that required by an equivalent single-phase system.

But there are other excellent reasons for using three-phase a.c. systems, which include:

• 'smoother' energy delivery (see below), compared with single-phase a.c.
• three-phase machines are self-starting; single-phase machines are not.
• three-phase machines are simpler, smaller, lighter and more efficient than single-phase machines of equivalent rating.
• three-phase machines run more smoothly and quietly than single-phase machines.

The waveforms shown in Figure 23.1 compare the rates at which energy is supplied (i.e. the 'power') to a

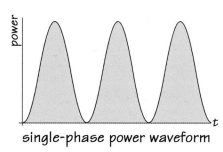
single-phase power waveform

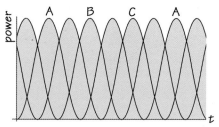
three-phase power waveform

Figure 23.1

An Introduction to Electrical Science, Waygood, ISBN 9780415810029, 2013. © Taylor & Francis

load by a single-phase and by a three-phase supply. The grey areas below each waveform represent the power 'supplied' by each system (in these examples, we are assuming the load is purely resistive in each case): for the three-phase system, the power supplied is close to being constant whereas, for the single-phase system, the power is supplied as a series of pulses.

Generating three-phase voltages

A three-phase alternator's armature consists of three, independent, windings that are physically displaced from each other by 120°. Accordingly, the voltages (labelled **A**, **B** and **C** in the diagram) which are induced into each winding, each reach their peak values, in sequence, 120 electrical degrees apart – as illustrated in Figure 23.2.

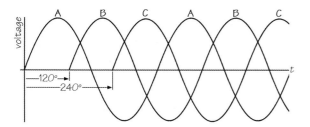

Figure 23.2

The sequence in which each voltage reaches peak value (**A-B-C**) is called the '**phase sequence**' of the system. Should the voltages reach their peak values in *reverse* sequence (**A-C-B**) – e.g. should the alternator run backwards – then we describe the system as having a 'negative phase sequence', but this topic is beyond the scope of this text.

As already explained, practically all high-voltage a.c. transmission and distribution systems are three-phase systems. For low-voltage a.c. distribution systems, single-phase supplies are required for residences while three-phase supplies may be required for small workshops, etc. This is achieved as illustrated in Figure 23.3 of a typical low-voltage, three-phase, distribution system. The three conductors, labelled **a**, **b** and **c**, are each energised at a potential of 230 V with respect to the fourth, neutral, conductor, with each potential displaced by 120°. As we shall learn, the potential differences between conductors **a-b**, **b-c** and **c-a** are 400 V.

The electricity network company tries to 'balance' their loads (i.e. apply a similar electrical load to each line) as far as possible, by supplying adjacent consumers from alternate lines, as illustrated in Figure 23.3. The significance of these connections will become clear later in this chapter.

This chapter is intended to supply us with an overview of three-phase a.c. systems, and is *not* intended to cover the subject in-depth.

Three-phase connections and terminology

Delta and star connections

The two most common three-phase connections are known as '**delta**' and '**star**'. In the United States and Canada, the 'star' connection is more commonly known as a '**wye**' connection.

A **delta** connection is shown in Figure 23.4, for a supply and for a load. Essentially, the three supply

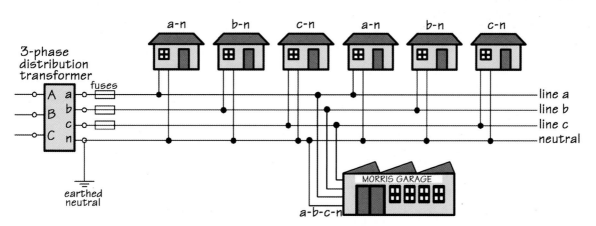

Figure 23.3

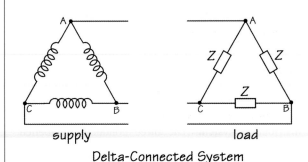

supply   load

Delta-Connected System

Figure 23.4

windings (as well as three identical load impedances) are connected is series with each other, and the external connections are made from points **A**, **B** and **C**, between each supply winding (e.g. alternator windings or transformer secondary windings) or load impedance.

A **star** connection is shown in Figure 23.5, for a supply and for a load made up of three identical impedances. The common point of connection for a star connection is called its '**star point**'; at the supply, the star point is normally connected to earth for the purpose of stabilising the three voltages **A-N**, **B-N** and **C-N,** and it is from this point that a neutral connection is made.

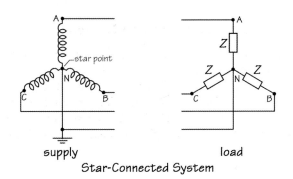

supply   load

Star-Connected System

Figure 23.5

There is absolutely no reason why a delta-connected supply cannot feed a star-connected load, or why a star-connected supply cannot feed a delta-connected load, providing the voltage requirements are met.

There are various other three-phase connections, but these are all beyond the scope of this text.

Delta-connected supplies are connected to their loads using *three* conductors, and are known as

'**three-phase**, **three-wire systems**', which accounts for why high-voltage pole-mounted distribution lines carry three conductors, and high-voltage transmission towers carry multiples of three conductors.

Star-connected supplies are connected to a common point, called a 'star point', which is normally then connected to earth, providing the **neutral point** of the system. Star-connected supplies are connected to their loads using *four* conductors, one of which is a neutral conductor, and are known as '**three-phase, four-wire, systems**'. Typically, in the UK at least, low-voltage distribution systems are three-phase, four-wire systems, which is why you will typically see four vertically mounted conductors on low-voltage distribution poles (any additional conductors are used for supplying street lighting, etc.).

> Delta and star configurations not only apply to alternator windings, but also to a three-phase transformer's primary and secondary windings, and to the way in which three-phase loads are connected.

Phase and line

The individual windings of a delta- or star-connected alternator or transformer, and their corresponding load impedances (as illustrated in Figures 23.6 and 23.7) are called '**phases**', whereas the conductors that interconnect three-phase supplies and their loads are called '**lines**'.

* In a three-phase, three-wire, system, there are three '**line conductors**'.
* In a three-phase, four-wire, system, there are three '**line conductors**' and a '**neutral**' conductor.

> Generally speaking, the **phases** of any three-phase machine, including transformers, are inaccessible, whereas the lines and their terminals are easily accessible. This is because the phases are normally enclosed within the machine or transformer tank.

Figures 23.6 and 23.7 showing three-phase loads should make this clear. Don't become confused over the 'crossed over' line conductors. The only reason for doing this is to make the vertical sequence **A-B-C** match the clockwise sequence **A-B-C**.

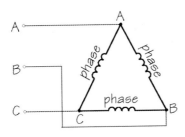

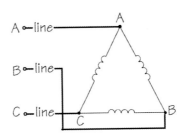

Figure 23.6

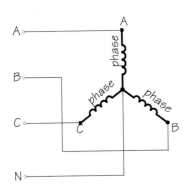

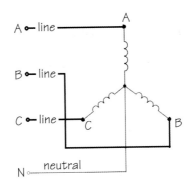

Figure 23.7

Vitally important!

Unfortunately, in the field (as well as in some textbooks), the term 'phase' is frequently misused which leads to confusion. For example, 'line conductors' are sometimes referred to as 'phase conductors' or, simply, as 'phases'. *This is completely incorrect*, and you **must** learn to use the correct terminology if you want to avoid becoming confused!

A single-phase supply can be obtained from a three-phase system very easily, simply by connecting the single-phase load between any pair of line conductors or, alternatively, between any line conductor and a neutral conductor – depending on the level of voltage required for the single-phase load.

Phase and line voltages and currents

The potential difference appearing across any phase is called a '**phase voltage**' (E_p), and the current flowing through any phase is called a '**phase current**' (I_p).

The potential difference appearing between any pair of line conductors is called a '**line voltage**' (E_L), and the current flowing through a line conductor is called a '**line current**' (I_L).

Identifying line conductors and terminals

Three-phase **line** conductors and terminals (*not* 'phases'!) are identified in various ways, according to relevant national electrical standards. At the time of writing this textbook, in the UK's electricity supply industry, lines are still identified using the colours **red-yellow-blue**. For commercial and industrial installations these colours have been replaced by **brown-black-grey**, in accordance with EU harmonisation requirements. Alternative systems of identifying lines include the use of *letters* (e.g. **A-B-C**), *numerals* (e.g. **1-2-3**), or a *combination of the two* (e.g. L_1-L_2-L_3).

Whichever system is used, the order, or sequence, in which the colours, letters or numerals are listed match the phase sequence of the three-phase system. For example, the letters, **A-B-C**, would indicate normal, or 'positive', phase sequence, whereas the letters, **A-C-B**, would indicate reverse, or 'negative', phase sequence. We will *not* be considering a negative phase sequence situation in this chapter.

In this chapter, we will be using the letters **A-B-C** to identify line conductors or line terminals. Upper-case letters can then be used to indicate high voltages, while lower-case letters can be used to indicate lower voltages, whenever we need to differentiate between the two (such as for transformers). This is also in accordance with the relevant British Standard for the marking of transformer connections.

> It's important to understand that this system of colours, letters or numbers is used to identify **line conductors** (or **line terminals**), *not* phases. The phases themselves, are not individually identified although, if necessary, they can be identified in terms of their line terminals; for example, 'phase A-B', 'phase B-C' and 'phase C-A'.
>
> Many textbooks talk about '**phase A**', '**phase B**' and '**phase C**'. This makes no sense whatsoever, and simply leads to confusion.

Sense arrows and double-subscript notation

Line and phase voltages can be identified using 'sense arrows', which we met in an earlier chapter on *series, parallel and series-parallel circuits*, combined with what is termed '**double-subscript notation**'. This system helps us determine the sense, or direction, in which a voltage or a current is acting *at any given instant*, and allows us to easily construct three-phase phasor diagrams.

The directions of voltage and current sense arrows are always drawn as illustrated in the schematic diagrams in this section. You will notice that sense arrows that represent line currents *always* point towards the load.

Double-subscript notation takes the following forms:

$$\overline{E}_{AB} \text{ or } \overline{E}_{AN}$$

These examples are read as: *'the potential at point A with respect to point B'*, and as: *'the potential at point A with respect to the neutral'*. So the first subscript letter

defines the line whose potential is being measured, and the second subscript defines the point of reference.

Double-subscript notation applies to both **line voltages** and **phase voltages**, and they *always* follow the phase sequence, A-B-C-A, so they're written as:

$$\overline{E}_{AB}, \overline{E}_{BC}, \text{and } \overline{E}_{CA}$$
$$\overline{E}_{AN}, \overline{E}_{BN}, \text{and } \overline{E}_{CN}$$

Double-subscript notation can also be used with **phase currents**, and takes the following form:

$$\overline{I}_{AB} \text{ or } \overline{I}_{AN}$$

Double-subscript notation for currents only applies to *phase* currents, and the above examples are read as *'the current flowing through the load, from terminal A towards terminal B'* or as *'the current flowing through the load, from terminal A towards the neutral terminal'*.

For currents, too, double-subscript notation *always* follows the phase sequence, A-B-C-A, so they're always written:

$$\overline{I}_{AB}, \overline{I}_{BC}, \text{and } \overline{I}_{CA}$$
$$\overline{I}_{AN}, \overline{I}_{BN}, \text{and } \overline{I}_{CN}$$

For **line currents**, we use '**single-subscript notation**', which takes the following form:

$$\overline{I}_{A}$$

This example is always read as, *'the current flowing in line A from the supply towards the load'*.

Figure 23.8 illustrates how this system of sense arrows and double-/single-subscript notation is used to identify the potential differences and currents for four-wire star-connected load.

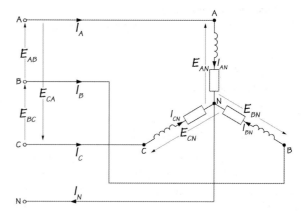

Figure 23.8

Figure 23.9 illustrates how sense arrows and double-/single-subscript notation is used to identify the potential differences and currents for three-wire delta-connected load.

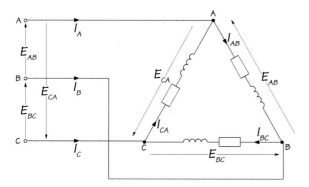

Figure 23.9

Do not get confused by the way in which the line conductors have been drawn, in Figures 23.8 and 23.9, with line B crossing over line C. This has been done for no other reason than neatness and consistency! The reason for doing this is simply to ensure that the vertical sequence of the line conductors follow the order A-B-C, and that each terminal of the star or delta connection is labelled A-B-C in a clockwise sequence.

Three-phase phasor diagrams

There are two '**Golden Rules**' which we must *always* follow whenever we want to construct a three-phase **phasor diagram**. These are:

1 *Always draw a phasor diagram as it applies to the **load**, never to the supply.*
2 *Always draw the phasor diagram in the following order:*
   a phase voltages
   b line voltages
   c phase currents
   d line currents.

Balanced star-connected load

So, let's draw a phasor diagram for the following star-connected load. We will assume a 'balanced' load: that is, each phase has an identical impedance and phase angle and, so, draws an identical phase current. *Unbalanced loads are well beyond the scope of this text.*

Of course, the phase current could either lag or lead its corresponding phase voltage, depending on whether the load is resistive-inductive or resistive-capacitive but, in the following example, we'll assume that each load is resistive-inductive, with the phase current lagging by ϕ degrees.

The first thing we must do, then, is to draw a schematic diagram of a star-connected load, and insert the **sense arrows**, labelled using **double-/single-subscript notation**, for *all* voltages and currents. Don't forget, the sequence of letters used in double-subscript notation *must* follow in the sequence: A-B-C-A (Figure 23.10).

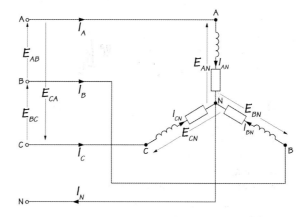

Figure 23.10

So, following the second of our two 'golden rules', we start by drawing the **phase voltages**, and it makes sense to choose $\bar{E}_{AN}$ as the reference phasor which, as usual, is drawn along the horizontal, positive, axis (Figure 23.11).

Figure 23.11

Next, we draw the second phase voltage, $\bar{E}_{BN}$, lagging the reference phasor, by 120° (Figure 23.12).

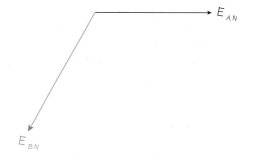

Figure 23.12

To complete drawing the phase voltages, we draw $\bar{E}_{CN}$, lagging the reference phasor by 240° (Figure 23.13).

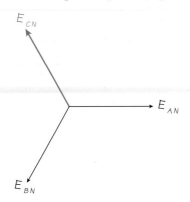

Figure 23.13

Having now drawn all the phase voltages, the next step is to construct each of the line voltages: $\bar{E}_{AB}$, $\bar{E}_{BC}$ and $\bar{E}_{CA}$. We'll start by drawing line voltage, $\bar{E}_{AB}$. So let's refer back to the schematic diagram to find out how this line voltage relates to the phase voltages (Figure 23.14).

Line voltage, $\bar{E}_{AB}$, is the voltage between terminal **A** and terminal **B**. So it's made up of the phase voltage acting *from* terminal **A** *to* terminal **N** and the phase voltage acting *from* terminal **N** *to* terminal **B** – i.e. phase voltage $\bar{E}_{AN}$ and phase voltage $\bar{E}_{NB}$ (which is in the *opposite* direction to sense arrow, $\bar{E}_{BN}$). These, of course, are *phasor* quantities, so they must be added vectorially.

> Imagine you are *walking* from point A to point B. Your route will take you from point A to point N and, then, from point N to point B. The 'distance' you have walked is the sum of A to N *plus* N to B. You (vectorially) add the voltages in exactly the same sequence!

If we now refer back to the partly completed phasor diagram, we can readily identify voltage phasor $\bar{E}_{AN}$; *but there is no phasor labelled $\bar{E}_{NB}$!* However, we can easily obtain it, simply by *reversing the voltage phasor $\bar{E}_{BN}$* (Figure 23.15).

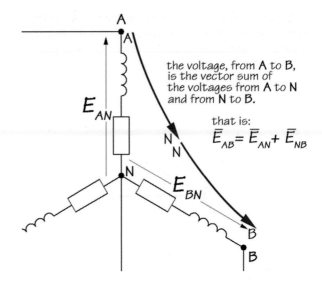

the voltage, from A to B, is the vector sum of the voltages from A to N and from N to B.

that is:
$$\bar{E}_{AB} = \bar{E}_{AN} + \bar{E}_{NB}$$

Figure 23.14

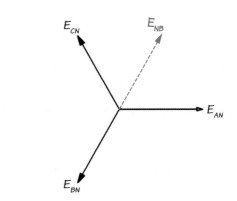

Figure 23.15

So, now, all we have to do is to **vectorially-add** $\bar{E}_{AN}$ and $\bar{E}_{NB}$ to obtain line voltage $\bar{E}_{AB}$ (Figure 23.16):

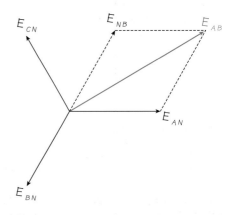

Figure 23.16

We can repeat this exercise line voltage, $\bar{E}_{BC}$, where (referring back to our schematic diagram and its sense arrows) (Figure 23.17):

$\bar{E}_{BC} = \bar{E}_{BN} + \bar{E}_{NC}$ (where $\bar{E}_{NC}$ is simply phase voltage $\bar{E}_{CN}$ reversed)

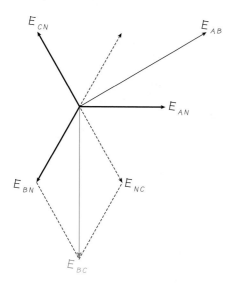

Figure 23.17

... and, again, for line voltage, $\bar{E}_{CA}$ (Figure 23.18), where:

$\bar{E}_{CA} = \bar{E}_{CN} + \bar{E}_{NA}$
(where $\bar{E}_{NA}$ is simply phase voltage $\bar{E}_{AN}$, reversed)

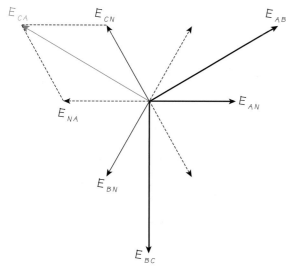

Figure 23.18

Let's clean up the phasor diagram, by removing the 'reversed' phase voltages, which we only needed to construct the line voltages (Figure 23.19).

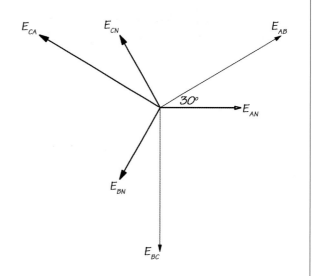

Figure 23.19

It's obvious, from the voltage phasor diagram, that *the line voltages are larger than the phase voltages*, and they *lead the phase voltages by 30°*. But *how much larger are the line voltages*, compared to the phase voltages? Well, we'll worry about that in a moment!

In the meantime, let's now finish the phasor diagram, by constructing the **current phasors**.

In a star-connected system, the phase and line currents are identical, so when we sketch the phase currents, we are *also* sketching the line currents. As with the previous example, we'll assume that the phase currents lag the phase voltage by ϕ degrees, so we start by drawing the phase current, $\bar{I}_{AN}$ (which is exactly the same as the line current $\bar{I}_A$), lagging (clockwise) the phase voltage, $\bar{E}_{AN}$, by a phase angle of ϕ degrees, behind its corresponding phase-voltage, $\bar{E}_{AN}$ (Figure 23.20).

To complete the phasor diagram, we simply repeat this process for phase/line currents, $\bar{I}_{BN}$ ($\bar{I}_B$) and $\bar{I}_{CN}$ ($\bar{I}_C$) (Figure 23.21).

Relationship between line and phase voltages

By simple observation, it's quite obvious from the completed phasor diagram that, for a **star-connected system**, the *line voltages are larger than the phase voltages* – but the question is, *how much larger?* Of course, we *could* draw the phasor diagram to scale and, then, measure and compare the two (in fact, you might like to do that with the phasor diagram in Figure 23.21 which has been drawn to scale).

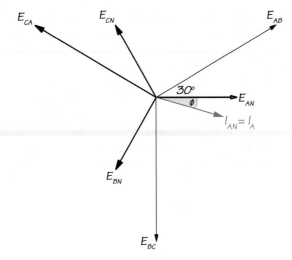

Figure 23.20

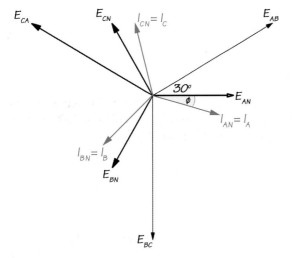

Figure 23.21

However, we can, instead, apply simple trigonometry to the problem, as follows.

Referring back to the completed phasor diagam, let's compare the length of phase voltage $\bar{E}_{AN}$ to the line voltage $\bar{E}_{AB}$ (Figure 23.22).

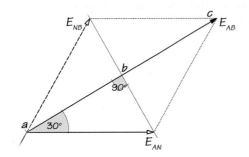

Figure 23.22

If we draw a line between the tips of the phase voltage phasors $\bar{E}_{AN}$ and $\bar{E}_{NB}$, it will bisect the line voltage phasor, $\bar{E}_{AB}$, at point **b**. The phase voltage $\bar{E}_{AN}$ and the distance **a–b** then form a right-angled triangle, so we can find the length **a–b** as follows:

$$\cos 30° = \frac{\text{adjacent}}{\text{hypotenuse}} = \frac{\text{distance } ab}{\bar{E}_{AN}}$$

$$\text{distance } ab = \bar{E}_{AN} \cos 30°$$

The line voltage, $\bar{E}_{AN}$, is equal to the distance **a–c**, or **twice** the distance **a–b**, so:

$$\bar{E}_{AB} = \text{distance } ac = 2 \times \text{distance } ab = 2 \times \bar{E}_{AN} \cos 30°$$
$$= 2 \times 0.866\, \bar{E}_{AN}$$
$$= 1.732\, \bar{E}_{AN}$$

By an amazing coincidence (!), **1.732** just happens to be the **square root of 3**! So, for a **wye-connected** system:

$$\bar{E}_L = \sqrt{3}\ \bar{E}_P$$

As the **supply voltages** are independent of any load, this important relationship *always* holds true for voltages in a star-connected system.

Worked example 1   What is the line voltage for a three-phase, four-wire, system having a phase voltage of 230 V?

Solution

$$\bar{E}_L = \sqrt{3}\ \bar{E}_P = \sqrt{3} \times 230 \approx 400\ \text{V (Answer)}$$

The worked example, above, explains what we mean when we describe the UK's low-voltage supply as being a '400/230-V system'. The '400 V' refers to its line voltage, while the '230 V' refers to its phase voltage.

Balanced delta-connected load

Now let's draw a phasor diagram for the following delta-connected load. Again, we will assume a 'balanced' load: that is, each phase has an identical impedance and phase angle and, so, draws an identical phase current. *Unbalanced loads are well beyond the scope of this text.*

We'll again assume that each load is resistive-inductive, this time with a phase angle of φ degrees, lagging – i.e. each phase current will lag its corresponding phase voltage by φ degrees (any angle between 0° and –90°).

As for the previous, star-connected load, the first thing we must do is to sketch a schematic diagram of the delta-connected load, and insert **sense arrows** into the schematic diagram for all voltages and currents, and to **label these sense arrows** using double-/single-subscript notation (Figure 23.23).

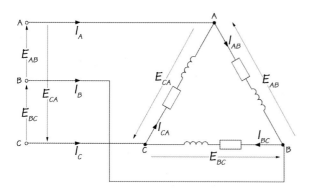

Figure 23.23

So, following the second of our two 'golden rules', we start by drawing the **phase voltages** and, this time, it makes sense to choose $\bar{E}_{AB}$ as the reference phasor which, as usual, is drawn along the horizontal, positive, axis. To save time (now that we know how to do so), we'll also draw the other two phase voltages, $\bar{E}_{BC}$ and $\bar{E}_{CA}$, which lag the reference phasor by 120° and 240°, respectively (Figure 23.24).

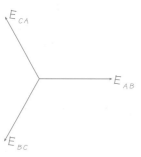

Figure 23.24

For a delta-connected system, of course, the line voltages are identical to the phase voltages. So the phasor diagram in Figure 23.24 represents both the phase *and* the line voltages.

Next, we'll start to construct the phase and line currents; starting, as always, with the phase currents $\bar{I}_{AB}, \bar{I}_{BC}$, and $\bar{I}_{CA}$.

The first step, then, is to draw the phase current, $\bar{I}_{AB,}$ which we are told to assume lags the corresponding phase current by an angle of φ degrees (Figure 23.25).

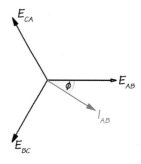

Figure 23.25

Again, we can save time by drawing the other two phase currents, $\bar{I}_{BC}$ and $\bar{I}_{CA}$, at 120° and 240°, respectively, from the first (Figure 23.26).

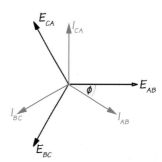

Figure 23.26

The next step, is to construct the line currents, starting with $\bar{I}_A$. To understand how to do this, we must first look at junction **A** in our schematic diagram (Figure 23.27).

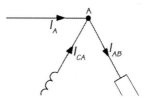

Figure 23.27

If we apply Kirchhoff's Current Law to junction **A**, it can be expressed as follows:

The phasor sum of the currents approaching the juction equals the phasor sum of the currents leaving that junction. That is:

$$\overline{I}_A + \overline{I}_{CA} = \overline{I}_{AB}$$

Rearranging this equation, in terms of $\overline{I}_A$:

$$\overline{I}_A = \overline{I}_{AB} - \overline{I}_{CA}$$

Which is exactly the same thing as,

$$\overline{I}_A = \overline{I}_{AB} + \overline{I}_{AC}$$

If we now examine the phasor diagram, we see that we already have a phase current, $\overline{I}_{AB}$, but unfortunately we *don't* have a phase current $\overline{I}_{AC}$! However, that's easily rectified, by simply *reversing phasor* $\overline{I}_{CA}$ (Figure 23.28).

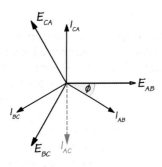

Figure 23.28

We now simply vectorially add the phase currents, $\overline{I}_{AB}$ and $\overline{I}_{AC}$, to obtain the line current, $\overline{I}_A$ (Figure 23.29).

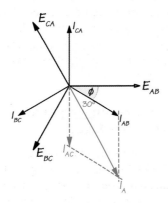

Figure 23.29

As you can see, the line current, $\overline{I}_A$, *lags* the phase current, $\overline{I}_{AB}$, by 30°. The line current is also greater than the phase current which, of course, we already knew simply by inspection. If we were to conduct a similar analysis to that we did for the line and phase voltages in a star-connected system, we would find that the line current is $\sqrt{3}$ larger than the phase current.

To complete drawing the line voltages, we apply exactly the same procedure – i.e. apply Kirchhoff's Current Law to junctions **B** and **C** of the schematic diagram:

$$\overline{I}_B = \overline{I}_{BC} + \overline{I}_{BA} \quad \text{(where } \overline{I}_{BA} \text{ is simply } \overline{I}_{AB}, \text{ reversed).}$$

$$\overline{I}_C = \overline{I}_{CA} + \overline{I}_{CB} \quad \text{(where } \overline{I}_{CB} \text{ is simply } \overline{I}_{BC}, \text{ reversed).}$$

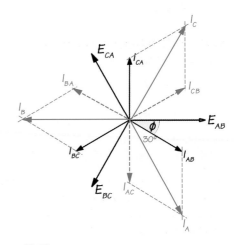

Figure 23.30

Finally, let's clear up the phasor diagram, by removing all the construction lines (Figure 23.31).

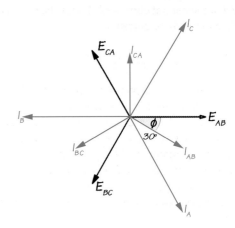

Figure 23.31

In this particular example, we were told that the phase current *lagged* the corresponding phase voltage by ϕ degrees (as indicated by the shaded sector). The *actual* phase angle, of course, depends entirely on the ratio of resistance to impedance of the loads and, of course, it could either lead or lag the corresponding phase voltage, depending on whether the actual load is an *R-L* load or an *R-C* load.

So, for a **delta-connected** system:

$$\overline{I}_L = \sqrt{3}\,\overline{I}_P$$

Summary for balanced star and delta systems

Three-phase system supplying a balanced star-connected load

$$\overline{E}_L = \sqrt{3}\,\overline{E}_P \qquad \overline{I}_L = \overline{I}_P$$

Three-phase system supplying a balanced delta-connected load:

$$\overline{E}_L = \overline{E}_P \qquad \overline{I}_L = \sqrt{3}\,\overline{I}_P$$

Behaviour of currents in balanced three-phase systems

In this chapter, we are only concerned with **balanced loads**. A 'balanced' load is one in which each phase is identical in all respects: each having the same impedance and phase angle.

Neutral current in a balanced-star load?

In a star-connected load, the three phase currents, $\overline{I}_{AN}, \overline{I}_{BN}$ and $\overline{I}_{CN}$, each converge on the star point, and return to the supply, via the neutral conductor. So we can say that the neutral current $(\overline{I}_N)$ must be the phasor sum of the three phase currents:

$$\overline{I}_N = \overline{I}_{AN} + \overline{I}_{BN} + \overline{I}_{CN}$$

Now, let's see what happens if we vectorially add the three phase currents in a **balanced** star-connected system. To add *three* phasors, we start by adding *any two* and, then, *add the resultant to the third*. In the example in Figure 23.32, $\overline{I}_{NC}$ is the phasor sum of $\overline{I}_{AN}$ and $\overline{I}_{BN}$, and is *equal and opposite* (and will, therefore, *cancel*) $\overline{I}_{CN}$.

So, the phasor sum of the three phase currents is **zero**! That is:

$$\overline{I}_N = \overline{I}_{AN} + \overline{I}_{BN} + \overline{I}_{CN} = 0$$

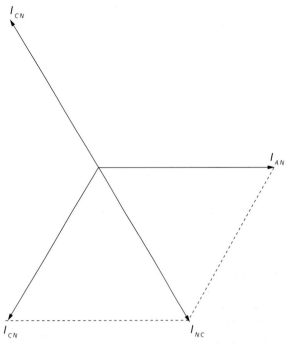

Figure 23.32

So, *what's the point in having a neutral conductor?*

Well, in practise, star-connected loads are **rarely perfectly balanced** and, if the three-phase load is unbalanced, even slightly, there *will* be a neutral current – albeit a small one (in comparison with the phase currents).

Real three-phase loads are rarely perfectly balanced, so there is practically *always* a neutral return current and, therefore, the need for a neutral. An exception to this are three-phase loads such as motors, which are *always* balanced because each of their phase windings is identical; so, if a motor has a star-connected field winding, there is *no* requirement for a neutral.

A failure of the neutral conductor supplying an unbalanced load will result in any neutral current having to return through the line conductors, resulting in *unbalanced phase voltages* (*not* line voltages, they are independent of load) appearing across the load – however, any further discussion of this topic is beyond the scope of this chapter.

Where is the return current in a balanced-delta load?

It's probably instinctive to understand why no neutral current should flow back to the supply from a balanced, **star-connected** load. However, many students are confused as to how three line conductors can supply current *to* a **delta-connected** load, if there is no *return* conductor (equivalent to a neutral) for current to flow back to the supply!

The answer, of course, is that these line currents are *not* constant-value currents; they are both *out of phase* with each other, and *vary in value and direction*. Hopefully, the following explanation should make this clear.

In Figure 23.33**a**, we show the waveforms representing the three line currents, each displaced by 120°, and flowing through lines **A**, **B** and **C**. We have randomly chosen three instants in time, **x**, **y** and **z**, in order to examine what is happening to the currents at each of these particular moments.

- At **instant x** (occurring at a displacement angle of 30°), the current in lines **A** and **C** are both +5 A. The positive sign indicates that these currents are acting in the positive direction (i.e. flowing *towards* the load), while the current in line **B** is –10 A. The negative sign indicates that this current is acting in the negative direction (*away* from the load, *back towards the supply*).
- At **instant y** (occurring at a displacement angle of 120°), the current in line **A** is +8.66 A, so is flowing *towards* the load. The current in line **B** is passing through the horizontal axis and, so, is zero. The current in line **C** is –8.66 A, so is flowing *away* from the load.
- At **instant z** (occurring at a displacement angle of 210°), the currents in line **A** and **C** are both –5 A and, so, are both flowing *away* from the load. The current in line **B** is +10 A, so is flowing *towards* the load.

If we now transfer this data to the schematic diagram (Figure 23.33**b**), you can see that:

- At **instant x**, a total of 10 A is flowing along lines **A** and **C** *towards the load*, while a total of 10 A is flowing *back to the supply*, along line **B**.

- At instant **y**, a current of 8.66 A is flowing along line **A** *towards the load*, while a current of 8.66 A is flowing along **C** *back to the supply*. No current is flowing through line **B**.
- At instant **z**, a current of 10 A is flowing along line **B** *towards the load*, while a total of 10 A is flowing through lines **A** and **B**, *back to the supply*.

Power in three-phase systems

You will recall that power in a single-phase circuit is given by:

$$P = \bar{E}\,\bar{I}\cos\phi$$

where:

$$P = \text{power (watts)}$$
$$\bar{E} = \text{voltage (volts)}$$
$$\bar{I} = \text{current (amperes)}$$
$$\cos\phi = \text{power factor}$$

As already explained, in this chapter we are only going to consider **balanced loads** (i.e. with each phase being identical in all respects).

For a three-phase load, the **total power** will be *the sum of the power developed by each phase*. So, the total power in a three-phase *balanced* load is given by:

$$P = 3\times(\bar{E}_P\bar{I}_P\cos\phi)$$

Unfortunately, in practise, we *cannot* measure phase voltages or phase currents *directly* because, usually, the phases are not easily accessible. For example, we cannot measure the phase voltages or currents

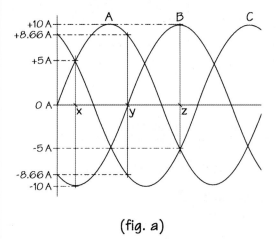

(fig. a)

Figure 23.33

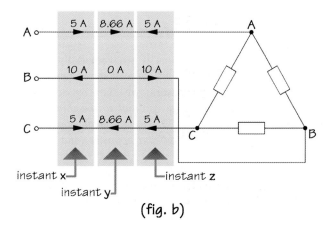

(fig. b)

for a three-phase motor because its phase windings are *inside* the casing of the machine and, therefore, *inaccessible*. However, we can *always* measure *line* voltages and *line* currents because they are 'outside' the machine and, therefore, easily accessible.

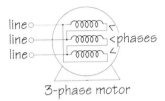

3-phase motor

Figure 23.34

So, it would be much more practical if we could develop an equation that gave us the total power of a three-phase system *in terms of line values* rather than in terms of phase values.

The explanation is as follows:

For a delta-connected system:	**For a star-connected system**
$\bar{E}_P = \dfrac{\bar{E}_L}{\sqrt{3}}$ and $\bar{I}_P = \bar{I}_L$	$\bar{E}_P = \bar{E}_L$ and $\bar{I}_P = \dfrac{\bar{I}_L}{\sqrt{3}}$
$P = 3\bar{E}_P\bar{I}_P\cos\phi$	$P = 3\bar{E}_P\bar{I}_P\cos\phi$
$= 3(\dfrac{\bar{E}_L}{\sqrt{3}})\times\bar{I}_L\times\cos\phi$	$= 3\bar{E}_L\times(\dfrac{\bar{I}_L}{\sqrt{3}})\times\cos\phi$
$= 1.732\,\bar{E}_L\,\bar{I}_L\cos\phi$	$= 1.732\,\bar{E}_L\,\bar{I}_L\cos\phi$
$= \sqrt{3}\,\bar{E}_L\,\bar{I}_L\cos\phi$	$= \sqrt{3}\,\bar{E}_L\,\bar{I}_L\cos\phi$

So, as you can see, it doesn't really matter whether you are dealing with a balanced star-, or a balanced delta-connected, exactly the same equation applies:

$$P = \sqrt{3}\,\bar{E}_L\,\bar{I}_L\cos\phi$$

where:

$$P = \text{power (watts)}$$
$$\bar{E}_L = \text{line voltage (volts)}$$
$$\bar{I}_L = \text{line current (amperes)}$$
$$\cos\phi = \text{power factor}$$

Once again, a reminder that the equations used in this section only apply for balanced loads.

By extension, we can show the equations for **apparent power** (in volt amperes) and **reactive power** (in reactive volt amperes):

$$\text{apparent power} = \sqrt{3}\,\bar{E}_L\,\bar{I}_L$$
$$\text{reactive power} = \sqrt{3}\,\bar{E}_L\bar{I}_L\sin\phi$$

And, of course, the same relationship exists between these three quantities in three-phase circuits as it did for single-phase circuits, i.e.

$$(\text{apparent power})^2 = (\text{true power})^2$$
$$+ (\text{reactive power})^2$$

Summary of important three-phase relationships

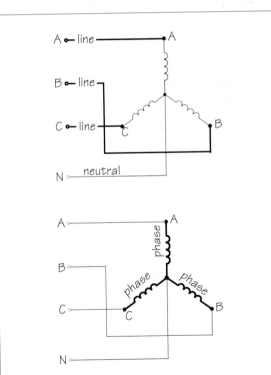

Figure 23.35

For a **star-connected** system (Figure 23.35):

$$\bar{E}_{line} = \sqrt{3}\,\bar{E}_{phase} \text{ and } \bar{I}_{line} = \bar{I}_{phase}$$

For a **balanced** load:

$$\bar{I}_{neutral} = 0$$

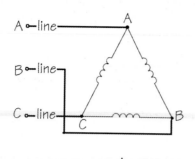

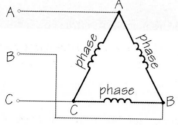

Figure 23.36

For a **delta-connected** system (Figure 23.36):

$$\bar{E}_{line} = \bar{E}_{phase} \text{ and } \bar{I}_{line} = \sqrt{3}\,\bar{I}_{phase}$$

For both **star** *and* **delta** connections, with a *balanced* load:

$$\text{true power} = 3\,\bar{E}_p \bar{I}_p \cos\phi \text{ or}$$
$$\text{true power} = \sqrt{3}\,\bar{E}_L \bar{I}_L \cos\phi$$

and

$$\text{apparent power} = 3\,\bar{E}_p \bar{I}_p \text{ or}$$
$$\text{apparent power} = \sqrt{3}\,\bar{E}_L \bar{I}_L$$

(**true power** is measured in **watts**; **apparent power** in **volt amperes**)

Worked example 2 Complete the missing data in the schematic diagram shown as Figure 23.37.

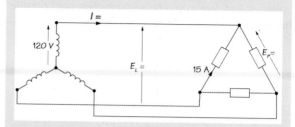

Figure 23.37

Solution

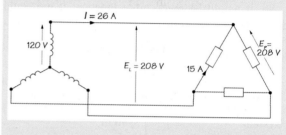

Figure 23.38

Worked example 3

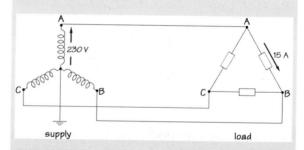

Figure 23.39

In the schematic shown in Figure 23.39, calculate each of the following:

a line voltage
b line current
c transformer's phase current
d load's phase voltage.
e the load's apparent power.

Solution

a   We know the supply transformer's secondary phase voltage $(\bar{E}_P)$ is 230 V so, for a star connection:

$$\bar{E}_L = \sqrt{3}\ \bar{E}_S = \sqrt{3}\times 230 \simeq 400\ \text{V (Answer a.)}$$

b   We know that the load's phase current is 15 A so, for a delta connection:

$$\bar{I}_L = \sqrt{3}\ \bar{I}_P = \sqrt{3}\times 15 \simeq 26\ \text{A (Answer b.)}$$

c   For a star connection, the phase current is the same as the line current, so the supply's phase current:

$$= \text{line current} = 26\ \text{A (Answer c.)}$$

d   For a delta connection, the phase voltage is the same as the line voltage, so the load's line voltage:

$$= \text{phase voltage} = 400\ \text{V (Answer d.)}$$

e   apparent power $= \sqrt{3}\ \bar{E}_L \bar{I}_L = \sqrt{3}\times 400\times 26$
$$\simeq 18\ \text{kV}\cdot\text{A (Answer e.)}$$

Worked example 4   In the above example, what is the power of the load, assuming that it has a power factor of (a) 0.5, and (b) 0.75?

Solution

a   For a power factor of 0.5:

$$P = \sqrt{3}\ \bar{E}_L \bar{I}_L \cos\phi$$
$$= \sqrt{3}\times 400\times 26\times 0.5$$
$$\simeq 9\ 000\ \text{W or 9 kW (Answer a.)}$$

b   For a power factor of 0.75,

$$P = \sqrt{3}\ \bar{E}_L \bar{I}_L \cos\phi$$
$$= \sqrt{3}\times 400\times 26\times 0.75$$
$$\simeq 13\ 500\ \text{W or 13.50 kW (Answer b.)}$$

$$\bar{I}_L = \sqrt{3}\ \bar{I}_L = \sqrt{3}\times 31 \simeq 53.7\ \text{A}$$

Finally . . .

Now that you have completed this chapter, are you able to achieve the objectives or learning outcomes listed at the beginning of this chapter?

Ask yourself, 'Can I . . .'

1 briefly explain the major advantages of three-phase a.c. systems, compared with single-phase a.c. systems.
2 identify delta- and star-connected systems.
3 distinguish between 'phases' and 'lines'.
4 briefly explain how lines are identified.
5 sketch a balanced, three-phase, delta-connected load, showing all voltage and current 'sense arrows'
6 construct a three-phase phasor diagram for a balanced star- and delta-connected load, showing all line and phase voltages and currents.
7 for a star-connected balanced load, define the relationships between its line and phase voltages and currents.
8 for a delta-connected balanced load, define the relationships between its line and phase voltages and currents.
9 express the total power of a balanced star- or delta-connected three-phase load, in terms of both their phase values and their line values.
10 Solve three-phase problems for balanced loads.

Online resources
The companion website to this book contains further resources relating to this chapter. The website can be accessed via the following link:
www.routledge.com/cw/waygood

Index

Locators in **bold** refer to figures/diagrams

An Introduction to Electrical Science, Waygood, ISBN 9780415810029, 2013. © Taylor & Francis